Greenhouse Warming: Abatement and Adaptation

Norman J. Rosenberg
William E. Easterling III
Pierre R. Crosson
Joel Darmstadter
Editors

Proceedings of a Workshop held in Washington, D.C. June 14–15, 1988

Workshop Sponsors

Resources for the Future

National Climate Program Office, National Oceanic and Atmospheric Administration

Carbon Dioxide Research Division, U.S. Department of Energy

U.S. Department of Agriculture

U.S. Environmental Protection Agency

International Federation of Institutes for Advanced Study

RESOURCES FOR THE FUTURE
Washington, D.C.

Edited proceedings of a workshop held in Washington, D.C., on June 14–15, 1988, organized by the Climate Resources Program of Resources for the Future and sponsored by Resources for the Future, the National Climate Program Office of the National Oceanic and Atmospheric Administration, the Carbon Dioxide Research Division of the U.S. Department of Energy, the U.S. Department of Agriculture, the U.S. Environmental Protection Agency, and the International Federation of Institutes for Advanced Study.

Resources for the Future publishes the papers in this volume to facilitate their wide and public availability. While extensively edited, they have not been subjected to the formal review and editing procedures of books from Resources for the Future. The publisher assumes responsibility neither for their accuracy and objectivity nor for the views expressed therein, which are those of the authors. These views should not be attributed to Resources for the Future.

Second printing, 1991

Library of Congress Cataloging-in-Publication Data

Greenhouse warming: abatement and adaptation / Norman J. Rosenberg
. . . [et al.].
p. cm.
Proceedings of a workshop held in Washington, D.C. June 14–15, 1988; workshop sponsors, Resources for the Future . . . [et al.]
Bibliography: p.
ISBN 0-915707-50-0
1. Greenhouse effect, Atmospheric—Congresses. 2. Climatic changes—Congresses. I. Rosenberg, Norman J., 1930–
II. Resources for the Future.
QC912.3.G737 1989 89-8483
363.7—dc20 CIP

A brochure describing the Climate Resources Program and a list of its publications are available from the Climate Resources Program, Resources for the Future, 1616 P Street, N.W., Washington, D.C. 20036.

RESOURCES FOR THE FUTURE

RESOURCES FOR THE FUTURE (RFF) is an independent nonprofit organization that advances research and public education in the development, conservation, and use of natural resources and in the quality of the environment. Established in 1952 with the cooperation of the Ford Foundation, it is supported by an endowment and by grants from foundations, government agencies, and corporations. Grants are accepted on the condition that RFF is solely responsible for the conduct of its research and the dissemination of its work to the public. The organization does not perform proprietary research.

RFF research is primarily social scientific, especially economic. It is concerned with the relationship of people to the natural environmental resources of land, water, and air; with the products and services derived from these basic resources; and with the effects of production and consumption on environmental quality and on human health and well-being. Grouped into four units—the Energy and Natural Resources Division, the Quality of the Environment Division, the National Center for Food and Agricultural Policy, and the Center for Risk Management—staff members pursue a wide variety of interests, including forest economics, natural gas policy, multiple use of public lands, mineral economics, air and water pollution, energy and national security, hazardous wastes, the economics of outer space, and climate resources. Resident staff members conduct most of the organization's work; a few others carry out research elsewhere under grants from RFF.

Resources for the Future takes responsibility for the selection of subjects for study and for the appointment of fellows, as well as for their freedom of inquiry. The views of RFF staff members and the interpretations and conclusions of RFF publications should not be attributed to Resources for the Future, its directors, or its officers. As an organization, RFF does not take positions on laws, policies, or events, nor does it lobby.

Contents

List of Figures and Tables

FIGURES

TABLES

Preface

The papers presented in this volume grew out of a workshop organized by the Climate Resources Program at Resources for the Future (RFF). That program, initiated in 1987, is part of RFF's Energy and Natural Resources Division. The program has two major objectives:

1. To identify the ways in which climate impacts upon agriculture, forestry, and water resources and to incorporate climatic information into scientific and socioeconomic analyses of natural-resource-based industries, with the aim of improving their management and their resilience to climatic stresses such as drought, flood, extreme heat and cold; and
2. To study the prospects for future climatic change and the far-reaching physical and socioeconomic impacts that may follow in the wake of such change, with the aim of providing information on policies and practices that can be effective for government and the private sector in delaying and abating climatic change and in coping with such change as may be unavoidable.

The second of these objectives was addressed by the Workshop on Controlling and Adapting to Greenhouse Warming held in Washington, D.C., on June 14–15, 1988, and is the subject of this volume.

The workshop was funded by Resources for the Future, with additional support from the National Climate Program Office of the National Oceanic and Atmospheric Administration, the Carbon Dioxide Research Division of the U.S. Department of Energy, the U.S. Department of Agriculture, and the U.S. Environmental Protection Agency. The workshop logistics were handled with great skill by William Easterling, assisted by RFF staff members Debra Montanino, K Storck, Janet Ekey-Ostrov, and Angela Blake.

Preparation of these proceedings involved RFF's Publications office. Dorothy Sawicki was project editor; Joan Engelhardt was responsible for the graphics and overall design; and Rachel Cox, Barbara de Boinville, and Samuel Allen provided technical editing. Correspondence with authors, coordination of manuscript flow, and numerous aspects of manuscript preparation were the responsibility of Angela Blake. Scientific review and editorial assistance were provided by Mary S. McKenney. Their very competent assistance, which we gratefully acknowledge, made preparation of this volume possible.

Norman J. Rosenberg
Senior Fellow and Director,
Climate Resources Program
Energy and Natural Resources Division
Resources for the Future

Contributors

Sandra S. Batie
Professor of Agricultural Economics
Virginia Polytechnic Institute and State University
Blacksburg, Virginia

Ian Burton
Director, International Federation of Institutes for Advanced Study
Toronto, Ontario, Canada

Chester L. Cooper
Advisor on International Programs
Resources for the Future
Washington, D.C.

Pierre R. Crosson
Senior Fellow
Resources for the Future
Washington, D.C.

Joel Darmstadter
Senior Fellow
Resources for the Future
Washington, D.C.

William E. Easterling III
Fellow, Climate Resources Program
Resources for the Future
Washington, D.C.

Jae Edmonds
Technical Leader, Economic Programs
Pacific Northwest Laboratory
Washington, D.C.

Kenneth D. Frederick
Senior Fellow
Resources for the Future
Washington, D.C.

Robert W. Fri*
President and Senior Fellow
Resources for the Future
Washington, D.C.

Peter H. Gleick
Director, Global Environment Program
Pacific Institute
Berkeley, California

Gjerrit P. Hekstra
Directoraat-General voor de Milieuhygiene
Ministry of Housing, Physical Planning and Environment
Leidschendam, The Netherlands

N. S. Jodha
Head, Mountain Farming Systems Division
International Centre for Integrated Mountain Development
Kathmandu, Nepal

D. Gale Johnson*
Professor of Economics
University of Chicago
Chicago, Illinois

Michael Oppenheimer*
Senior Scientist
Environmental Defense Fund
New York City

Martin L. Parry
Professor of Geography
University of Birmingham
Birmingham, United Kingdom

Paul R. Portney
Senior Fellow and Director,
Center for Risk Management
Resources for the Future
Washington, D.C.

*Panelist (see chapter 14 in this volume) at the Workshop on Controlling and Adapting to Greenhouse Warming.

Roger R. Revelle
Professor of Science and Public Policy
University of California, San Diego
La Jolla, California

Norman J. Rosenberg
Senior Fellow and Director,
Climate Resources Program
Resources for the Future
Washington, D.C.

Milton Russell*
Professor of Economics and Senior Fellow,
Energy, Environment, and Resources Center,
University of Tennessee; Senior Economist,
Oak Ridge National Laboratory
Oak Ridge, Tennessee

Stephen H. Schneider
Head, Interdisciplinary Climate Systems Section
National Center for Atmospheric Research
Boulder, Colorado

Roger A. Sedjo
Senior Fellow
Resources for the Future
Washington, D.C.

Herman H. Shugart
William W. Corcoran Professor
of Environmental Sciences
University of Virginia
Charlottesville, Virginia

Allen M. Solomon
Senior Researcher
International Institute for Applied Systems Analysis
Laxenburg, Austria

*Panelist (see chapter 14 in this volume) at the Workshop on Controlling and Adapting to Greenhouse Warming.

Note from the editors

Several of the bibliographic entries in our reference lists were in press or forthcoming at the time of original publication of this volume. This second printing (February 1991) affords us the opportunity to provide, for the convenience of readers, the complete facts of publication for many of those references. They are as follows:

Crosson, Pierre. 1989. "Climate Change and Mid-Latitudes Agriculture: Perspectives on Consequences and Policy Responses," *Climatic Change* vol. 15, pp. 51–73.

Easterling, W. E. 1989. "Farm-Level Adjustments to Climatic Change by Illinois Corn Producers," in J. Smith and D. Tirpak, eds., *The Potential Effects of Global Climate Change on the United States* (Washington, D.C., U.S. Environmental Protection Agency, Office of Policy, Planning, and Evaluation).

Hanchey, J. R., K. E. Schilling, and E. Z. Stakhiv. 1987. "Water Resources Planning Under Climate Uncertainty," in *Preparing for Climate Change*. Proceedings of the First North American Conference on Preparing for Climate Change: A Cooperative Approach (Washington, D.C., Government Institutes).

Ingram, H. M., H. J. Gortner, and M. K. Landy. 1990. "The Political Agenda," pp. 421–443 in P. E. Waggoner, ed., *Climate Change and U.S. Water Resources* (New York, Wiley).

Jones, P. D., P. M. Kelley, C. M. Goodess, and T. R. Karl. 1989. "The Effect of Urban Warming on the Northern Hemisphere Temperature Average," *Journal of Climate* vol. 2, pp. 285–290.

Martin, Ph., N. J. Rosenberg, and M. S. McKenney. 1989. "Climatic Change and Evapotranspiration: Simulation Studies of a Wheat Field, a Forest, and a Tall Grass Prairie," *Climatic Change* vol. 14, pp. 117–151.

Mearns, L. O., P. H. Gleick, and S. H. Schneider. 1990. "Climate Forecasting," pp. 87–137 in P. E. Waggoner, ed., *Climate Change and U.S. Water Resources* (New York, Wiley).

Mearns, L. O., S. H. Schneider, S. L. Thompson, and L. R. McDaniel. 1990. "Analysis of Climate Variability in General Circulation Models: Comparison with Observations and Changes in Variability in 2 × CO_2 Experiments." *Journal of Geophysical Research* vol. 95, pp. 20469–20490.

Peterson, D. F., and A. D. Keller. 1990. "Irrigation," pp. 269–306 in P. E. Waggoner, ed., *Climate Change and U.S. Water Resources* (New York, Wiley).

Rind, D., R. Goldberg, and R. Ruedy. 1989. "Change in Climate Variability in the 21st Century," *Climatic Change* vol. 14, pp. 5–38.

Rosenberg, N. J., B. A. Kimball, Ph. Martin, and C. F. Cooper. 1990. "From Climate and CO_2 Enrichment to Evapotranspiration," pp. 151–175 in P. E. Waggoner, ed., *Climate Change and U.S. Water Resources* (New York, Wiley).

1

Introduction

The Editors

On June 14 and 15, 1988, Resources for the Future conducted a workshop called Controlling and Adapting to Greenhouse Warming. The workshop was held at the National Academy of Sciences in Washington, D.C., at a time of year when its citizens ruefully prepare for the departure of the normally pleasant spring and the onset of the normally oppressive summer. The summer of 1988 turned out (perhaps with the help of the greenhouse effect?) to be even hotter than usual, both climatically and politically—but more on that later.

Resources for the Future's sponsorship of the workshop was augmented by contributions from a number of agencies of the federal government: the National Climate Program Office of the National Oceanic and Atmospheric Administration, the Carbon Dioxide Research Division of the U.S. Department of Energy, the U.S. Department of Agriculture, and the Environmental Protection Agency. The International Federation of Institutes for Advanced Study (IFIAS) was also a cosponsor.

The purpose of the workshop was to provide a forum for natural and social scientists—individually and in teams—to review opportunities for controlling and/or diminishing the rate of greenhouse gas emissions and their accumulation in the atmosphere, and opportunities for adjustment to whatever degree of warming is unavoidable.

An audience of about 150 academics, U.S. and Canadian officials, representatives of major industries, consultants, and members of nongovernmental research and advocacy groups participated in discussions of the formal papers and responded to comments made by a panel of distinguished individuals with broad experience in research, government, and environmental policy.

This volume is a compilation of papers presented at the workshop. At the time of the workshop most of the papers were in rough draft. The versions printed here as chapters contain considerably more detail than could be presented at the workshop. The papers have been heavily edited and revised since they were first presented, and have been subjected to extensive review within Resources for the Future. In editing the papers we have tried to minimize duplication wherever possible, to add the requisite scientific detail so that our readers need take none of the statements on faith alone, and to annotate the papers so that both the lay reader and the specialist are directed to the primary sources of information used by the authors.

The chapters in the volume follow almost precisely the order in which the papers were presented at the workshop. They are divided into three parts: Background (chapters 2 through 6), Natural Resource Sectors (chapters 7 through 10), and Perspectives (chapters 11 through 14).

Leading off part I, chapter 2 by Stephen H. Schneider and Norman J. Rosenberg summarizes the current state of knowledge and the uncertainties regarding the greenhouse effect, its physical mechanisms, and its climatic implications. The authors also describe what is known concerning the direct effects of the rising carbon dioxide (CO_2) concentration in the atmosphere on plant growth and water use.

In chapter 3 Joel Darmstadter and Jae Edmonds deal with the relationship between economic development and greenhouse gas emissions. Adapting the findings of an economic-energy CO_2 model, they present a reference-case projection of global CO_2 emissions by the year 2050. In addition, they examine critically two alternative scenarios—one that would maintain emissions at the current rate and one at half that rate—and discuss some of the vital questions raised in pursuing such low CO_2 futures: problems of interfuel substitution, exploitation of conservation potentials, policy initiatives, and institutional changes.

In chapter 4 Gjerrit P. Hekstra explains the physical linkages between greenhouse warming, climatic change, and sea-level rise. He reports on the wide range of rise that analysts have predicted, on the consequences for coastal ecosystems, agriculture, and other enterprises, and on the demographic, economic, and social impacts that may follow. Hekstra proposes policies for adaptation to sea-level rise should it become necessary.

Chapter 5 by Pierre R. Crosson addresses the question of how governments and the international community will decide how much global warming is too much and considers the obstacles to an international agreement to enforce the limit. In chapter 6 Paul R. Portney examines the greenhouse warming issue by using analogies and contrasts with environmental health problems that are usually studied in the framework of risk assessment and risk management.

The second set of papers, grouped under Natural Resource Sectors, includes four chapters, each one dealing with the possible consequences of greenhouse warming on a particular natural resource sector. William E. Easterling III, Martin L. Parry, and Pierre R. Crosson deal in chapter 7 with impacts of climate change on agriculture and possible responses thereto, stressing that future world agriculture must be considered in all its complexity and diversity—from the highly mechanized, capitalized, and information-based operation of the most advanced farmers to the small-scale, labor-intensive, risk-averse agriculture in developing countries. They explore as well how farmers' future regional comparative advantages will be determined by the pace and extent of climate changes, the rate at which adaptations can be made, and the manner in which environmental costs are handled on farms.

In chapter 8 Roger A. Sedjo and Allen M. Solomon examine the potential effects of greenhouse-induced climate change on forests. They also consider the many complex issues that must be addressed before massive afforestation can be used as a tool to reduce the accumulation of CO_2 in the atmosphere in order to moderate the rate of climate change. One emphasis is on managed forests.

In chapter 9 Sandra S. Batie and Herman H. Shugart deal with the vulnerabilities of unmanaged forests and other unmanaged ecosystems to the effects of a changed climate. They note that deriving policies to protect these systems against climate change is particularly challenging because many of the social values the systems provide are difficult to quantify (as is species diversity, for example) and raise difficult issues of intergenerational equity. Consequently, for these systems traditional benefit–cost analysis is not appropriate as a management tool. Batie and Shugart nonetheless emphasize that the economic way of thinking can provide valuable guidelines for policies to protect these systems. As an alternative to benefit–cost analysis, they propose a "safe minimum standard" approach as a basis for developing policy.

Kenneth D. Frederick and Peter H. Gleick, in chapter 10, discuss the impacts on regional hydrology that could result from greenhouse-induced climate change, and the alternatives for dealing with these impacts. They consider four broad approaches for adapting to climate-induced changes in water supplies: development of new infrastructure for storing and transporting water, improved management of existing facilities, demand management (including water pricing and marketing), and technological change.

In the final set of papers, called Perspectives, N. S. Jodha examines how developing countries perceive the greenhouse warming–climate change issue and its relevance to the many urgent problems these countries now face. He argues in chapter 11 that the unique vulnerabilities of such countries, and experience in dealing with those vulnerabilities, might enable these nations to adapt to future climate change more readily, in some ways, than the developed countries.

Ian Burton in chapter 12 puts the climate change issue in the context of those events and phenomena now transforming our planet that the scientific community calls global change—acid precipitation, deforestation, desertification, diminishing biodiversity, and others. Climate change is only one of this set; yet climate change may be affected by all of these earth-transforming processes and may in turn affect them. Burton describes a program now being developed, and centered in the social sciences, whose objective is an understanding of the human dimensions of global change.

In chapter 13 Roger R. Revelle's broad perspective provides additional commentary on the questions of abatement and adaptation. In the course of his discussion he returns to issues and uncertainties raised in earlier chapters. On the abatement question, he emphasizes the great uncertainties remaining about the role of the ocean as a carbon sink and as a factor determining global atmospheric circulation. He also explores the possible expansion of the boreal forest in response to spontaneous warming in the high latitudes as another effect that might reduce the rate and extent of CO_2 accumulation in the atmosphere. Revelle deals also with the impacts of climate change on the need for irrigation in the United States, and provides additional thoughts on adaptation in the developing countries.

Chapter 14 summarizes comments made by a panel reacting to the papers presented at the workshop, and introduces other ideas on the questions of abatement and adaptation. The panel was chaired by Chester L. Cooper and included D. Gale Johnson, Michael Oppenheimer, Milton Russell, and Robert W. Fri. In his report, Cooper ties together the panel presentations, the discussion they engendered, and the formal papers to highlight policy issues surrounding the abatement–adaptation debate.

The reader will find, not only in chapter 14 but throughout this volume, that a wide range of informed, yet divergent and even conflicting, views are expressed. A full airing of these views—unanimous or not—is our purpose.

* * * * *

Nearly a year will have passed between the workshop and the publication of this book of proceedings. But given the need for coordination of subject matter, the review process, technical and stylistic editing, and all the other complicated steps involved in publication, only a slap-dash volume could have been produced much faster. We knew when we chose this publication process that the greenhouse issue was hot and getting hotter and that some of the views expressed and positions taken at the workshop might be altered by subsequent events. None of us would have predicted, however, the tremendous surge in public attention to the greenhouse issue that this past year has witnessed. The surge was due in part to a reaction to the hot summer of 1988 in North America; to the severe drought in the northern plains states and the corn belt; to a prominent scientist's declaration that the emergence of the greenhouse warming signal from the normal "noise" of climatic variation is a near certainty (Hansen, 1988); to the declaration by the Toronto Conference on the Changing Atmosphere (also in June 1988) that "humanity is conducting an unintended, uncontrolled, globally pervasive experiment whose ultimate consequences could be second only to a global nuclear war" (Conference on the Changing Atmosphere, 1988); and to the call at this same meeting for a 20 percent reduction in fossil fuel emissions of CO_2 by 2005 and further reductions in subsequent years. Public interest has been further fueled by (or perhaps reflected in) the introduction in the 100th U.S. Congress of S. 2667—the National Energy Policy Act of 1988—and a similar bill in the House of Representatives, aimed at steering the United States and other nations in directions pointed out by the report of the Brundtland Commission (World Commission on Environment and Development, 1987) and the Toronto Conference Statement. While no action was taken on S. 2667 in 1988, the bill has been reintroduced as S. 305 in the 101st Congress.

In view of all the ensuing activity, how relevant is this proceedings volume one year after the workshop? Is it likely to remain relevant for some time to come? Only the reader can judge this objectively. At least five critical issues were addressed at our workshop: Will greenhouse warming actually happen? If so, what will be its regional consequences? How do we decide how much warming is too much, and what options exist for abating or delaying it and for adapting to it if necessary? What are the policy steps needed to facilitate abatement and adaptation? What are the impediments to development of needed policies?

According to a recent report (Kerr, 1989), 1988 was the year of greatest global surface mean temperature on record. So perhaps the first question is being answered. However, according to a different source (Hanson et al., 1989), the temperature record of at least the contiguous United States from 1895 to 1987 shows no evidence of mean surface warming. Beyond that, there are concerns that the data used in calculating trends in global mean surface temperature may be flawed, for a number of reasons. Among them is the "urban heat island" effect, whereby the growth of population centers around weather stations has raised temperatures for reasons unrelated to greenhouse warming (see Karl et al., 1988; Jones et al., 1989). We must assume that the jury is still out on whether or not the greenhouse effect is already manifest. Nonetheless, the workshop was, and this book is, predicated on the assumption that if emissions of radiatively active trace gases continue at current or accelerated rates, or even at somewhat lower rates than today, greenhouse warming will increase and climate will change. Endorsement of this view by the electronic and print media and by environmental activists, legislators, and others does not *necessarily* weaken that assumption.

On the second question, our understanding of the regional climatic consequences of greenhouse gas emissions is not significantly better at this writing than it was in June 1988. Schneider and Rosenberg, in chapter 2, report that the best judgment of experts is that real progress in finding answers to this question may take another ten to fifty years of intensive research.

In October 1988 the World Meteorological Organization and the United Nations Environmental Programme convened an organizational meeting in Geneva, Switzerland, of the Intergovernmental Panel on Climate Change. This new entity is charged with assessing the state of scientific knowledge of climate change and its impacts, and of the full range of responses needed to diminish the rate and extent of change and to adapt as necessary to unavoidable change. A number of other national and regional efforts are now under way with essentially the same objectives.

The issues dealt with in these proceedings—options for abatement and adaptation, policies needed to facilitate abatement and adaptation, and impediments to development of such policies—are just those with which these intensified assessment efforts must deal. We hope that our work contributes part of what is needed for these efforts to succeed.

REFERENCES

Conference on the Changing Atmosphere: Implications for Global Security. 1988. Conference Statement, Toronto, Ontario, Canada, June 27–30.

Hansen, James E. 1988. "The Greenhouse Effect: Its Impacts on Current Global Temperature and Regional Heat Waves." Testimony before the U.S. Senate Committee on Energy and Natural Resources, 100th Congress, June 23, 1988.

Hanson, K., G. A. Maul, and T. R. Karl. 1989. "Are Atmospheric Greenhouse Effects Apparent in Climatic Record of the Contiguous U.S. (1895-1987)?" *Geophysical Research Letters* vol. 16, pp. 49–52.

Jones, P. D., P. M. Kelley, C. M. Goodess, and T. R. Karl. 1989. "The Effect of Urban Warming on the Northern Hemisphere Temperature Average," *Journal of Climate*, in press.

Karl, T. R., H. F. Diaz, and G. Kukla. 1988. "Urbanization: Its Detection and Effect in the United States Climate Record," *Journal of Climate* vol. 1, pp. 1089–1123.

Kerr, R. A. 1989. "1988 Ties for Warmest Year," *Science* vol. 243, p. 891.

World Commission on Environment and Development. 1987. *Our Common Future* (New York, Oxford University Press).

Part I

Background

2

The Greenhouse Effect: Its Causes, Possible Impacts, and Associated Uncertainties

Stephen H. Schneider
Norman J. Rosenberg

The earth's climate changes. It is vastly different now from what it was 100 million years ago, when dinosaurs dominated the planet and tropical plants thrived at high latitudes; it is different from what it was 18,000 years ago, when ice sheets covered much more of the Northern Hemisphere than they do now. In the future it will surely continue to evolve. In part, the evolution will be driven by natural causes, such as small variations in the earth's orbit. But future climatic change will probably have another important source as well: human activities. We may already be feeling the climatic effects of having polluted the atmosphere with gases such as carbon dioxide (CO_2).

There is no doubt that the concentration of carbon dioxide in the atmosphere has been rising; it is roughly 25 percent higher now than it was a century ago. It is also broadly accepted that when the carbon dioxide concentration rises, the temperature at the earth's surface must rise too. Carbon dioxide is relatively transparent to visible sunlight, but it absorbs long-wavelength, infrared radiation emitted by the earth. Hence it tends to trap heat near the surface (see figure 2-1). This is known as the greenhouse effect, and its existence and basic mechanisms are not questioned by atmospheric scientists. For example, it explains the very hot temperatures on Venus (whose thick atmosphere is mostly carbon dioxide) as well as the frigid conditions on Mars (whose largely carbon dioxide atmosphere is very thin).

What is questioned is the precise amount of warming and the regional pattern of climatic change that can be expected on the earth from the anthropogenic increase in the atmospheric concentration of carbon dioxide and other greenhouse gases. (The cumulative effect of chlorofluorocarbons, nitrogen oxides, ozone, and other trace greenhouse gases could be comparable to that of carbon dioxide over the next century—see figure 2-2). It is the regional patterns of changes in temperature, precipitation, and soil moisture that will determine what impact the greenhouse effect will have on natural ecosystems, agriculture, and water supplies.

HOW MUCH MIGHT CLIMATE CHANGE?

A number of researchers have attempted to model the possible climatic impacts of carbon dioxide. Most of them have followed the same approach: they give the model an initial increase of carbon dioxide (usually doubling the atmospheric concentration), allow it to run until it reaches a new thermal equilibrium, and then compare the new climate to the control climate, that is, the climate state of the model for present conditions.

Figure 2-2, for example, shows the results of a simple, one-dimensional, globally averaged climatic model's response to a scenario of the cumulative effect of increases in trace greenhouse gases (alternatively referred to as radiatively active or radiatively important trace gases) projected over the next fifty or so years by Ramanathan et al. (1985). The sensitivity of such one-dimensional models is typically half that of more comprehensive three-dimensional models that include more feedback processes (discussed below). Note the very large uncertainty represented by the error bars on figure 2-2. These are meant to account for uncertainties in the anthropogenic and biogeochemical cycle factors which control the sources and sinks that control the trace gas concentrations. In addition to these uncertainties in composition, there are additional uncertainties arising from internal processes in climatic models which cannot be fully validated. Roughly speaking, the uncertainty in modeling the sensitivity of the globally averaged surface temperature increase to any given trace gas increase scenario is another factor of two.

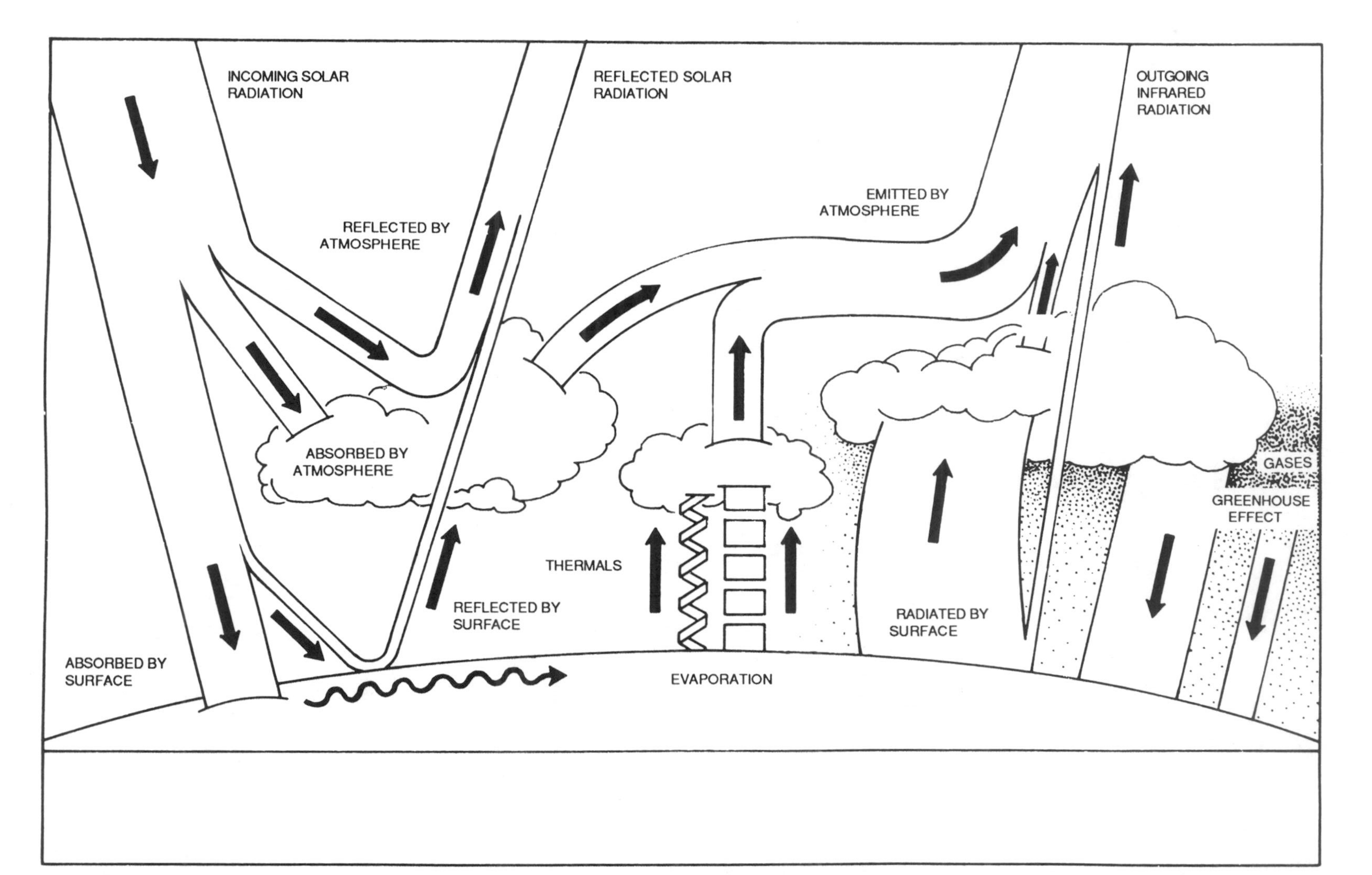

Figure 2-1. Schematic illustration of the earth's radiation and energy balances. The greenhouse effect is well established. It arises because the earth's atmosphere tends to trap heat near the surface. Carbon dioxide, water vapor, and other trace greenhouse gases are relatively transparent to the visible and near-infrared wavelengths that carry most of the energy of sunlight, but they absorb more efficiently the longer, infrared (IR) wavelengths emitted by the earth. Hence an increase in the atmospheric concentration of greenhouse gases tends to warm the surface by downward reradiation of IR, as shown. (*Source:* Adapted, with permission, from S. H. Schneider, "Climate Modeling," *Scientific American* vol. 256, p. 78. © 1987 by Scientific American, Inc. All rights reserved.)

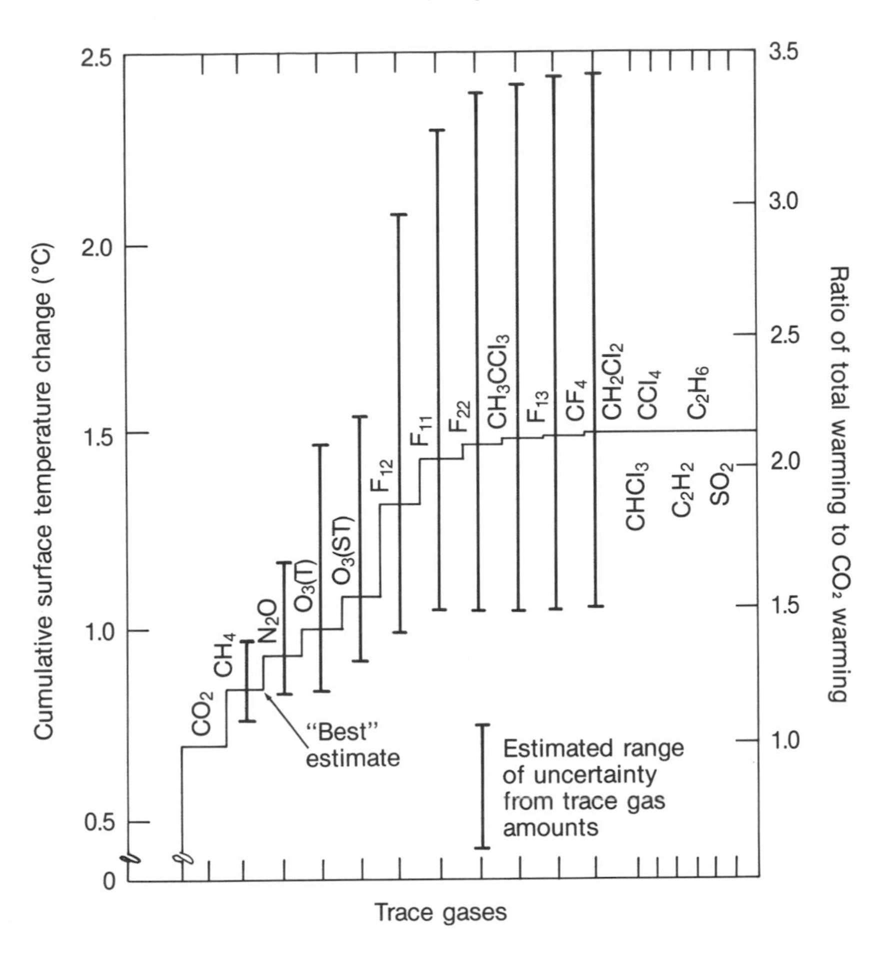

Figure 2-2. Cumulative surface warming from assumed trace gas scenario. (Period fifty years from 1980 levels.) (*Source:* Adapted, with permission, from V. Ramanathan et al., "Trace Gas Trends and Their Potential Role in Climate Change," *Journal of Geophysical Research* vol. 90, pp. 5547–5566. © 1985 by the American Geophysical Union.)

Modeling the Climatic System

The climate system involves the interaction of biota, air, sea, ice, and land components, with solar radiation providing nearly all of the energy that drives the system. Variations in the gaseous and particulate constituents of the atmosphere, along with changes in the earth's position relative to the sun, act to vary the amount and distribution of sunlight received by the system. The unreflected portion of that energy drives the atmospheric circulation, which in turn is linked by means of wind stress and heat transfer to the circulation of the oceans. The atmosphere and oceans are both influenced by the extent and thickness of the ice covering the land and sea as well as by the land surface itself. Since each of these components has a different range of response times, the whole system must be viewed as continuously evolving, with some parts of the system lagging behind or leading other parts.

The system also contains feedback loops between the interacting components. These amplify (positive feedback) or damp (negative feedback) perturbations. For example, any increase in the area of polar ice or snow cover will cause more of the incoming solar radiation to be reflected, leaving less to be absorbed by the surface. This will result in a positive feedback loop, wherein lowered surface temperature favors a further increase in ice and snow cover, assuming that the availability of moisture for snowfall remains adequate. However, it is quite possible that increasing snow cover and the associated coldness of a continental interior gradually would limit the overlying atmosphere's ability to import moisture into the region. This scenario would eventually result in decreased amounts of snowfall and would limit further growth of the snow cover in a negative feedback loop.

Climate prediction, like most other forecasts of complex systems, is essentially a process of extrapolation (see Land and Schneider, 1987). We attempt to determine the future behavior of the climate system from knowledge of its past behavior and present state using basically two approaches. One employs empirical statistical methods, such as regression equations, with past and present observations to obtain the most probable extrapolation in time. The other approach performs the extrapolation using "first principles"—equations believed to represent the physical, chemical, and biological processes that govern the behavior of the climate system on the pertinent temporal and areal scales. The latter approach is what is usually meant by the term *climate modeling*.

Since the statistical approach is dependent on historical data, it is obviously limited in its ability to predict climatic conditions that have not been previously observed or are caused by different processes. The statistical method cannot answer "What if?" questions, such as the potential effects of rapidly increased atmospheric carbon dioxide on climate. Thus, the more promising approach to climate prediction for conditions, or forcings, different from the present is modeling. In that case, the validation of such models becomes a chief concern, as we will discuss later.

Climate models vary in their spatial resolution, that is, in the number of dimensions they simulate and the amount of spatial detail they include. An example of an extremely simple model is one that calculates only the average temperature of the earth, independent of time, as an energy balance arising from the earth's average reflectivity and the average "greenhouse" properties of the atmosphere. Such a model is zero dimensional: it collapses the real temperature distribution on the earth to a single point, or global average. By contrast, three-dimensional climate models simulate the way temperature varies with latitude, longitude, and altitude. The most sophisticated of these include dynamical effects and are known as general circulation models (GCMs). They predict the evolution over time not only of temperature but also of humidity, wind speed and direction, soil moisture, sea ice, and other climatic variables modeled over three dimensions in space. General circulation models are usually more comprehensive than simpler models in terms of the physical, chemical, or biological detail they include, but they are also much more expensive to design and interpret.

Climate Sensitivity and Scenarios

Uncertainty about parameterizations of important feedback mechanisms like clouds, soil moisture, and sea ice is one reason the ultimate goal of climate modeling—reliably forecasting details of the future of key variables such as temperature and rainfall patterns on a regional basis—is not yet realizable. Another source of uncertainty that is external to the models themselves is human behavior. To forecast, for example, what impact carbon dioxide emissions will have on climate, one would need to know how much carbon dioxide is going to be emitted and how that emission will be distributed or removed by the physical, chemical, and biological processes of the carbon cycle.

What the climate models often can do well is analyze the sensitivity of the climate to various uncertain or unpredictable variables. In the case of the carbon dioxide problem, one could construct a set of plausible economic, land-use, technological, and population-growth scenarios to project CO_2 or other trace gas growth and employ a model to evaluate the climatic consequences of each scenario. Climatic factors whose correct values are uncertain, such as cloud-feedback parameters, could be varied over a plausible range of values. The results would indicate which of the uncertain factors is most important in sensitizing the climate to a carbon dioxide buildup; one could then focus research on those factors. The results would also give some idea of the range of climatic futures to which natural ecosystems and societies may be forced to adapt and the rates at which change would occur. How to respond to such information, of course, is a political issue.

Mearns, Gleick, and Schneider (forthcoming) prepared table 2-1 to provide impact assessment specialists with ranges of climate changes that reflect their interpretation of state-of-the-art modeling results. These projections were based on their analysis of available results and provide what are believed to be plausible estimates about the direction and magnitude of some important anthropogenically forced climate changes over the next fifty years or so—a typical estimate for an equivalent doubling of carbon dioxide—together with a simple high, medium, or low level of confidence for each variable. (Equivalent doubling means that point where carbon dioxide and other trace greenhouse gases have a radiative effect equivalent to doubling the preindustrial value of carbon dioxide from about 280 parts per million by volume [ppmv] to 560 ppmv.) As another measure of the nature of the uncertainties, a rough estimate is included on table 2-1 of the time that may be necessary to achieve a widespread scientific consensus on the direction and magnitude of the change. In some cases, such as the magnitude and direction of changes in sea level and global annual average temperature and precipitation, such a consensus has virtually been reached. In other cases, such as changes in the extent of cloud cover, time-evolving patterns of regional precipitation, and the daily, monthly, and interannual variance of many climatic variables, the large uncertainties surrounding present projections will be reduced only with considerably more research.

Let us consider, for example, the first row on table 2-1, temperature change. The global average change of +2 degrees Celsius (°C) to +5°C is typical of most national and international assessments of an equivalent doubling of greenhouse gases, neglecting transient delays (for example, Bolin et al., 1986). The neglect of transients means that the range given is based on the assumption that trace gases have increased over a long enough period for the climate to come into equilibrium with the increased concentration of greenhouse gases. In reality, the large heat capacity of the oceans will delay realization of most of the equilibrium warming by perhaps many decades (see Schneider and Thompson, 1981). This implies that at any specific time when we reach an equivalent CO_2 doubling (by say 2030), the actual global temperature increase may be considerably smaller than the +2°C to +5°C listed in table 2-1. However, this "unrealized warming" (see, for example, Hansen et al., 1985) will eventually be experienced when the

Table 2-1. Projections of Climatic Changes Likely to Occur as the Result of an Equivalent Doubling of Atmospheric CO_2

Phenomenon	Probable global annual average change	Distribution of change		Interannual variability[a]	Significant transients	Confidence of projections		Estimated time for research that leads to consensus (years)
		Regional average	Change in seasonality			Global average	Regional average	
Temperature	+2°C to +5°C	−3°C to +10°C	Yes	Down?	Yes	High	Medium	0–5
Sea level	+0.1 m to 1.0 m	[b]	No	[c]	Unlikely	High	Medium	5–20
Precipitation	+7% to +15%	−20% to +20%	Yes	Up?	Yes	High	Low	10–50
Solar radiation	−10% to +10%	−30% to +30%	Yes	[c]	Possible	Low	Low	10–50
Evapotranspiration	+5% to +10%	−10% to +10%	Yes	[c]	Possible	High	Low	10–50
Soil moisture	—[c]	−50% to +50%	Yes	[c]	Yes	[c]	Medium	10–50
Runoff	Increase	−50% to +50%	Yes	[c]	Yes	Medium	Low	10–50
Severe storms	[c]	[c]	[c]	[c]	Yes	[c]	[c]	10–50

Note: Equivalent doubling means that point where carbon dioxide and other trace greenhouse gases have a radiative effect equivalent to doubling the pre-industrial value of carbon dioxide.

Source: Adapted, by permission of the American Association for the Advancement of Science, from L. Mearns, P. H. Gleick, and S. H. Schneider, "Prospects for Climate Change," in Paul Waggoner, ed., *Climate Change and U.S. Water Resources* (New York, Wiley for AAAS, forthcoming), table I.

[a] Inferences based on preliminary results of Rind et al. (forthcoming).

[b] Sea-level increases at approximately the global rate except where local geological activity prevails.

[c] No basis for quantitative or qualitative forecast.

climate-system thermal response catches up to the greenhouse-gas forcing.

Forecasts of regional- and watershed-scale changes in temperature, evaporation, and precipitation are most germane to impact assessment. But, as table 2-1 suggests, such regional forecasts are much more uncertain than global equilibrium projections. Regional temperature ranges given in table 2-1 are much larger than global changes and even allow for some regions of negative change. For example, surface temperature increases in higher northern latitudes are up to several times larger than the global average response, at least in equilibrium (see figures 2-3 and 2-4). Because of the importance of regional and local impact information, techniques need to be developed to evaluate the smaller-scale effects of large-scale climatic changes. For example, Gleick (1987) employs a regional hydrology model driven by large-scale climate change scenarios from various GCM inputs.

Figures 2-5 and 2-6 show equilibrium precipitation changes from CO_2-doubling GCM runs using models from three different laboratories—Geophysical Fluid Dynamics Laboratory (GFDL), Goddard Institute for Space Studies (GISS), and National Center for Atmospheric Research (NCAR). Although the detailed differences in precipitation evident on the maps are considerable, increased evaporation from the warmer summertime land temperatures creates substantial summertime soil moisture decreases in midlatitude midcontinents. Figure 2-7, from Kellogg and Zhao (1988), shows this effect for five different equilibrium GCM runs in which CO_2 doubled. Although details differ from one GCM to another, such dryness is certainly a plausible scenario from these separate calculations.

Unfortunately, all the results on figures 2-3 through 2-7 are for equilibrium CO_2 doublings. In reality, different latitudes would approach equilibrium at different rates, primarily because they include different amounts of land and land warms up faster than ocean. In addition, well-mixed parts of oceans warm more slowly than shallow-mixed layers. Hence, during the transient phase before equilibrium is reached, the warming and other climatic effects induced by the enhanced greenhouse effect could well display time-evolving worldwide patterns significantly different from those inferred on the basis of equilibrium simulations. Furthermore, the social and ecological impacts of climatic changes would probably be greatest during times of rapid change, before equilibrium is reached and before human beings have had a chance to identify and adapt to their new environment.

To represent the transient phase adequately, one would need a time-evolving scenario of trace gas increase to force a three-dimensional model of the atmosphere coupled to a three-dimensional model of the ocean that includes the effects of horizontal and vertical heat transport. A handful of such coupled models have been run, but none that analyzes regional changes over the next century has yet been published. The coupled models are still too uneconomical to be run routinely, and they are not yet trustworthy enough for one to rely on only one or two experimental simulations. Once they have been improved, and the next generation of faster computers becomes available, one will be able to state more confidently how the climatic anomalies caused by rising levels of greenhouse gases might be distributed over space and time.

Changes in Variability

Even more uncertain than changes in regional climatic means, but perhaps most important, are estimates for changes in measures of climatic variability, such as the frequency and magnitude of severe storms and aggravated heat waves and the reduced probabilities of damaging frosts (see Parry and Carter, 1985; Mearns et al., 1984). For example, some modeling evidence suggests that hurricane intensities will increase with climatic warming (Emanuel, 1987). Such issues are just now beginning to be evaluated using equilibrium climate-model results, and they will, of course, also have to be studied for realistic transient cases to be of maximum value to impact assessors.

Variability of temperature and precipitation, and other combinations of climatic factors, can have important implications for the environmental impact of trace gas increases. Since biological entities often exhibit nonlinear behavior (for example, nonlinear responses to frost or heat stress), change in the probability of extreme events can be significant. Table 2-2, from Mearns et al. (1984), uses the example of a change in the probability of extreme heat waves—five or more days in a row with maximum temperature above a 95 degree Fahrenheit (°F) or 100°F threshold—at these U.S. locations if only mean temperature were to increase by 1.7°C (3°F) and there were no changes to the standard deviation of daily temperature maximums nor to the autocorrelation of daily temperature variations. Tables 2-3 and 2-4 show how the GISS GCM standard deviations of surface temperature and precipitation change in response to a greenhouse gas increase scenario and in their control run (current conditions) at several U.S. locations (Rind et al., forthcoming). Observations appropriate to these grid boxes are also given in tables 2-3 and 2-4. Despite considerable noise in the GISS results, qualitatively, at least, they imply that standard deviations of surface temperatures are more likely to decrease than increase as a result of warming caused by radiatively active trace gases—a result which supports the empirical analysis of Mearns et al. (1984, figure 1) using natural climatic forcing factors instead of anthropogenic ones.

Other measures of climatic variability, such as blocking frequency (that is, the likelihood of extended periods of blocked, or persistent, regional climatic anomalies), have

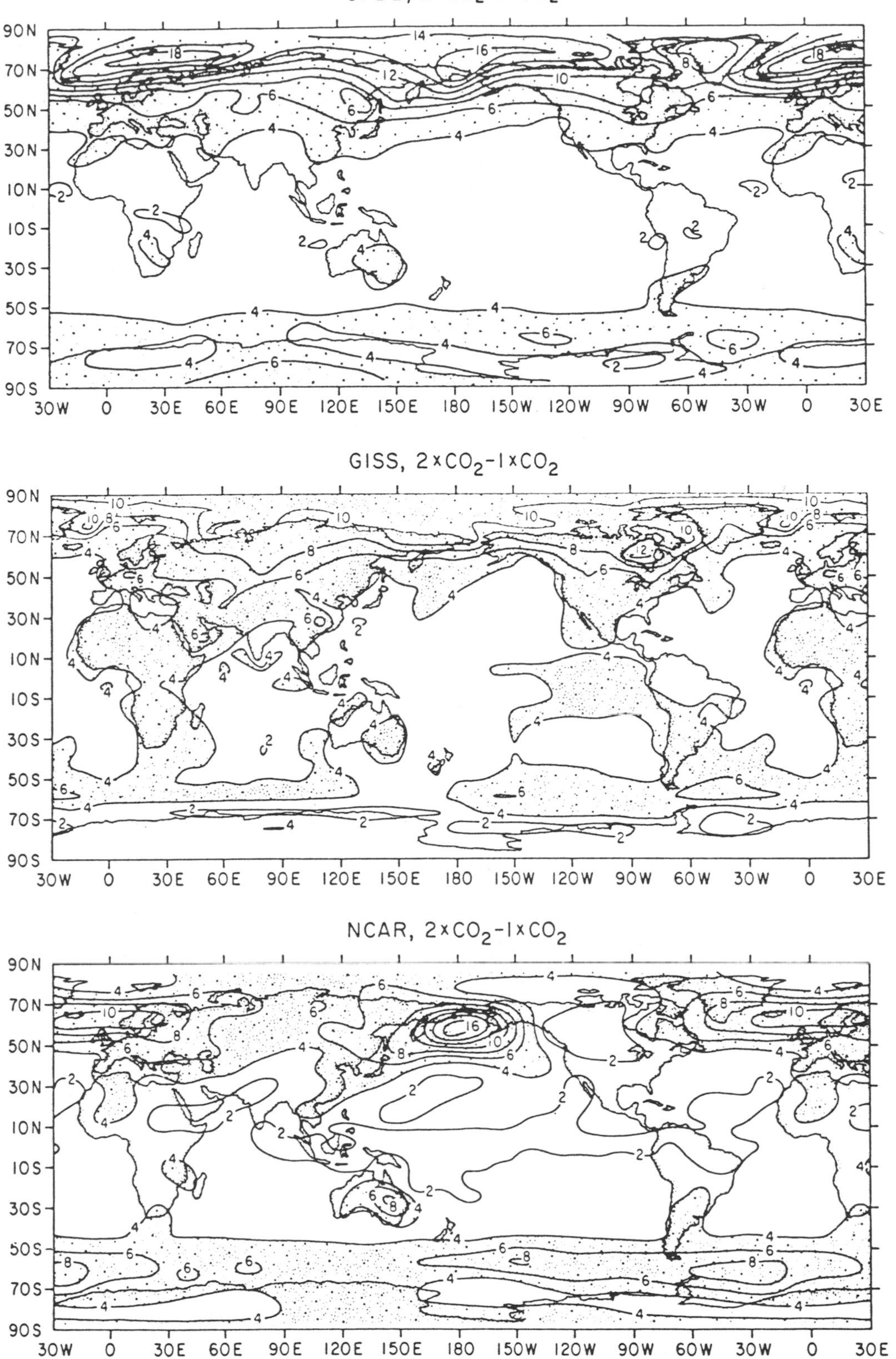

Figure 2-3. Geographical distribution of the surface air temperature change (in degrees Celsius), under conditions of an equivalent doubling of CO_2 and other greenhouse gases minus the preindustrial carbon dioxide concentration (2 × CO_2) − (1 × CO_2) by 2030 for winter (December, January, and February) simulated with general circulation models (GCMs) of (top) the Geophysical Fluid Dynamics Laboratory (GFDL), (middle) the Goddard Institute for Space Studies (GISS), and (bottom) the National Center for Atmospheric Research (NCAR). Stippling indicates a decrease in precipitation rate. For details of which specific GCM versions were chosen, see Schlesinger and Mitchell (1987). (*Source:* Reprinted, with permission, from Schlesinger and Mitchell, 1987, p. 760. © 1987 by the American Geophysical Union.)

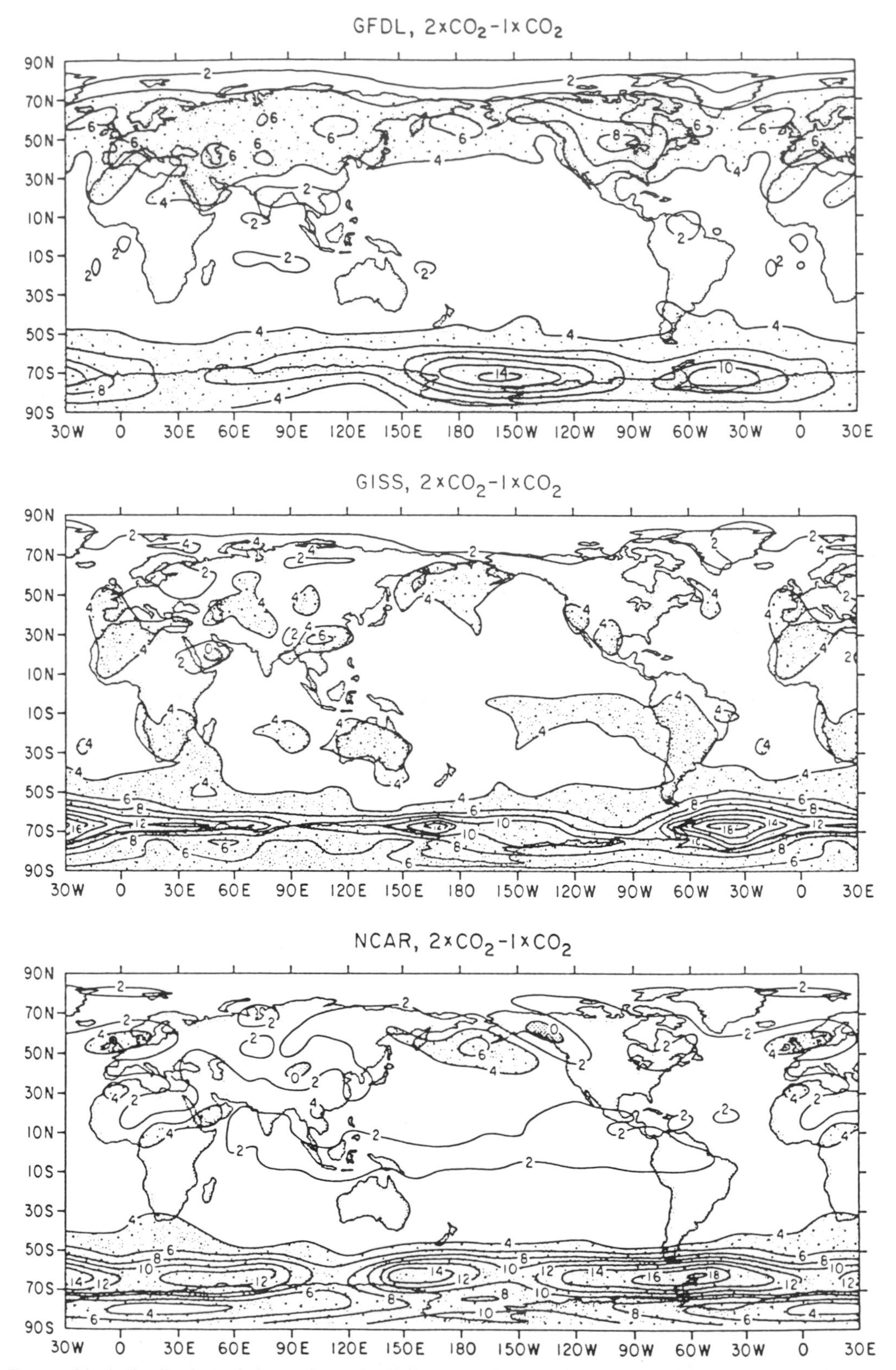

Figure 2-4. Geographical distribution of the surface air temperature change (in degrees Celsius), under conditions of (2 × CO_2) − (1 × CO_2) for summer (June, July, and August) simulated with general circulation models (GCMs) of (top) the Geophysical Fluid Dynamics Laboratory (GFDL), (middle) the Goddard Institute for Space Studies (GISS), and (bottom) the National Center for Atmospheric Research (NCAR). Stippling indicates a decrease in precipitation rate. For details of which specific GCM versions were chosen, see Schlesinger and Mitchell (1987). (*Source:* Reprinted, with permission, from Schlesinger and Mitchell, 1987, p. 760. © 1987 by the American Geophysical Union.)

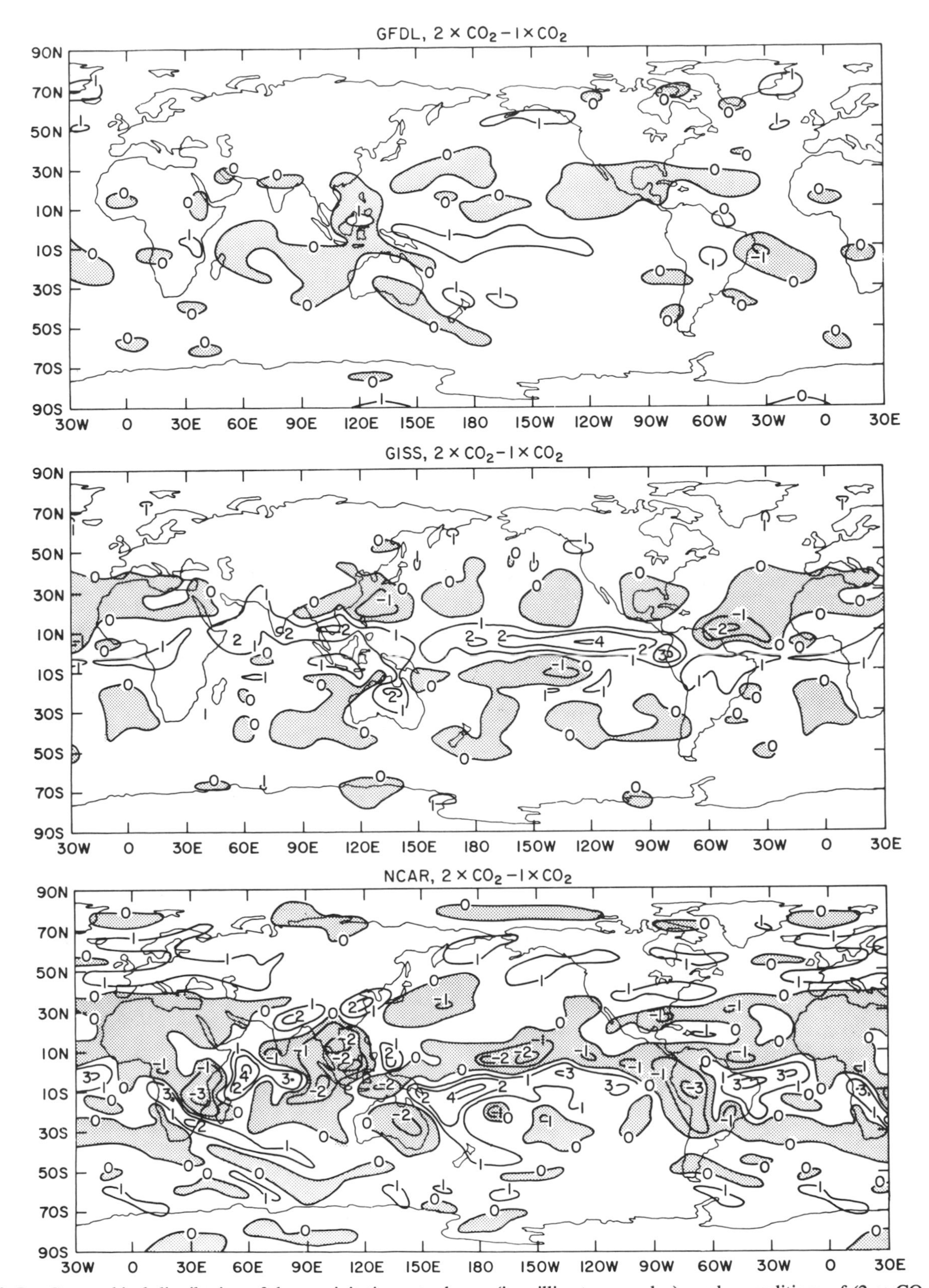

Figure 2-5. Geographical distribution of the precipitation rate change (in millimeters per day), under conditions of $(2 \times CO_2) - (1 \times CO_2)$, for winter (December, January, and February) simulated with general circulation models (GCMs) of (top) the Geophysical Fluid Dynamics Laboratory (GFDL), (middle) the Goddard Institute for Space Studies (GISS), and (bottom) the National Center for Atmospheric Research (NCAR). Stippling indicates a decrease in precipitation rate. For details of which specific GCM versions were chosen, see Schlesinger and Mitchell (1987). (*Source:* Reprinted, with permission, from Schlesinger and Mitchell, 1987, p. 760. © 1987 by the American Geophysical Union.)

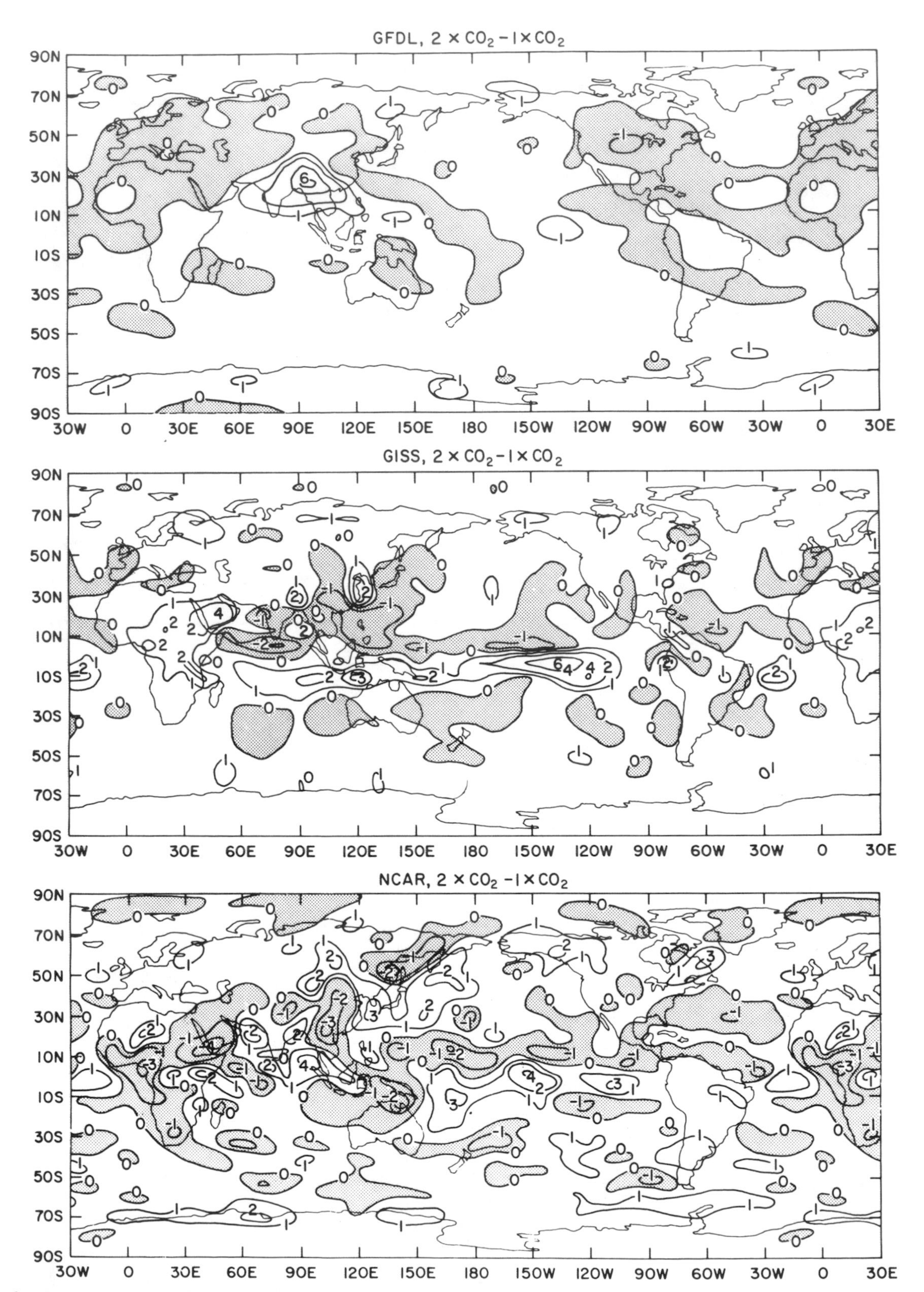

Figure 2-6. Geographical distribution of the precipitation rate change (in millimeters per day), under conditions of $(2 \times CO_2) - (1 \times CO_2)$, for summer (June, July, and August) simulated with general circulation models (GCMs) of (top) the Geophysical Fluid Dynamics Laboratory (GFDL), (middle) the Goddard Institute for Space Studies (GISS), and (bottom) the National Center for Atmospheric Research (NCAR). Stippling indicates a decrease in precipitation rate. For details of which specific GCM versions were chosen, see Schlesinger and Mitchell (1987). (*Source:* Reprinted, with permission, from Schlesinger and Mitchell, 1987, p. 760. © 1987 by the American Geophysical Union.)

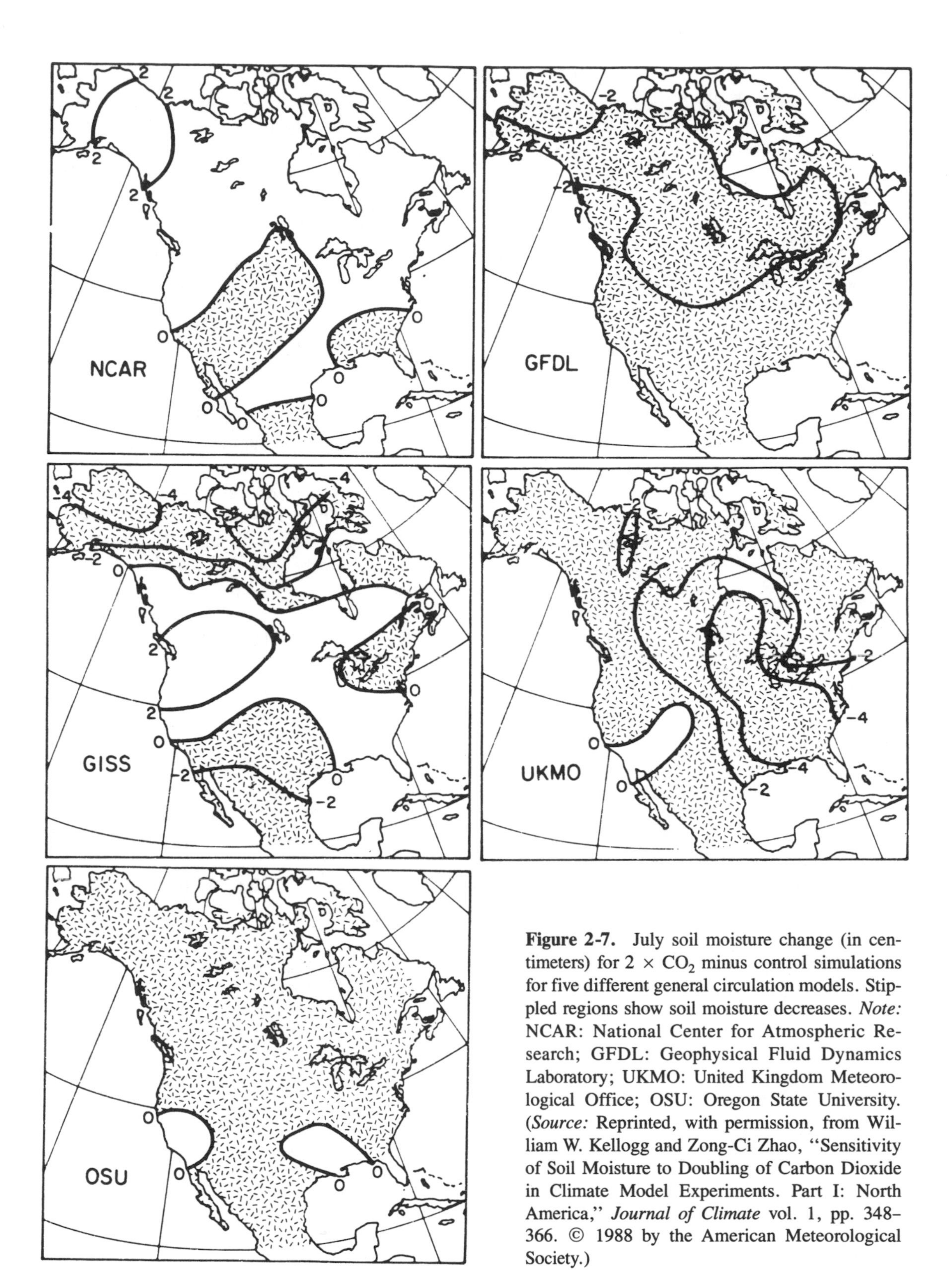

Figure 2-7. July soil moisture change (in centimeters) for 2 × CO_2 minus control simulations for five different general circulation models. Stippled regions show soil moisture decreases. *Note:* NCAR: National Center for Atmospheric Research; GFDL: Geophysical Fluid Dynamics Laboratory; UKMO: United Kingdom Meteorological Office; OSU: Oregon State University. (*Source:* Reprinted, with permission, from William W. Kellogg and Zong-Ci Zhao, "Sensitivity of Soil Moisture to Doubling of Carbon Dioxide in Climate Model Experiments. Part I: North America," *Journal of Climate* vol. 1, pp. 348–366. © 1988 by the American Meteorological Society.)

Table 2-2. Changing Odds of a Heat Wave in Selected Cities if Mean Temperature Increases by 3°F (1.7°C)

City	Maximum temperature	Odds now (%)	Odds if +3°F (%)
Washington, D.C.	95°F	17	47
Des Moines, Iowa	95°F	6	21
Dallas, Texas	100°F	38	68

Note: A heat wave is five or more days in a row at or above the maximum temperature indicated in the first column.

Source: Reprinted, with permission, from Mearns et al. (1984).

been analyzed in a preliminary way (Bates and Meehl, 1986). The results of the NCAR GCM with CO_2 doubled are not very significant with respect to the northern hemispheric scale frequency of blocking events, but the locations of anomaly patterns do seem to shift significantly.

Although analysis of the variability of GCM results is very important for impact assessment, it is just beginning to become part of the research agenda of climate modelers.

Model Validation

Perhaps the most perplexing question about climate models is whether they can ever be trusted enough to provide grounds for altering social policies, such as a decision to embark on new dam construction or to govern the rates and amounts of carbon dioxide emissions. How can models so fraught with uncertainties over details be verified? There are actually several methods. None of them is sufficient on its own, but together they can provide significant (albeit circumstantial) evidence of a model's credibility.

The first method is to check the model's ability to simulate today's climate. The seasonal cycle is one good test because the temperature changes involved are large—several times larger, on a hemispheric average, than the change from an ice age to an interglacial period. General circulation models do remarkably well at mapping the seasonal cycle, which strongly suggests that their thermal sensitivity to large-scale radiative forcing is probably accurate to a rough factor of two. The seasonal test, however, does not indicate how well a model simulates slow processes, such as changes that involve deep ocean circulation or ice cover; these may have important long-term effects and are important to predict transient response reliably. Figures 2-8a through 2-8d offer another example of the performance of a particular GCM (Chervin, 1986) in a seasonal test, in this case the ability to simulate the large seasonal signal of changes in daily variability in the Great Plains. This model does remarkably well for this location (Mearns et al., 1989).

A second method of verification is to isolate individual physical components of the model, such as its parameterizations, and test them against real data from the field. For example, one can check whether the model's parameterized cloudiness matches the level of cloudiness appropriate to a particular grid box. The problem with this test is that it cannot guarantee that the complex interactions of many individual model components are properly treated. The model may be good at predicting average cloudiness but bad at representing cloud feedback. In that case the simulation of the overall climatic response to, say, increased carbon dioxide is likely to be inaccurate. A model should be able to reproduce, to better than, say, 25 percent accuracy, the flows of thermal energy among the atmosphere,

Table 2-3. Standard Deviation (S.D.) of Current Daily Temperature (°C), Modeled S.D. Under Current Conditions, and Changes in Standard Deviation (Δ S.D.) Predicted for the Future in Four Regions of the United States

Month	Location	Observed S.D.	Current S.D.	2010s Δ S.D.	2030s Δ S.D.	~2060 Δ S.D.
Jan.	South Georgia Plains	4.81	8.15	0.61	−1.19	−0.83
	Southeast	4.53	6.90	−0.14	−1.14	−0.23
	West Coast	3.63	5.86	−0.61	0.05	−0.16
	Great Lakes	4.97	5.79	0.44	−0.33	−0.44
April	South Georgia Plains	3.72	5.77	−0.57	−0.27	−0.80
	Southeast	3.71	5.50	−0.65	−1.61	−1.24
	West Coast	2.59	4.29	0.77	0.60	0.33
	Great Lakes	4.65	6.15	−0.51	−0.26	−1.39
July	South Georgia Plains	1.74	2.56	0.54	−0.19	0.18
	Southeast	1.50	2.34	0.14	−0.22	−0.24
	West Coast	2.40	3.56	0.03	0.54	0.28
	Great Lakes	2.38	3.02	−0.48	−0.84	−0.14
Oct.	South Georgia Plains	3.79	5.16	1.16	0.97	1.35
	Southeast	3.59	5.21	−0.54	−0.25	−0.73
	West Coast	3.15	6.51	−0.55	−0.30	−0.80
	Great Lakes	4.09	5.46	−0.37	0.91	−0.06

Source: Rind et al. (forthcoming).

Table 2-4. Standard Deviation (S.D.) of Current Daily Precipitation (mm/day), Modeled S.D. Under Current Conditions, and Changes in Standard Deviation (Δ S.D.) Predicted for the Future in Four Regions of the United States

Month	Location	Observed S.D.	Current S.D.	2010s Δ S.D.	2030s Δ S.D.	~2060 Δ S.D.
Jan.	South Georgia Plains	1.08	2.80	0.05	0.05	1.68
	Southeast	4.35	4.62	−1.20	−1.35	−0.85
	West Coast	3.23	4.55	−0.18	0.34	0.13
	Great Lakes	2.23	4.06	−1.07	−0.94	−0.50
April	South Georgia Plains	2.51	3.26	0.94	1.99	1.17
	Southeast	4.35	3.85	0.95	−0.15	0.81
	West Coast	1.41	2.76	0.07	1.02	−0.12
	Great Lakes	3.85	3.29	−0.43	−0.31	0.44
July	South Georgia Plains	2.79	3.08	−0.10	−0.09	0.36
	Southeast	4.13	3.31	0.28	0.29	0.11
	West Coast	0.57	1.53	0.44	0.24	0.71
	Great Lakes	3.68	2.48	−0.06	0.72	0.35
Oct.	South Georgia Plains	2.75	1.79	0.52	0.34	0.00
	Southeast	3.77	3.88	0.72	−0.15	−0.28
	West Coast	1.86	2.69	1.20	−0.63	1.34
	Great Lakes	3.58	2.26	0.52	0.76	0.95

Source: Rind et al. (forthcoming).

the underlying surface, and space (see figure 2-1). Together, these three elements comprise the well-established greenhouse effect on earth, and they constitute a formidable and necessary test for all climate models. Such a test is an example of physical verification of model components. Most models do well at this test.

For determining overall, long-term simulation skill there is a third method: checking the model's ability to reproduce the very different climates of the ancient earth (see Barron, 1985; Imbrie et al., 1984; Schneider, 1987) or even of other planets. Paleoclimatic simulations of the Mesozoic era, the glacial/interglacial cycle, and other extreme past climates are intrinsically interesting as exercises in understanding the coevolution of the earth's climate with living things. As verification checks on climate models, however, the simulations are also crucial to estimating the earth's climatic and biological future.

Overall validation of climatic models thus depends on a constant appraisal and reappraisal of model performance in the three broad categories described above. Moreover, it is important to check model response to century-long forcings such as the 25 percent increase in carbon dioxide and other trace greenhouse gases observed since the beginning of the industrial revolution. Indeed, most climatic models are sensitive enough to predict that a warming trend on the order of 1°C should have occurred over the past century. The precise model "forecast" of the past 100 years also depends upon how the modeler accounts for such factors as changes in the solar constant and volcanic dust veils (see Schneider and Mass, 1975; Bryson and Dittberner, 1976; Gilliland and Schneider, 1984; or Hansen et al., 1981). Indeed, as figure 2-9 (Hansen et al., 1981) shows, the typical model prediction that we should already have experienced about 1°C of warming is broadly consistent but somewhat larger than that observed over the past 100 years. Possible explanations for the discrepancy include the following (see Gilliland and Schneider, 1984):

1. Climate models' sensitivity to trace gas greenhouse increases has been overestimated by about a factor of two.
2. Modelers have not properly accounted for known competitive external forcings such as volcanic dust veils, natural stratospheric aerosols, or changes in solar energy output.
3. Modelers have not accounted for other external forcings that might have occurred, such as regional-scale tropospheric aerosols associated with human agricultural and industrial activity.
4. Modelers have not properly accounted for the very large heat capacity of the oceans, which could slowly take more of the added greenhouse effect heating than present models suggest, thereby further delaying the signal.
5. Sensitivity of the present generation of models and observed climatic trends could both be correct, but since the models are typically run for the equivalent of carbon dioxide doubling and the actual world has experienced only a quarter of this increase, nonlinear processes have been properly modeled and have produced a sensitivity appropriate for doubling but not for 25 percent increase.

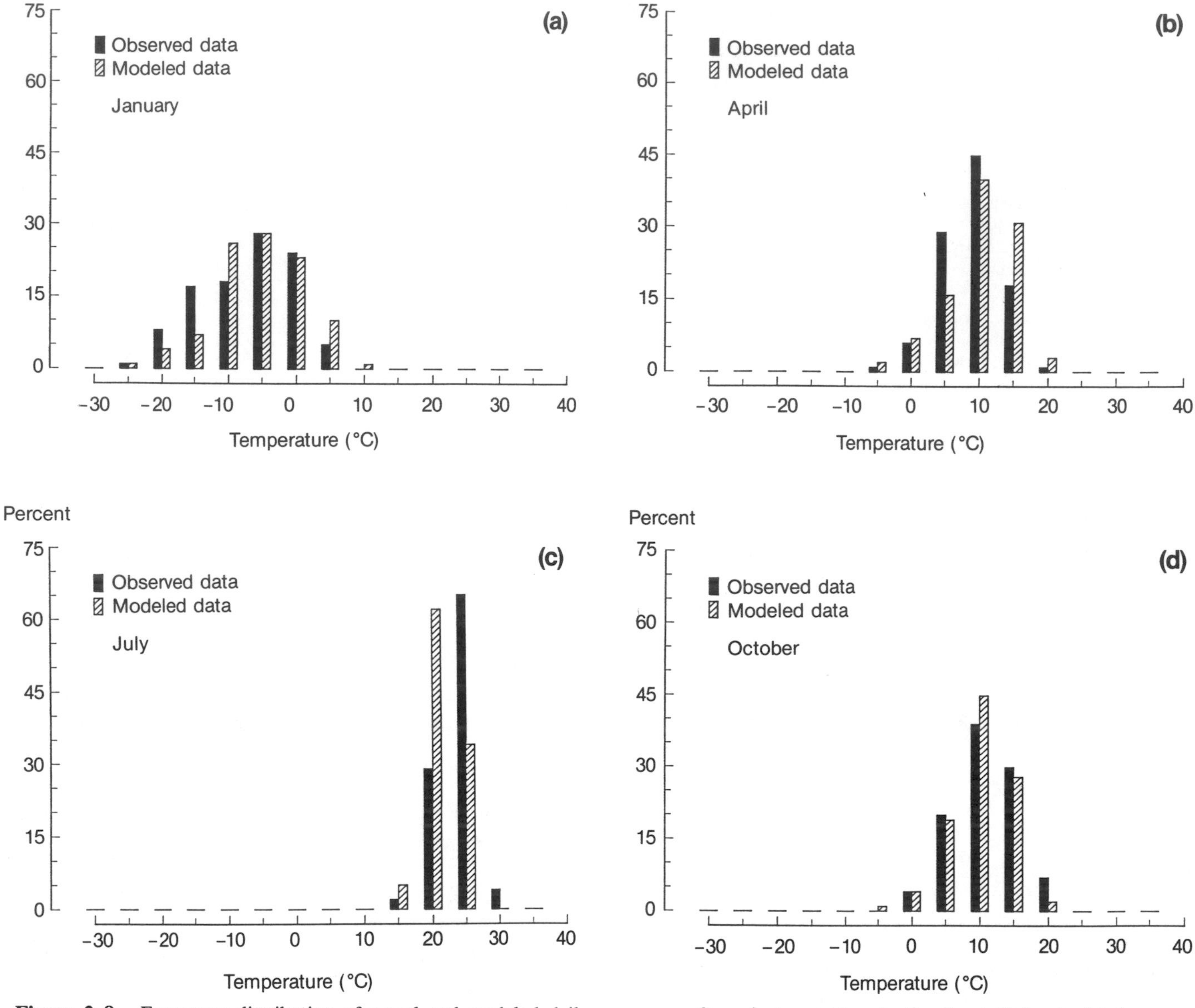

Figure 2-8. Frequency distribution of actual and modeled daily average surface air temperature in the Great Plains in (a) January, (b) April, (c) July, and (d) October. (*Source:* Mearns et al., 1989.)

6. Estimates of global climatic warming from the incomplete, nonhomogeneous network of thermometers has mis-estimated the actual warming that has occurred.

Despite these possible explanations for the factor of two or so difference between observed global temperature trends in the past century and those anticipated by most state-of-the-art GCMs, such rough agreement to a factor of two is still quite close, considering the remaining uncertainties. The reason that most climatologists do not yet proclaim the observed temperature increase to have been caused by the greenhouse effect (in other words, proof that a greenhouse effect signal is clearly detected in the climatic record) is that there is still a significant probability that the observed trend and the predicted warming are chance occurrences. Other factors, such as solar constant variations or stratospheric dust, simply have not yet been adequately accounted for—except over the past ten years or so, when adequate instrumentation has been measuring them. Nevertheless, the empirical test of global temperature trend versus model predictions certainly is consistent to a rough factor of two. Taken together with other circumstantial evidence (such as the excellent performance of most climatic models on the very tough seasonal cycle test, the ability of models to reproduce vastly different ancient paleoclimates, the ability of models to reproduce very hot conditions of

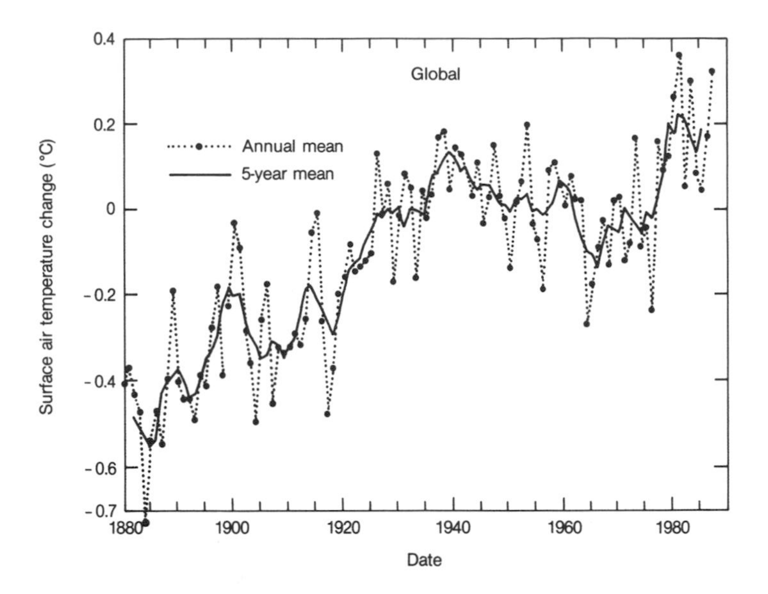

Figure 2-9. Global surface air temperature calculated from land stations over the past century. (*Source:* Reprinted, with permission, from J. Hansen et al., "Climate Impact of Increasing Atmospheric Carbon Dioxide," *Science* vol. 213, pp. 957–966. © 1981 by the AAAS.)

Venus and cold conditions of Mars, and the ability of the present generation of models to capture many of the important and diverse features of the present distribution of climates on earth), the case for the occurrence of rapid climatic change over the next several decades is strong. Of course, another decade or two of observations of climatic trends should lead to signal-to-noise ratios sufficiently obvious that almost all scientists will agree whether present estimates of climatic sensitivity to increasing trace gases have been predicted reasonably well or not. In essence, while scientists debate, the real climate system continues to perform the experiment for us.

BIOSPHERIC RESPONSES TO CLIMATE CHANGE: SOME EXAMPLES

The ways in which vegetation is distributed throughout the world's managed and unmanaged ecosystems depend upon climate and soil conditions. Soil formation, according to Jenny (1941), depends on a number of factors—among them the nature of the parent material, the length of time during which soil formation takes place, the vegetation, and the climate. Clearly, then, climate is critical in determining what grows where.

To illustrate the sensitivity of vegetation to climatic conditions, we present but two examples—one from a paleoclimatic reconstruction and the other based on modern records.

Paleoclimatic Example

The advance and retreat of the glaciers in North America were accompanied by geographic shifts in the distribution of dominant tree species. Figure 2-10 is taken from the work of Bernabo and Webb (1977), who used pollen analysis to reconstruct the changing climate of North America during the past 11,000 years. Spruce, adapted to cooler climates, and oak, adapted to warmer climates, moved northward as the glacier ice retreated. Eleven thousand years before the present (B.P.), eastern North Carolina was the site of a maximum concentration of oak pollen. Spruce was then found in northern Illinois and Indiana and as far south as Virginia. Today oak is to be found up to about the Canadian border, and spruce is concentrated in central Canada reaching to the tundra.

Modern Example

The Great Plains of North America, once equated with the "great American desert," is one of the world's major breadbaskets. Wheat production there has increased during this century from about twelve bushels per acre (0.81 tons per hectare) in 1925 to about thirty bushels per acre (2.02 tons per hectare) today, largely as a result of improved production technologies. New cultivars have been bred, fertilizer use has increased, and mechanization has facilitated seedbed preparation, application of fertilizers and herbicides, and harvesting.

Yet the upward yield trend has been interrupted a number of times since 1925. The first very serious drop in yields, shown in figure 2-11a (National Academy of Sciences, 1976), was clearly related to the 1930s drought which extended over most of the Great Plains states. The next such reduction occurred during the 1950s. The importance of the 1950s drought, which affected primarily the southwestern portion of the Great Plains, is more ambiguous. In fact, the onset of yield depression appears to have anticipated the drought, although yields remained depressed through the drought period. The overall impact in terms of

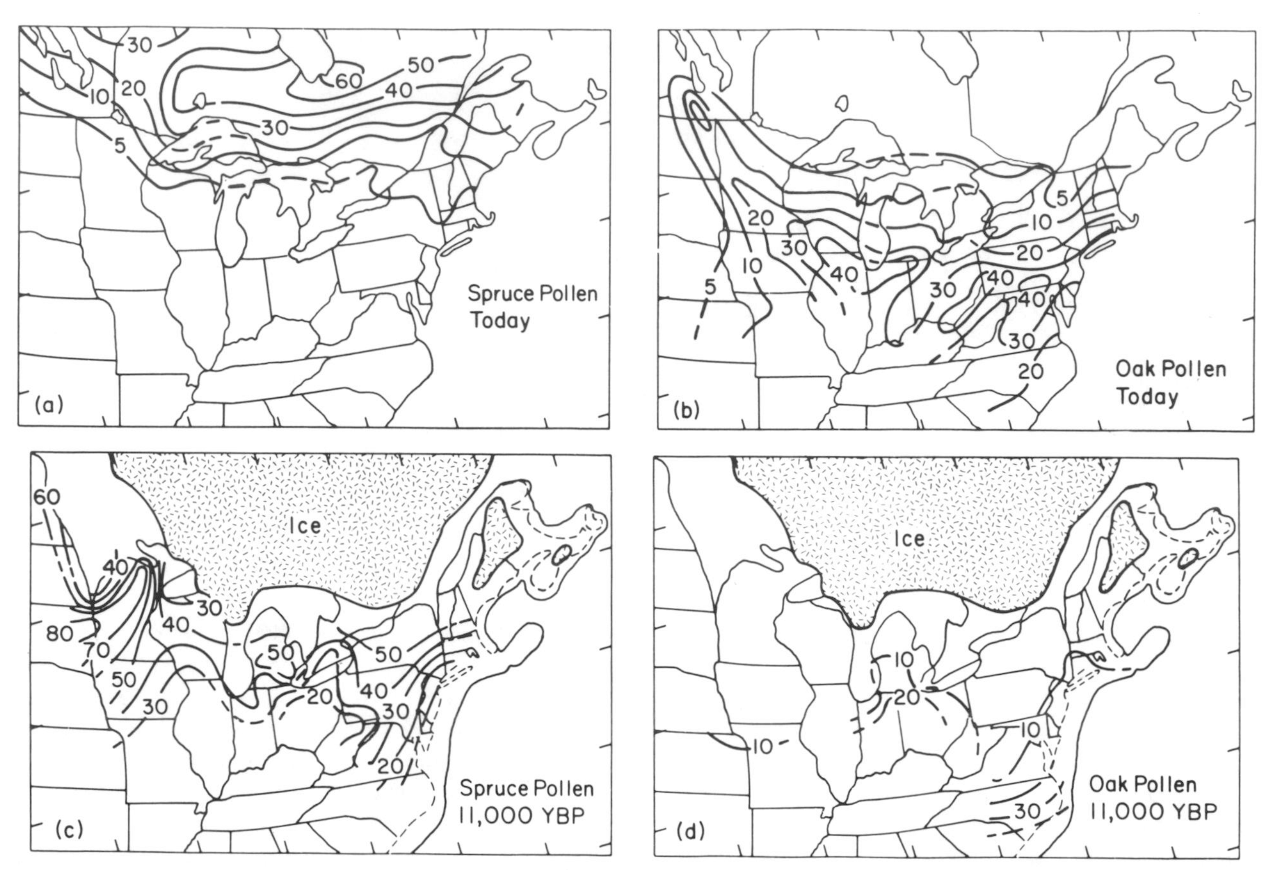

Figure 2-10. The relative fraction of spruce pollen (a and c) and oak pollen (b and d) in eastern North America today and at the end of the last Pleistocene (Ice Age) 11,000 years before the present (YBP). (*Source:* Reprinted, with permission, from J. C. Bernabo and Thompson Webb III, "Changing Patterns in the Holocene Pollen Record of Northeastern North America: A Mapped Summary," *Quaternary Research* vol. 8, pp. 64–96. © 1977 by Academic Press.)

lost production and social and economic disruptions was less severe than in the 1930s. Figure 2-11a ends with 1969. The record of Great Plains wheat yields is extended from 1969 to 1986 in figure 2-11b. Wheat yields had reached an apparent plateau of about thirty bushels per acre although yields appear to have risen somewhat in the first half of the 1980s. A serious drop in yield occurred during the period 1974–76. Elsewhere, Rosenberg (1986) has detailed the specific sequence of events that led to this sharp loss in absolute per acre yield. A short but intense drought explains much of the reduction in the Great Plains wheat crop during the mid-1970s.

Warrick and Bowden (1981) have proposed a "lessening hypothesis" to explain the apparent decrease in the vulnerability of Great Plains wheat production to successive droughts. They suggest that the many government programs put into effect after the 1930s have tended to insulate the wheat growers, agribusinesses, and dependent rural and urban people from the most severe economic stresses of drought. This view may be correct. However, as figure 2-12 from an atlas of the Palmer Moisture Anomaly Index (Karl and Knight, 1985) shows, the geographical distribution and severity of the three major twentieth-century droughts have differed greatly. Their duration also differed considerably: of the three, the drought of 1974–76 was also the shortest. However, observe the map of the Palmer Drought Severity Index (related to, but not identical with, the anomaly index) for early September 1988, shown in figure 2-13. Interestingly, the geographical distribution of drought effects in 1988 was quite similar to that in 1936. Whether the drought of 1988 will continue into 1989 or reoccur with equal severity within the next few years is, of course, unknown.

No drought in the Great Plains since the 1930s has been as severe and protracted nor has posed so serious a test of the region's productive, economic, and societal capacity to cope with drought. As there is no reason to believe that a drought of equal severity cannot occur again, the question is not simply academic. More to the point, however, is the fact that some current state-of-the-art GCMs predict that, as the result of greenhouse warming, droughts may become more frequent, more severe, or longer lasting, or all three, in the midlatitudes of Northern Hemisphere continents (for example, see figure 2-7; Manabe and Wetherald, 1986; Schlesinger and Mitchell, 1985; or Rosenberg, 1987).

POSSIBLE IMPACTS OF THE GREENHOUSE EFFECT ON NATURAL RESOURCES: SOME EXAMPLES

As shown above, the various global climatic models and other surrogate means of predicting climatic change do not always yield consistent results. Our review and analysis of these predictions lead to an ordering in terms of the decreasing likelihood of predicted climatic phenomena, to wit: a global warming and precipitation increase, warming in the high latitudes, midsummer drying in the midlatitudes of Northern Hemisphere continents, increased monsoonal rainfall, and alterations in soil-water storage and runoff as a result of changes in evapotranspiration. Here we list and briefly describe a few of the possible impacts of projected climatic changes upon specific natural resource sectors. We illustrate each possible change by reference to a specific impact. In other chapters of this volume, these and other possible impacts of greenhouse warming are dealt with in considerably greater detail.

Warming in the High Latitudes/Changes in the Boreal Forest

There is relatively strong agreement among GCM predictions that, at least in equilibrium, warming will be greatest in the high latitudes (see figures 2-3 and 2-4). Year-round warming and recessions of the permafrost might open large areas of tundra to colonization by boreal tree species. At the southern margins of the boreal forests, competition with more temperate species or with grasses, or both, might cause a northward retreat of the forest. Changes in species composition or biomass production are also considered likely (see Sedjo and Solomon, chapter 8 in this volume, and Batie and Shugart, chapter 9 in this volume). However, the climate could change more rapidly than the forests could respond. This would put the forests out of equilibrium with the changing climate, making specific time-evolving perturbations difficult to accommodate.

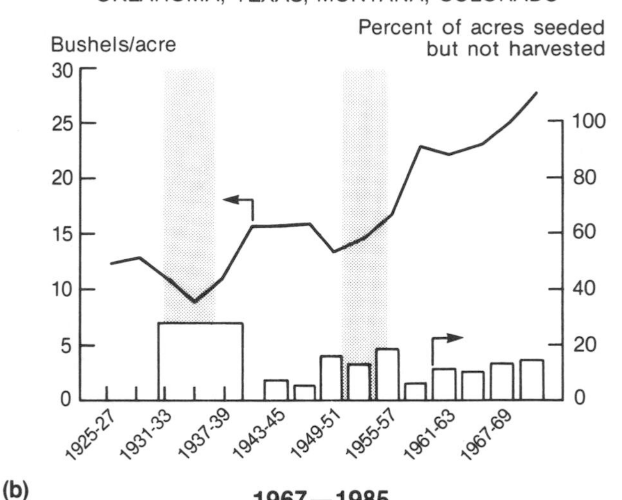

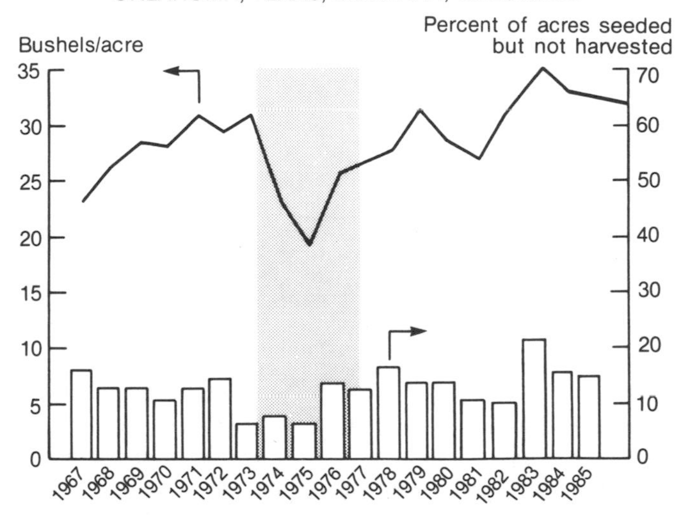

Figure 2-11. Average wheat yields and acreage abandonment in the Great Plains states for (a) 1925–69 and (b) 1967–85. (*Sources:* [a] reprinted, with permission, from National Academy of Sciences, *Climate and Food,* © 1976 by National Academy Press, Washington, D.C.; [b] adapted, with permission, from C. Ford Runge, ed., "The Future of the North American Granary, © 1986 by Iowa State University Press, Ames.)

Midsummer Drying in Midlatitudes/Changes in Midwestern Corn Belt

Maize is a tropical grass that grows best in a temperature range of about 30°C to 33°C (Loomis and Williams, 1963). In very general terms one can say that the northern limit of corn culture in North America today is determined by length of growing season; the western limit is determined by soil-water availability; and the southern limit, by the occurrence of excessively high temperatures during the reproductive period. In the United States, corn has no eastern limit except that imposed by topography and economic advantages favoring other crops.

Most analyses of the impacts of global warming on the corn belt are based on climate change scenarios that assume increasing temperature with or without small changes in precipitation. Such analyses (for example, Newman, 1982; Blasing and Solomon, 1982; Decker and Achutuni, 1988) predict a northeastern movement of the corn belt into

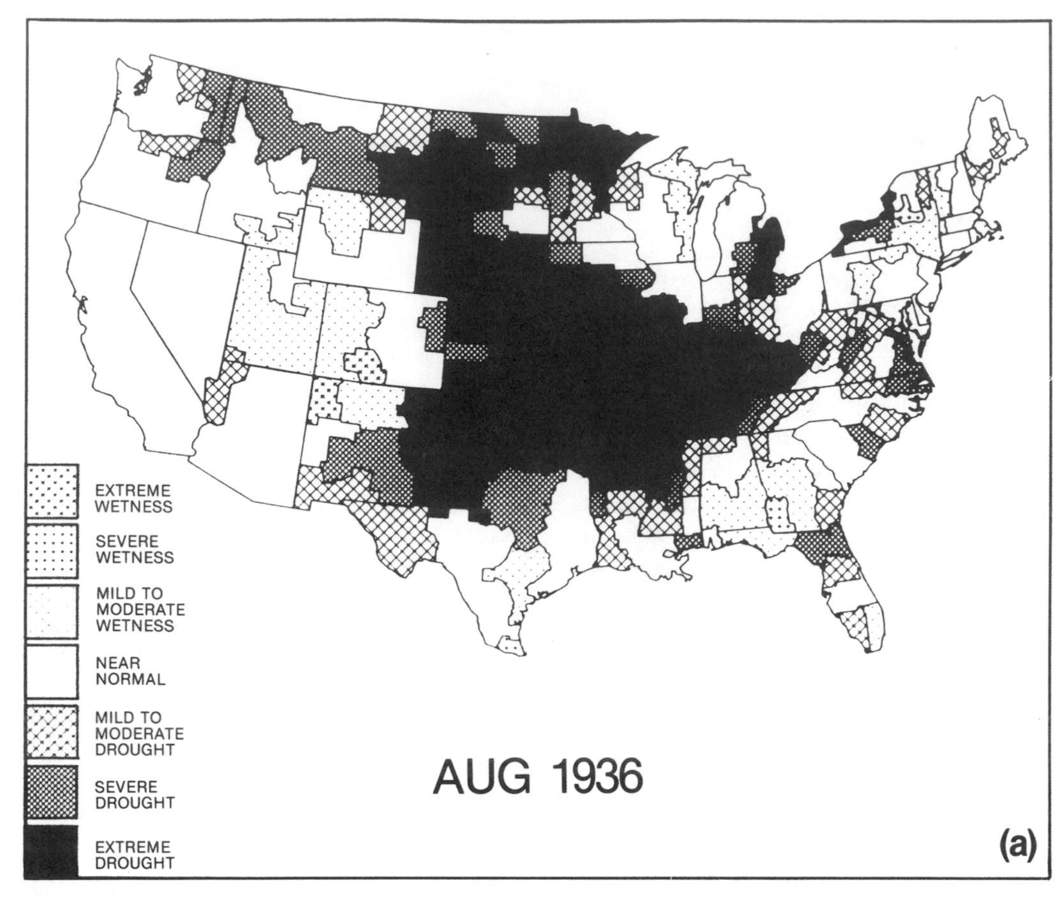

Figure 2-12. Palmer Moisture Anomaly Index (PMAI) for the contiguous United States in (a) August 1936, (b) August 1956, and (c) August 1976. (*Source:* Karl and Knight, 1985.)

northern Minnesota, Wisconsin, and Michigan, and into southern Ontario, as these regions warm and their growing seasons lengthen. Other crops such as sorghum and wheat or double cropped wheat-soybean or wheat-sorghum systems could possibly gain in importance in the western and southern margins as the current corn belt recedes. Winter wheat could also move north into present spring wheat regions. How pests and pathogens will react to such changes is speculative. Easterling, Parry, and Crosson (chapter 7 in this volume) explore these questions more fully and propose alternatives for analysis of the impacts of climatic change on U.S. agriculture.

Northward Advance of Monsoonal Rainfall/Greening of the Sahel

GCMs predict that, in a greenhouse-warmed world, the intertropical convergence zones (ITCZs) will advance further northward into Africa and Asia during summer because land surfaces could warm up more in the pre-monsoon season than would the nearby oceans. This would enhance the ocean continent pressure gradient that drives monsoon circulation. Of course, such a conclusion is based on GCM equilibrium experiments and could be altered in a realistic transient calculation.

If such an advance occurs, regions such as the African Sahel will receive more rain. Pasturage would be improved and longer lasting. Rain-fed agriculture might advance further northward and, in the zones where it is currently practiced, might become less risky. If storage is available for that water, greater rainfall in the Sahel could increase the availability of water for irrigation. More reliable monsoon rainfall in India would certainly be welcome, but this prospect must be balanced against the potential for more catastrophic floods.

Actual evaporation would also increase in such regions for two reasons. First, there would be more water to evaporate, and the soil surface would be wetter for long periods of time. Second, temperatures would rise in the tropics (although less there than in the higher latitudes), and evaporation rate is directly related to air and surface temperatures. During the monsoon season, however, greater humidity and increased cloudiness would moderate evaporation. During the nonrainy season, evaporation rates are

Figure 2-12 (continued)

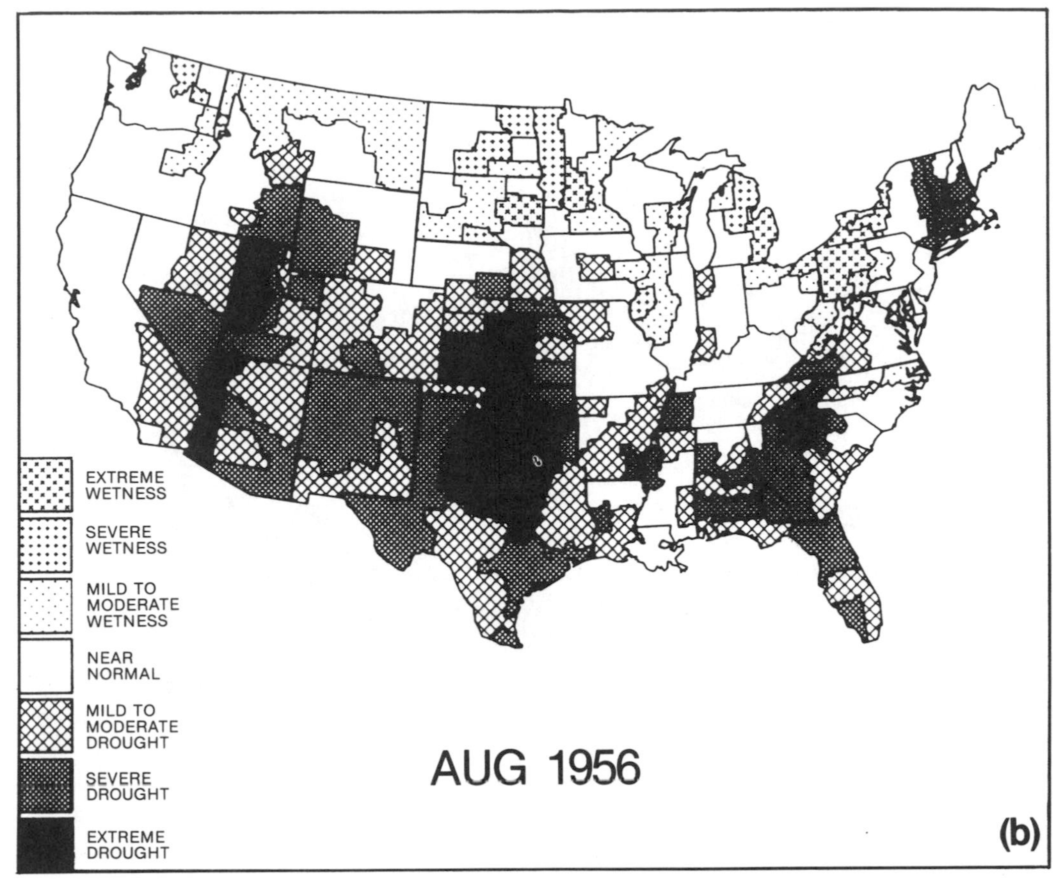

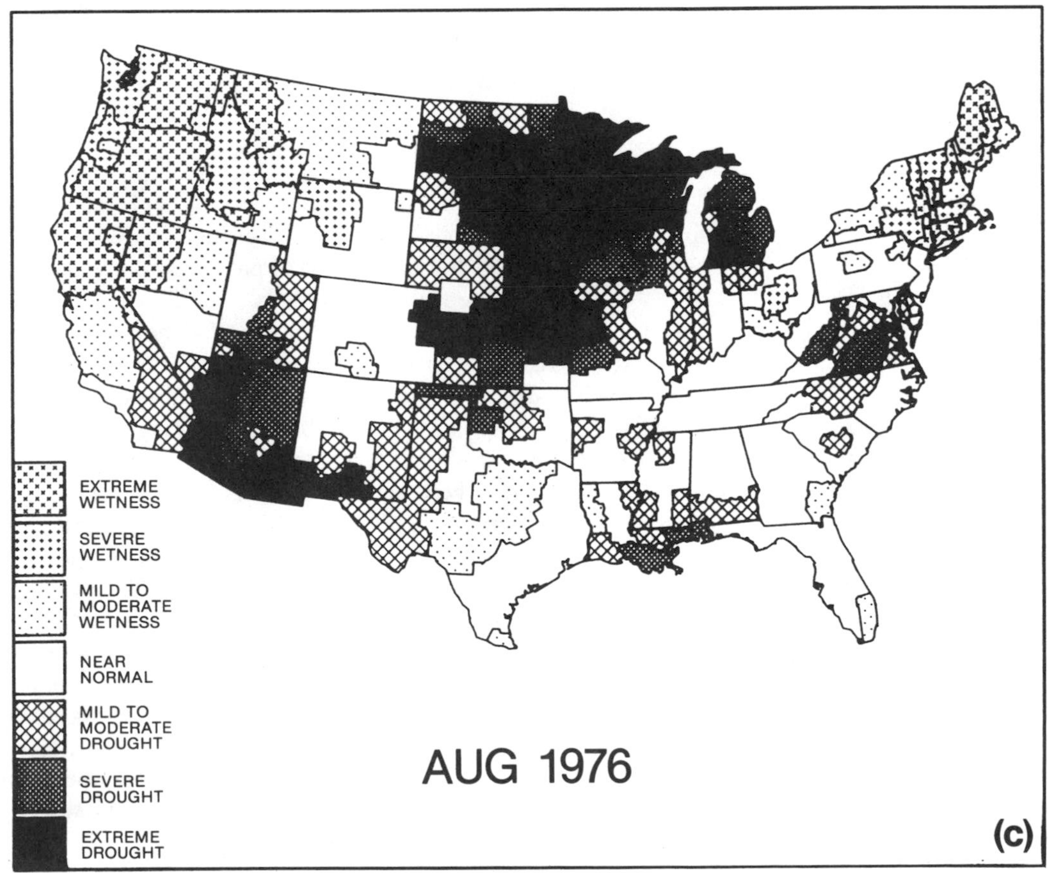

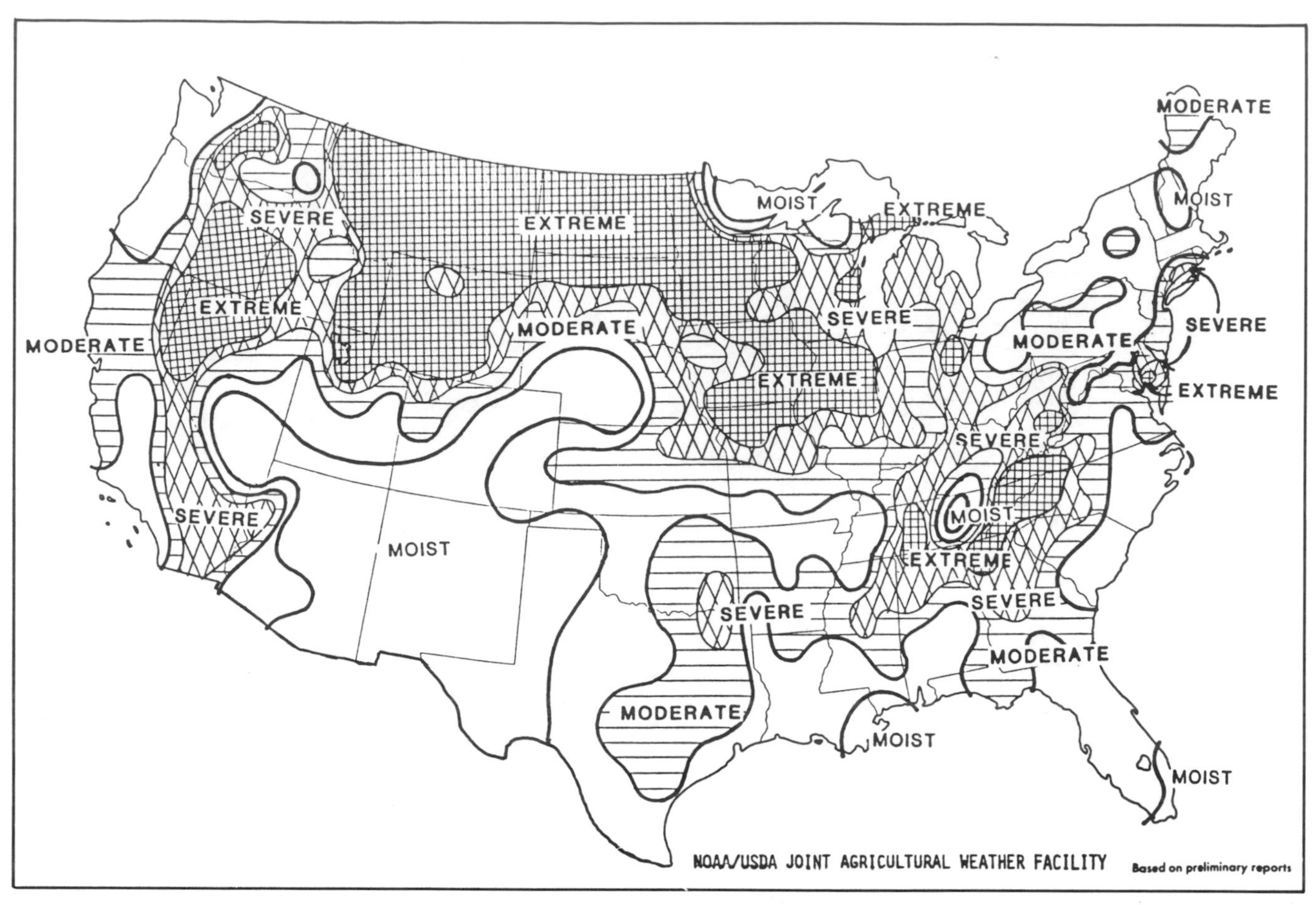

Figure 2-13. Palmer Drought Severity Index for the contiguous United States as of September 3, 1988. (*Source:* U.S. Department of Commerce and U.S. Department of Agriculture, 1988.)

likely to increase as a result of higher temperatures—but by amounts that are not easily predicted, as the following section will show.

Changes in the Hydrologic Cycle/Availability of Water Supplies

Intensification of the hydrologic cycle on a global average basis is predicted by all GCMs as an outcome of surface warming from the greenhouse effect. Evaporation in the models increases globally by perhaps 7 to 12 percent for an effective doubling of the preindustrial CO_2 concentration. Since the atmosphere cannot store large quantities of water vapor, precipitation will return the excess to the surface. But this increase in precipitation will not be uniformly distributed: some regions will get more and some less than at present. The interannual, seasonal, and daily variability may also change, as could storm frequency and intensity and the duration of rainfall events within a given season. Further, the proportions of the precipitation deposited in the form of rain, snow, and dew may change.

Even if all the above-mentioned uncertainties concerning precipitation in a climate-changed world were removed, it would still be difficult to predict just how the hydrology of any particular region would be affected. The amounts of water that are stored in soil for the support of crops and the amounts that run off into streams and reservoirs are determined by the excess of precipitation over evapotranspiration, and uncertainties regarding evapotranspiration in a climate-changed world are also great.

For the most part, analysts considering the potential impacts of climatic change on water resources have treated evapotranspiration as a function of temperature change alone. Revelle and Waggoner in a 1983 study by the National Academy of Sciences calculated that a 2°C temperature increase would decrease runoff in the Colorado River Basin by 29 percent, a 10 percent decrease in precipitation would decrease runoff by 11 percent, and a combination of these climatic changes would reduce runoff by 40 percent. A 10 percent increase in precipitation would not offset a 2°C increase; runoff in this case would still decrease by 18 percent.

Gleick (1987) calculated the water balance of the Sacramento Basin in a similar fashion, considering changes in temperature of plus or minus 2°C and plus or minus 4°C and changes in precipitation of plus or minus 10, 20, 40, 60, and 80 percent. He found that runoff in winter increases because more of the precipitation falls in the form of rain and less as snow and because the snow melts earlier in the spring. In summer, his simulation shows a greatly decreased runoff because of higher temperatures, even with as much as a 20 percent increase in precipitation. It may turn out, if such analyses are correct, that the greatest impact of greenhouse warming on natural resources will occur because of changes in the seasonality and amounts of precipitation and of evapotranspiration.

Since the possible changes in evapotranspiration are so profound and the process depends not only on temperature but also on other climatic factors that may change and on the direct responses of plants to elevated carbon dioxide concentration in the atmosphere, this subject is explored below in greater detail.

BIOLOGICAL RESOURCES AND DIVERSITY

There is great, and justified, concern today that the rate at which biological resources are being lost is increasing rapidly. Large areas of tropical forest in Amazonia have already been converted to agriculture and ranching, and more are threatened with conversion (Malingreau and Tucker, 1988). Such forests are particularly rich in the numbers and diversity of plant and animal species. Tropical forests are also being heavily exploited in Southeast Asia, Africa, and Central America (Repetto, 1988). The question we introduce here is whether prospective greenhouse warming will alter in any way current trends in species loss.

Tropical deforestation must add significant amounts of CO_2 to total annual CO_2 emissions into the atmosphere, but how much is uncertain. Detwiler and Hall (1988) have found estimates to range from less than 1 gigaton (Gt) to more than 2 Gt of carbon per annum (1 Gt is equal to 10^{15} grams). They believe that the smaller estimates are more likely. Methane emissions following deforestation are also measurable (Crutzen et al., 1979). Deforestation, therefore, contributes to greenhouse warming.

Most models show relatively minor temperature changes in store for the tropical regions (see figures 2-3 and 2-4), but precipitation could change markedly (see figures 2-5 and 2-6). The extent to which these possible changes are likely to be beneficial or detrimental to individual species and to the ecological stability of these regions is not yet apparent.

The tropical forests, boreal forests, and deserts of the world occupy vast and, in many instances, continuous areas. The risk of species loss due to climatic change alone is mitigated somewhat by their very vastness, especially if we believe that, at the outset, climatic change will be felt at the margins of such regions. However, isolated biological reserves supporting species already facing extinction may be exposed to rapidly changing conditions sufficiently different from the present that the species' destruction will be hastened (see Peters and Darling, 1985), particularly if the reserves are small or their conservation is not well enforced. In this respect, at least, climate change could indeed lead to reduced diversity. But this matter clearly requires considerably more analysis than we are able to give it here.

DIRECT EFFECTS OF CO_2 ENRICHMENT ON PLANT PHYSIOLOGICAL BEHAVIOR

One of the greenhouse gases, CO_2, directly affects plant growth and water use. Consequently, its increasing concentration in the atmosphere complicates our efforts to predict the future effects of greenhouse-induced climatic change on crops, forests, and unmanaged ecosystems.

Elevated CO_2 concentration in the atmosphere increases the rate of photosynthesis. Photosynthesis and respiration can be written as:

$$CO_2 + H_2O \leftrightarrows CH_2O + O_2$$

where the reaction to the right is the light-dependent photosynthesis that forms organic compounds and the leftward reaction is respiration, which forms CO_2 by the oxidation of organic compounds. The presence of elevated CO_2 concentrations increases the photosynthetic reaction rate, but has little effect on the respiration reaction rate.

Effects of CO_2 Concentration on Photosynthesis and Growth

There are three major categories of higher plants, each with its own photosynthetic mechanism: the C_3, C_4, and CAM groups. The C_3 photosynthetic pathway is relatively inefficient in that a considerable portion of the primary photosynthetic product is lost through a respiratory mechanism that is itself linked with photosynthesis (photorespiration). When air in growth chambers or greenhouses is enriched with carbon dioxide, the C_3 group in particular (small grains, legumes, root crops, most trees) responds with a significantly increased rate of photosynthetic fixation of carbon. The C_4 group (tropical grasses such as maize, millet, and sorghum) has evolved another photosynthetic pathway and a special anatomy which facilitates efficient photosynthesis. C_4 plants possess a CO_2-concentrating mechanism that nullifies the photorespiratory losses of carbon dioxide that account for the lower efficiency of C_3 plants. The C_4 plants make efficient use of CO_2 even at current ambient atmospheric levels. Consequently, their photosynthetic response to CO_2 enrichment is not as marked as in C_3 species. For the purposes of this discussion, the CAM group—usually succulent desert plants—can be neglected. A detailed explanation of the biochemistry of photosynthesis and photorespiration is given by Beer (1986).

Effects of CO_2 Concentration on Stomatal Behavior

When plants are subjected to elevated CO_2 concentrations, their stomates—the pores through which water vapor and carbon dioxide pass—close partially. The partial closure is related to CO_2-induced changes in photosynthesis (Raschke, 1986; Morison, 1987), but the controlling mechanism is poorly understood. Increased CO_2 concentration in the air around a leaf increases the substomatal CO_2 concentration. This concentration change probably triggers the processes that lead to stomatal closure. Whatever the cause, stomatal closure increases the resistance of the leaf to the passage of vapor into the atmosphere (the stomatal resistance). Partial stomatal closure impedes photosynthesis, but to a much smaller degree, for reasons that are explained in a number of sources (for example, Rosenberg et al., 1983) but need not concern us here.

The consequences of stomatal closure for stomatal resistance and transpiration have been well documented. Kimball and Idso (1983) reviewed nine CO_2 enrichment experiments involving forty-six observations on eighteen species including both C_3 and C_4 crops, herbaceous weeds, and woody plants in which either transpiration or stomatal resistance was measured. The doubling of CO_2 concentration led to an average reduction in transpiration rate of 34 percent. In another review of twenty-one experiments with fifty-five observations and eight C_3 and C_4 crop species, Cure (1985) computed an increase of 51 percent in stomatal resistance. In a third review of twenty-three reports covering eighty observations on twenty-five species, Morison (1987) found a 67 $\pm$ 14 percent increase in stomatal resistance for a doubling of CO_2 concentration from 330 to 660 ppmv. Thus, these reviews suggest that a doubling of CO_2 will increase stomatal resistance by about 57 $\pm$ 14 percent, on the average.

It has often been stated that the stomata of C_4 plants are more sensitive to CO_2 concentration than are those of C_3 plants. However, reviews of Kimball and Idso (1983), Cure (1985), and Morison (1987) fail to reveal any significant differences between C_3 and C_4 species. The impact of a doubled atmospheric CO_2 concentration on stomatal conductance should not be greatly different on the average for C_3 and C_4 species. In fact, the variation in stomatal resistance within a single species, due either to leaf age or past irrigation history, appears greater than variation occasioned by CO_2 enrichment between representative C_3 and C_4 species.

EFFECTS OF CO_2 ENRICHMENT ON NET PRIMARY PRODUCTIVITY

Will the increasing CO_2 concentration in the atmosphere and the climatic changes attendant upon the greenhouse effect alter the net productivity of managed and unmanaged ecosystems? One way to answer this question, perhaps the most direct way, would be to determine whether the terrestrial biosphere is already assimilating more carbon in response to the increased accumulation of atmospheric carbon over the last 100 years or so. Of the 5 to 6 Gt of carbon humans currently inject annually into the atmosphere in the form of CO_2, about 3 Gt remain. That which is drawn out of the atmosphere is believed to be captured by the oceans or the terrestrial biosphere or both. As Dahlman (1988) points out, there are many difficulties in determining whether overall global biotic activity is increasing in response to CO_2 enrichment and whether, as a result, the biosphere is sequestering some of that missing CO_2. Detwiler and Hall (1988) point out that, depending on the analysis, from -0.3 to $+2.8$ Gt of carbon are yet to be accounted for in attempts to balance the annual carbon cycle. And as the climate changes, carbon stored in high-latitude tundra or bogs could be released, further complicating the projection of CO_2 concentrations.

Direct CO_2 Effect

Because there is little direct evidence of change in biotic activity, and even that evidence is controversial, we are forced to rely on the results of growth-chamber and greenhouse studies. Based on a review of hundreds of such studies, Kimball (1983a, 1983b, 1986) has estimated that, on average, a doubling of CO_2 concentration will increase growth and yield by about 34 percent ($\pm$6 percent; 95 percent confidence band) in C_3 plants and about 14 percent ($\pm$10 percent) in C_4 plants. Most of these studies are for food and fiber crops, but a small number dealing with woody plants show similar biomass increases. Unless climatic changes or pests and pathogens somehow stimulated by climatic change interfere, we expect that CO_2 enrichment of the atmosphere should increase growth and yield in most annual plants and biomass accumulation in most perennial species.

Climatic Change Effects

It is important to recognize that CO_2 enrichment of the atmosphere will not occur in a static climatic situation. Changes in solar radiation reaching the earth's surface and in temperature, humidity, and windiness are predicted to occur as a result of greenhouse warming. Rosenberg et al. (1989) have reviewed the ways in which such specific changes may affect photosynthesis. Suffice it to say here that plant response to any change in one factor (say an increase in temperature) depends on whether the prevailing temperature is optimal, suboptimal, or superoptimal for that species, on how the temperature change will manifest itself in terms of diurnal and seasonal changes, and on how other climatic factors change synchronously. The contribution of an increase in temperature to moisture stress, for example, could be offset somewhat by an increase in humidity or cloudiness.

Evidence accumulated by Kimball (1985) does show some interesting interactions between CO_2 concentration and climatic factors. For example, in eighteen studies of plant growth under doubled CO_2 concentration, plants with adequate water supply increased growth by 43 percent. Plants grown under water stress produced less, of course, but the yield reduction was mitigated significantly by exposure to the higher CO_2 concentration. Thus, CO_2 could offset some of the detrimental effects of a drier climate, should that occur. Larger root volumes in response to greater photosynthetic activity could also contribute to this effect. Similarly, growth reduction due to salinity is offset somewhat by increased CO_2 concentration. There is also some evidence (Idso et al., 1987) that the positive benefits of elevated CO_2 concentration are enhanced by temperatures above about 19°C and are actually offset by temperatures below that threshold.

It is not our purpose to examine all of the possible climatic changes and how they would be likely to interact with CO_2 enrichment to alter plant growth. Rather, our purpose is to show that the range of possible outcomes is very great. Climatic changes in one region could negate any beneficial effects of CO_2 enrichment; in another region, climatic change might increase the opportunities for CO_2-enrichment effects to be expressed.

Effects on Evapotranspiration

Will the environmental changes described above affect evapotranspiration and, consequently, the availability of soil-water to support plant growth and the runoff into streams and reservoirs that serves so many essential purposes? The water that falls to earth in the form of rain or snow is eventually returned to the atmosphere by evaporation. The hydrologic equation can be written as:

$$P - E + \Delta SW + R + DS = 0 \tag{1}$$

where P is precipitation, E is evaporation, ΔSW is the change in water storage in the soil, R is runoff, and DS is seepage to depths below which the roots of plants extract water.

Both P and E are expected to increase in their global means as the greenhouse effect takes hold, but, as explained above, we are not yet able to specify with confidence how that increase will be manifested regionally or will evolve in time. Indeed, both increases and decreases are expected. Nonetheless, it is evident that changes in E will be of great importance, since these affect the availability of water to sustain plants and to supply the needs of irrigation, industry, and municipalities.

Most first-cut approaches to predicting the impacts of climate change on evaporation have been based on the assumption that it will increase in direct proportion to increasing temperature. Such an approach is logical, since the atmosphere's ability to hold water vapor (its saturation vapor pressure, or e_s) increases exponentially with temperature. The drying power of the atmosphere increases as the difference between e_s and the actual vapor pressure (e_a) increases.

Evaporation on land, however, is actually controlled by a number of atmospheric and soil factors. The atmospheric factors include the amount of energy that sunlight and winds from drier, warmer regions provide for evaporating water as well as the actual humidity and temperature of the air. Evaporation occurs freely from water surfaces, wet soils, and vegetation at a rate determined by atmospheric conditions. When soils are less than fully wet or when water passes from the soil through the vascular systems of plants into the air (transpiration), evaporation occurs at a reduced rate. In real fields and forests evaporation directly from the soil and transpiration by plants occur simultaneously, and since it is practically impossible to separate these processes, we term the whole process *evapotranspiration* (ET).

In view of the above, and bearing in mind both climatic changes and changes in plant growth and behavior due to CO_2 enrichment of the atmosphere, scientists at Resources for the Future (RFF) have developed a method to simulate how ET may change as a result of greenhouse warming and CO_2 enrichment (Martin et al., forthcoming). The analysis is based on a well-known model of potential evapotranspiration by Penman (1948) adapted by Monteith (1964). The Penman–Monteith (P–M) model is:

$$LE = \frac{s(Rn + S) + \rho_a Cp(e_s - e_a)/r_a}{s + \gamma[(r_a + r_c)/r_a]} \tag{2}$$

where LE = evapotranspiration in terms of the energy consumed in the process (latent heat flux)

Rn = net radiation (a fraction of the solar radiation, controlled in part by cloudiness)

S = soil heat flux (generally a small fraction of Rn)

e_s = saturation vapor pressure of water in the atmosphere, a function of temperature

e_a = actual vapor pressure, a function of weather conditions

r_a = aerial resistance to vapor diffusion from the evaporating surface to the atmosphere, controlled by plant height and roughness and by windiness

r_c = the resistance exerted by the full canopy of vegetation to the transfer of vapor into the atmosphere.

r_c is determined by the openness of the individual pores (stomates) on the leaves. High concentrations of CO_2 cause stomates to close, thereby increasing stomatal and canopy resistance. r_c is also a function of the number of stomates within the canopy, which increases with increasing leaf area. Leaf area may increase because of the stimulation of photosynthesis by elevated CO_2 concentrations in the atmosphere or may decrease because of substantial drying or other deleterious climatic change.

The remaining terms, s, ρ_a, Cp, and γ, are physical factors that are controlled by temperature to a degree, but that need not concern us here.

The P–M method permits analysis of the sensitivity of evapotranspiration (ET) to each of the climatic and plant factors that may be altered as a result of climatic change and CO_2 enrichment, individually and in combinations. It was calibrated by comparing its estimates to actual measurements of ET made in three typical ecosystems—a Nebraska wheat field, a Kansas grassland, and a Tennessee forest. Then climatic and plant factors were varied within credible ranges that might follow from greenhouse warming. These were established through an extensive review of the scientific literature.

Results of these simulations are shown in table 2-5 (adapted from Martin et al., forthcoming). Scenario A shows *LE* rates with no climatic change. A 3°C increase in temperature could increase *LE* by 17 to 24 percent on the average (scenario B). Other climatic and plant factors, if they change within the limits shown in scenario C, could increase or even decrease *LE* from rates with no climate change. Increased net radiation due to reduced cloudiness augments the effect of warming, and reduced net radiation moderates it (scenario D). The effects of higher and lower atmospheric humidity are shown to increase the range of

Table 2-5. Simulated Effects of Greenhouse-Induced Climate Change and CO_2-Enrichment-Induced Plant Responses on Evapotranspiration in a Wheat Field, a Grassland, and a Forest

		Change in					Wheat (10 days)		Grassland (7 days)		Forest (6 days)	
Line no.	Scenarios of change	*T* (°C)	*Rn* (%)	*e* (%)	r_s (%)	*LAI* (%)	*LE* (W m^{-2})	Change (%)	*LE* (W m^{-2})	Change (%)	*LE* (W m^{-2})	Change (%)
1	A. No climate change	0	0	0	0	0	423		233		250	
2	B. Temperature increase only	3	0	0	0	0	508	20	274	17	311	24
3	C. Single factor changes	0	10	0	0	0	449	6	250	7	268	7
4		0	−10	0	0	0	398	−6	216	−7	232	−7
5		0	0	10	0	0	395	−7	223	−4	230	−8
6		0	0	−10	0	0	451	7	243	4	271	8
7		0	0	0	40	0	373	−12	194	−17	211	−16
8		0	0	0	0	15	444	5	249	7	266	6
9	D. Temperature and net radiation change	3	10	0	0	0	536	27	292	25	330	32
10		3	−10	0	0	0	481	14	255	10	292	17
11	E. Temperature, net radiation, and humidity change	3	10	10	0	0	510	20	283	21	311	24
12		3	10	−10	0	0	562	33	302	29	350	40
13		3	−10	10	0	0	455	7	246	6	272	9
14		3	−10	−10	0	0	507	20	265	14	311	14
15	F. Scenario E plus stomatal resistance increase	3	10	10	40	0	453	7	238	2	265	6
16		3	10	−10	40	0	499	18	254	9	298	19
17		3	−10	10	40	0	404	−5	207	−11	232	−7
18		3	−10	−10	40	0	450	6	223	−4	264	6
19	G. Scenario F plus leaf index increase	3	10	10	40	15	478	13	256	10	284	13
20		3	10	−10	40	15	527	24	273	17	319	27
21		3	−10	10	40	15	427	1	223	−4	248	−1
22		3	−10	−10	40	15	476	12	240	3	284	13

Note: T = temperature, °C; R_n = net radiation; e = vapor pressure (humidity); r_s = stomatal resistance; LAI = leaf area index; LE = latent heat flux (evapotranspiration); and W m^{-2} = watts per square meter.

Source: Adapted from Martin et al. (forthcoming).

LE determined by the temperature increase and altered net radiation (scenario E). The worst-case scenario (line 12) shows increases from 29 to 40 percent above no climate change, depending on ecosystem. The best-case scenario (line 13) shows increases in *LE* of only 6 to 9 percent. If stomatal resistance were to increase by 40 percent (scenario F) the effects of the climate changes in scenario E would be dramatically reduced. Finally, the effect of greater leaf area, considered in scenario G, is shown to diminish, but not overcome, the moderating effects of an increase in stomatal resistance.

The analysis presented above illustrates the importance of considering all factors that may change because of greenhouse warming and CO_2 enrichment of the atmosphere in attempts to predict how natural ecosystems will respond to such changes. When this is done, a wider range of outcomes becomes available for consideration in the planning and policy development that will be needed to cope with the effects of greenhouse warming.

SUMMARY AND DISCUSSION

Despite the vast array of uncertainties in projecting future trace greenhouse gas emissions, evaluating the direct effects of CO_2 in stimulating photosynthesis and suppressing transpiration, calculating the time-evolving regional distribution of climatic anomalies, and even pinning down to better than a factor of two projected changes in global average temperature, a number of seemingly robust conclusions still emerge. First, the greenhouse effect is not a speculative, controversial hypothesis in the atmospheric sciences, but a well-validated principle that explains the hot climate of Venus, the cold climate of Mars, and the moderate conditions on our planet. There is a strong consensus of scientists that the increase in carbon dioxide, methane, chlorofluorocarbons, nitrogen oxides, and other trace greenhouse gases due to human activities will warm the climate by the middle of the next century from 1.0°C to 5.0°C and that global average temperatures could then be higher than experienced in the past million years of geologic history. There is also strong consensus that along with changes in temperature of this magnitude will come substantial regional anomalies in temperature, rainfall, evaporation, cloudiness, windiness, droughtiness, and so forth that could have substantial impact on agriculture, water supplies, human and animal health, biological diversity, energy demand, and sea level. While a scientific consensus is not possible over the likelihood of specific scenarios of time-evolving regional anomalies in temperature, rainfall, and so forth, nearly all knowledgeable scientists agree that substantial change is possible over the next several decades, and unprecedented change is possible beyond the middle of the next century. Since detailed predictions cannot easily be validated with current scientific tools, it is necessary to construct plausible scenarios of CO_2 increase, climatic response, biological response, and environmental and societal impacts. No single scenario should ever be taken literally, but a study of the sensitivity of various economic and ecological sectors to a range of plausible futures is a useful way—perhaps the only way—to estimate the potential distribution of costs and benefits from increasing trace greenhouse gases and to evaluate the utility and applicability of adjustments and adaptations.

Since all scenarios imply a distribution of costs and benefits depending on the specific magnitude and rates of climatic changes and their interactions with cropping patterns, standing forests, range animal distribution, hydrologic systems, and so forth, one point seems clearly evident: the more rapidly the climate changes, the more difficult it will be for natural ecosystems and human activities to anticipate and adapt to those changes. For example, it is known that the hardwood forests of the midwestern and northeastern United States were very much reduced in scale and displaced southward at the end of the last ice age some 11,000 years ago. At the same time, tree species of the boreal forests now located south of the tundra across Canada were largely present in the American Midwest and Northeast. Over the course of the subsequent 5,000 years, as the local climate warmed by some 10°C in response to the end of the last ice age, the forests advanced to their present positions (see figure 2-10). The average rate of climate change in the American Northeast, for example, was on the order of 2°C per thousand years (that is, 10°C warming over 5,000 years).

Global climate models project a warming in this region in response to increases in trace greenhouse gas over the next century ranging from 2°C to 10°C, a rate ten to fifty times faster than the average climate change from the waning of the last ice age to the present! Therefore, it is not so much the magnitude of climate change that concerns us, but the rates, for adaptation to rapid change that is not well forecast is obviously much more difficult than if we have a longer time to develop anticipatory strategies and to improve our skill in forecasting global changes.

How to respond to the prospect, or advent, of global climate change has two components, one scientific and one political. The scientific exercise of policy analysis involves estimating how a host of alternative energy, agricultural, water supply, population, and other policy areas would be affected by different rates of climate change or how climatic inputs can be managed by different policies for each of these sectors. The political component is, of course, policy choice, a value-laden activity. Reasonable people may accept the same sets of policy analyses and climatic projections yet differ on the implied policy choices. We feel now that the political process should recognize that the potentially rapid rates of global change believed by a strong consensus of scientists to be plausible over the next several decades demand serious consideration. We are led to conclude that a series of steps designed to slow the rate of growth of trace greenhouse gas emissions could buy time both to assess what global changes will occur and to devise adaptations to mitigate potential losses and take advantage of potential gains.

In particular, if such strategies involve sensible investments that are beneficial even in the absence of rapid climate change (what Schneider and Thompson (1985) have called the "tie-in strategy"), then a strong argument can be made for rapid implementation of such investments. Obvious examples of the tie-in strategy include the development and testing of crop strains and forest species better able to deal with a wide range of climate changes or CO_2 concentrations, development and testing of non-CO_2-producing energy systems, and a more rapid diffusion of energy-efficient technologies in both energy production and energy end-use sectors. Energy efficiency not only reduces the greenhouse effect but, among other benefits, reduces acid rain, acute air pollution in cities, and dependence on foreign energy supplies. More controversial, but also related, is the option to increasingly de-emphasize less efficient hydrocarbon fuels such as coal and to switch to natural gas, since the latter produces about 50 percent less CO_2 per unit of energy than does coal—oil being about halfway between the two. Burning methane also substantially reduces acid rain and smog in cities.

In summary, then, a detailed assessment of the climatic, biological, and societal changes that are evolving and should continue to occur into the next century cannot reliably be made with available scientific capabilities. Nevertheless, enough is known to suggest a range of plausible futures which include substantial redistributions of climatic resources with attendant impacts, both positive and negative, on natural resources and human well-being. The more rapidly climatic change evolves, the less easy it will be for natural and human systems to adapt. That this evolution may lead to an unprecedentedly altered climate sometime in the middle of the twenty-first century is plausible. To predict in detail the consequences of such unprecedented change is beyond the state of the art, since surprises are certain under such conditions. Nonetheless, we must try to predict the consequences systematically as far into the future as possible. To do so will require that technological, demographic, social, and economic futures be factored into analyses.

ACKNOWLEDGMENTS

Our discussion of the direct effects of CO_2 enrichment on plant physiological behavior draws heavily on Bruce A. Kimball's contribution to Rosenberg et al. (1989) and on a paper by Roger Dahlman (1988).

REFERENCES

Barron, E. J. 1985. "Climate Models: Applications for the Pre-Pleistocene," pp. 397–421 in A. Hecht, ed., *Historic and Paleoclimate Analysis and Modeling* (New York, Wiley).

Bates, G. T., and G. A. Meehl. 1986. "The Effect of CO_2 Concentration on the Frequency of Blocking in a General Circulation Model Coupled to a Simple Mixed Layer Ocean Model," *Monthly Weather Review* vol. 114, pp. 687–701.

Beer, S. 1986. "The Fixation of Inorganic Carbon in Plant Cells," pp. 3–11 in H. Z. Enoch and B. A. Kimball, eds., *Carbon Dioxide Enrichment of Greenhouse Crops* vol. 2, *Physiology, Yield, and Economics* (Boca Raton, Fla., CRC Press).

Bernabo, J. C., and Thompson Webb III. 1977. "Changing Patterns in the Holocene Pollen Record of Northeastern North America: A Mapped Summary," *Quaternary Research* vol. 8, pp. 64–96.

Blasing, T. J., and A. M. Solomon. 1982. *Response of the North American Corn Belt to Climatic Warming*. Publication 2134 (Oak Ridge, Tenn., Oak Ridge National Laboratory, Environmental Sciences Division).

Bolin, B., B. R. Döös, J. Jaeger, and R. A. Warrick, eds. 1986. *The Greenhouse Effect, Climatic Change, and Ecosystems*. SCOPE 29 (Chichester, U.K., Wiley).

Bryson, R. A., and G. J. Dittberner. 1976. "A Non-equilibrium Model of Hemispheric Mean Surface Temperature," *Journal of the Atmospheric Sciences* vol. 33, pp. 2094–2106.

Chervin, R. M. 1986. "Interannual Variability and Seasonal Climate Predictability," *Journal of the Atmospheric Sciences* vol. 43, pp. 233–251.

Crutzen, P. J., L. E. Heidt, J. P. Krasnec, W. H. Pollock, and W. Seiler. 1979. "Biomass Burning as a Source of Atmospheric Gases CO, H_2, N_2O, NO, CH_3CL, and COS," *Nature* vol. 282, pp. 253–256.

Cure, J. D. 1985. "Carbon Dioxide Doubling Responses: A Crop Survey," pp. 99–116 in B. R. Strain and J. D. Cure, *Direct Effects of Increasing Carbon Dioxide on Vegetation* DOE/ER-0238 (Washington, D.C., U.S. Department of Energy, Carbon Dioxide Research Division).

Dahlman, R. C. 1988. "Potential Mitigation of Rising Atmospheric CO_2 From Enhanced Biological Productivity." Paper presented at the symposium on Prospects for Mitigating Climate Warming by Carbon Dioxide Control, American Association for the Advancement of Science annual meeting, Boston, Mass., February 12.

Decker, W. L., and R. Achutuni. 1988. *The Use of Statistical Climate-Crop Models for Simulating Yield to Project the Impacts of CO_2-Induced Climate Change* DOE/ER-60444–H1 (Washington, D.C., U.S. Department of Energy, Carbon Dioxide Research Division).

Detwiler, R. P., and C. A. S. Hall. 1988. "Tropical Forests and the Global Carbon Cycle," *Science* vol. 239, pp. 42–50.

Emanuel, K. A. 1987. "The Dependence of Hurricane Intensity on Climate," *Nature* vol. 326, pp. 483–485.

Gilliland, R. L., and S. H. Schneider. 1984. "Volcanic, CO_2, and Solar Forcing of Northern and Southern Hemisphere Surface Air Temperatures," *Nature* vol. 310, pp. 38–41.

Gleick, P. H. 1987. "Regional Hydrologic Consequences of Increases in Atmospheric CO_2 and Other Trace Gases," *Climatic Change* vol. 10, pp. 137–160.

Hansen, J., and S. Lebedeff. 1987. "Global Trends of Measured Surface Air Temperature," *Journal of Geophysical Research* vol. 92, pp. 13345–13372.

Hansen, J., D. Johnson, A. Lacis, S. Lebedeff, P. Lee, D. Rind, and G. Russell. 1981. "Climate Impact of Increasing Atmospheric Carbon Dioxide," *Science* vol. 213, pp. 957–966.

Hansen, J., G. Russell, A. Lacis, I. Fung, and D. Rind. 1985. "Climate Response Times: Dependence on Climate Sensitivity and Ocean Mixing," *Science* vol. 229, pp. 857–859.

Idso, S. B., B. A. Kimball, M. G. Anderson, and J. R. Mauney. 1987. "Effects of Atmospheric CO_2 Enrichment on Plant Growth: The Interactive Role of Air Temperature," *Agricultural Ecosystem Environment* vol. 20, pp. 1–10.

Imbrie, J., J. D. Hays, D. G. Martinson, A. McIntyre, A. C. Mix, J. J. Morley, N. G. Pisias, W. L. Prell, and N. J. Shackleton. 1984. "The Orbital Theory of Pleistocene Climate: Support from a Revised Chronology of the Marine $^{18}0$ Record," in A. Berger, J. Imbrie, J. Hays, G. Kukla, and B. Saltzman, eds., *Milankovitch and Climate* pt. 1, NATO ASI Series, Series C: Mathematical and Physical Sciences, vol. 126 (Dordrecht, Holland, D. Reidel).

Jenny, H. 1941. *Factors of Soil Formation: A System of Quantitative Pedology* (New York, McGraw-Hill).

Karl, T. R., and R. W. Knight. 1985. *Atlas of Monthly Palmer Moisture Anomaly Indices (1931–1984) for the Contiguous United States* Historical Climatology Series 3-9 (Asheville, N.C., National Climate Data Center).

Kellogg, William W., and Zong-Ci Zhao. 1988. "Sensitivity of Soil Moisture to Doubling of Carbon Dioxide in Climate Model Experiments. Part I: North America," *Journal of Climate* vol. 1, pp. 348–366.

Kimball, B. A. 1983a. "Carbon Dioxide and Agricultural Yield: An Assemblage and Analysis of 430 Prior Observations," *Agronomy Journal* vol. 75, pp. 779–788.

———. 1983b. "Carbon Dioxide and Agricultural Yield: An Assemblage and Analysis of 770 Prior Observations," *Water Conservation Laboratory Report* no. 14 (Phoenix, Ariz., U.S. Water Conservation Laboratory).

———. 1985. "Adaptation of Vegetation and Management Practices to a Higher Carbon Dioxide World," pp. 185–204 in B. R. Strain and J. D. Cure, *Direct Effects of Increasing Carbon Dioxide on Vegetation* DOE/ER-0238 (Washington, D.C., U.S. Department of Energy, Carbon Dioxide Research Division).

———. 1986. "Influence of Elevated CO_2 on Crop Yield," pp. 105–115 in H. Z. Enoch and B. A. Kimball, eds., *Carbon Dioxide Enrichment of Greenhouse Crops* vol. 2, *Physiology, Yield, and Economics* (Boca Raton, Fla., CRC Press).

Kimball, B. A., and S. B. Idso. 1983. "Increasing Atmospheric CO_2: Effects on Crop Yield, Water Use, and Climate," *Agricultural Water Management* vol. 7, pp. 55–72.

Land, K. C., and S. H. Schneider. 1987. "Forecasting in the Social and Natural Sciences: An Overview and Analysis of Isomorphisms," *Climatic Change* vol. 11, pp. 7–31.

Loomis, R. S., and W. A. Williams. 1963. "Maximum Crop Productivity: An Estimate," *Crop Science* vol. 3, pp. 67–72.

Malingreau, J. P., and C. J. Tucker. 1988. "Large-Scale Deforestation in the Southeastern Amazon Basin of Brazil," *Ambio* vol. 17, pp. 49–55.

Manabe, S., and R. T. Wetherald. 1986. "Reduction in Summer Soil Wetness Induced by an Increase in Atmospheric Carbon Dioxide," *Science* vol. 232, pp. 626–628.

Martin, Ph., N. J. Rosenberg, and M. S. McKenney. Forthcoming. "Climatic Change and Evapotranspiration: Simulation Studies of a Wheat Field, a Forest, and a Tall Grass Prairie," *Climatic Change*.

Mearns, L. O., R. W. Katz, and S. H. Schneider. 1984. "Changes in the Probabilities of Extreme High Temperature Events with Changes in Global Mean Temperature," *Journal of Climate and Applied Meteorology* vol. 23, pp. 1601–1613.

Mearns, L. O., S. H. Schneider, S. L. Thompson, and L. R. McDaniel. 1989. "Analysis of Climate Variability in GCM Control Runs—Comparison with Observations." Draft manuscript.

Mearns, L. O., P. H. Gleick, and S. H. Schneider. Forthcoming. "Prospects for Climate Change," in Paul Waggoner, ed., *Climate Change and U.S. Water Resources* (New York, Wiley).

Monteith, J. L. 1964. "Evaporation and Environment," *Symposium of the Society for Experimental Biology* vol. 19, pp. 205–234.

Morison, James I. L. 1987. "Intercellular CO_2 Concentration and Stomatal Response to CO_2," pp. 229–251 in E. Zeiger, G. D. Farquhar, and I. R. Cowan, *Stomatal Function* (Stanford, Stanford University Press).

National Academy of Sciences. 1976. *Climate and Food: Climatic Fluctuation and U.S. Agricultural Production* (Washington, D.C., National Research Council, Board on Agriculture and Renewable Resources).

Newman, J. E. 1982. "Impacts of Rising Carbon Dioxide Levels on Agricultural Growing Seasons and Crop Water-Use Efficiencies," in *Environmental and Societal Consequences of a Possible CO_2-Induced Climate Change* DOE/EV/10019-8 (Washington, D.C., U. S. Department of Energy, Carbon Dioxide Research Division).

Parry, M. L., and T. R. Carter. 1985. "The Effect of Climatic Variations on Agricultural Risk," *Climatic Change* vol. 7, pp. 95–110.

Penman, H. L. 1948. "Natural Evaporation from Open Water, Bare Soil and Grass," pp. 120–145 in *Proceedings of the Royal Society, London Series A* no. 193.

Peters, R. L., and J. D. S. Darling. 1985. "The Greenhouse Effect and Nature Reserves," *Bio Science* vol. 35, pp. 707–717.

Ramanathan, V., R. J. Cicerone, H. B. Singh, and J. T. Kiehl. 1985. "Trace Gas Trends and Their Potential Role in Climate Change," *Journal of Geophysical Research* vol. 90, pp. 5547–5566.

Raschke, K. 1986. "The Influence of the CO_2 Content of the Ambient Air on Stomatal Conductance and the CO_2 Concentration in Leaves," pp. 87–102 in H. Z. Enoch and B. A.

Kimball, eds., *Carbon Dioxide Enrichment of Greenhouse Crops* vol. 2, *Physiology, Yield, and Economics* (Boca Raton, Fla., CRC Press).

Repetto, R. 1988. *The Forest for the Trees? Government Policies and the Misuse of Forest Resources* (Washington, D.C., World Resources Institute).

Revelle, R., and P. E. Waggoner. 1983. "Effects of a Carbon Dioxide-Induced Climate Change on Water Supply in the Western United States," pp. 419–432 in *Changing Climate* (Washington, D.C., National Academy of Sciences).

Rind, D., R. Goldberg, and R. Ruedy. Forthcoming. "Change in Climate Variability in the 21st Century," *Climatic Change*.

Rosenberg, N. J. 1986. "Climate, Technology, Climate Change, and Policy: The Long Run," pp. 93–127 in C. F. Runge, ed., *The Future of the North American Granary* (Ames, Iowa, Iowa State University Press).

———. 1987. "Drought and Climate Change: For Better or Worse," pp. 317–347 in D. A. Wilhite and W. E. Easterling, eds., *Planning for Drought: Toward a Reduction of Societal Vulnerability* (Boulder, Colo., Westview Press).

Rosenberg, N. J., B. L. Blad, and S. B. Verma. 1983. *Microclimate: The Biological Environment* (New York, Wiley).

Rosenberg, N. J., B. A. Kimball, C. F. Cooper, and Ph. Martin. 1989. "Climate Change, CO_2 Enrichment, and Evapotranspiration," in P. E. Waggoner, ed., *Climate and Water: Climate Variability, Climate Change, and the Planning and Management of U.S. Water Resources* (New York, Wiley).

Schlesinger, M. E., and J. F. B. Mitchell. 1985. "Model Projections of the Equilibrium Climatic Response to Increased Carbon Dioxide," chap. 4 in M. C. MacCracken and F. M. Luther, eds., *Projecting the Climatic Effects of Increasing Carbon Dioxide* DOE/ER-0237 (Washington, D.C., U.S. Department of Energy, Carbon Dioxide Research Division).

———. 1987. "Climate Model Simulations of the Equilibrium Climatic Response to Increased Carbon Dioxide," *Reviews of Geophysics* vol. 25, p. 760.

Schneider, S. H. 1987. "Climate Modeling," *Scientific American* vol. 256, no. 5, pp. 72–80.

Schneider, S. H., and C. Mass. 1975. "Volcanic Dust, Sunspots, and Temperature Trends," *Science* vol. 190, pp. 741–746.

Schneider, S. H., and S. L. Thompson. 1981. "Atmospheric CO_2 and Climate: Importance of the Transient Response," *Journal of Geophysical Research* vol. 86, pp. 3135–3147.

———. 1985. "Future Changes in the Global Atmosphere," pp. 397–430 in R. Repetto, ed., *The Global Possible: Resources, Development and the New Century* (New Haven, Conn., Yale University Press).

U.S. Department of Commerce and U.S. Department of Agriculture. 1988. *Weekly Weather and Crop Bulletin* vol. 75 (Washington, D.C.).

Warrick, R. A., and M. J. Bowden. 1981. "The Changing Impacts of Droughts in the Great Plains," pp. 111–137 in M. P. Lawson and M. E. Baker, eds., *The Great Plains: Perspectives and Prospects* (Lincoln, Nebr., University of Nebraska Press).

3

Human Development and Carbon Dioxide Emissions: The Current Picture and the Long-Term Prospects

Joel Darmstadter
Jae Edmonds

The scale of human activity has grown to the point that gaseous emissions produced by four areas of human endeavor—energy, agriculture, land use, and chemical manufacture—are changing the stock of gases in the atmosphere. That change, in turn, affects the global temperature and climate. Such gases are called greenhouse, or radiatively important, gases (RIGs). The nature and timing of future climate change depend on three things: the rate of emission of RIGs into the atmosphere, the capacity of removal mechanisms, and the interaction between atmospheric composition and the climate. This chapter will focus on the first of these three factors: the rate of emission of RIGs and its determinants. At the outset, we will briefly review the broad diversity of RIGs, but our particular emphasis will be on the relationship between energy use and the release of carbon dioxide (CO_2) from fossil fuels. Chapter 2 of this volume addresses the other two factors, as well as more general aspects of climate change and the greenhouse effect.

This chapter discusses four topics:

1. The section on present concentrations, sources, and sinks of greenhouse gas emissions by region and by human activity provides a brief listing of the greenhouse gases whose emissions are of interest, reviews the human and natural processes that give rise to these emissions, and discusses the disposition of the gases in the atmosphere.
2. The next section discusses long-term trends in CO_2 emissions from fossil fuels, examining in some detail the prospects for future CO_2 releases and providing a measure of future emissions uncertainty that might be experienced in the absence of any policy actions specifically designed to affect CO_2 emissions.
3. Alternative scenarios of future emissions are then introduced, as the following section examines some of the energy implications of emissions scenarios designed to stabilize or reduce the rate of global CO_2 emissions from fossil fuels.
4. The chapter concludes with a discussion of the feasibility of energy strategies to avert, mitigate, or delay climate change induced by the greenhouse effect. This final section examines some of the institutional factors that must be considered before emissions-reduction strategies can be designed.

PRESENT CONCENTRATIONS, SOURCES, AND SINKS OF GREENHOUSE GASES

It is generally assumed that, were it not for human activities, the composition of the atmosphere would be stable; that is, a balance would exist between natural emissions, natural sinks, and atmospheric processes. Without debating the validity of this assumption, and recognizing that it remains one of the critical uncertainties concerning the carbon cycle (Trabalka, 1985), we will here concern ourselves primarily with emissions generated by human activities, despite the fact that in some cases human perturbations occur against a background of vastly larger, but essentially balanced, natural rates of emission and reabsorption.

There are no firm rules for identifying which gases are to be considered RIGs and which are not. Most lists contain between ten and twenty gases. Somewhat more than half of these are emitted directly by human activities in sufficient quantities to affect the climate. The other gases either are by-products of atmospheric chemical processes or are so abundant in the atmosphere that human activities release amounts that are small in comparison to the atmospheric stocks. Seven that appear on most lists and which are emitted by human activities are the following: CO_2, methane (CH_4), nitrous oxide (N_2O), chlorofluorocarbons ($CFCl_3$ and CF_2Cl_2), nitrogen oxide (NO_x), and carbon monoxide (CO).

Appendix table 3-A-1 in this chapter assembles information about the sources, sinks, and trends in the atmospheric abundance of these gases. To put the human component in its broader perspective, we consider both human and natural emissions of these gases under "Sources" in this particular tabulation. Under the heading "Sinks," we consider three mechanisms for removal from the atmosphere: interactions with oceans, interactions with the land, and transformations to other atmospheric constituents via chemical reactions in the atmosphere. Under the heading "Atmospheric Trends/Characteristics," we review the current concentration, the rate of increase in that concentration, and the average residence time, or lifetime, of the gases in the atmosphere.

We note with interest that the average residence cycling time of carbon is on the order of 500 years, if removal of carbon to the deep ocean is the only sink for carbon considered. (The deep ocean is indeed the only permanent sink for CO_2.) This makes CO_2 the longest lived of the RIGs. N_2O and CFCs such as $CFCl_3$ and CF_2Cl_2 are also long-lived in the atmosphere, with residence times of 75 to 150 years. CH_4 is the most short-lived of the RIGs, with an atmospheric lifetime of about eight years. The short life of the average CH_4 molecule in the atmosphere is attributable to its chemical reactivity, and its lifetime can be affected by the atmospheric chemical environment that it encounters. The importance of CO as a RIG is derived from its ability to influence the lifetime of CH_4 in the atmosphere, since it reacts with hydroxyl ions that normally act as a sink for CH_4.

The uncertainty bounds on historical rates of man-made RIG emissions vary greatly. In general, the more controlled the environment under which the gas is produced, the better known is the rate of emission. For example, production rates of CFCs and fossil fuel CO_2 are better known than CO_2 and CO emissions related to land use. Uncertainty regarding man-made N_2O emissions is particularly high. The regional and temporal distribution of emissions is generally not well known.

Human Activities That Generate Emissions

The study of RIG emissions has grown to include not just a widening mix of gases, but also an increasingly broad spectrum of human endeavors. It is thus impossible adequately to forecast the emission of RIGs without understanding the diversity of human activities involved in their generation. Table 3-1 shows seven RIGs and their dominant human sources in the form of a matrix, partitioned by source. Five of the seven gases—CO_2, CO, CH_4, N_2O, and NO_x—have three interrelated sources: energy use, land-use changes, and agricultural practices. The two CFCs, $CFCl_3$ and CF_2Cl_2, are man-made chemicals with no natural emissions source.

The CFCs are radiatively active. They are not chemically active in the troposphere, and in the stratosphere they are broken down by ultraviolet radiation over long periods of time. CFCs are also an environmental concern because upon decomposition they liberate chlorine atoms that facilitate the destruction of ozone in the stratosphere. The 1987 Montreal Protocol to protect stratospheric ozone has already changed anticipated rates of future emissions. Actual rates of future emissions hinge critically on the success of the protocol's implementation, success in finding acceptable substitutes for CFCs, and the determination by policymakers as to whether even full implementation of the Montreal Protocol is sufficient to preserve the integrity of the ozone shield.

CO_2, CO, CH_4, N_2O, and NO_x have significant sources stemming from human activities. All have important natural sources as well. The three carbon compounds are components of the natural carbon cycle; they are also related chemically. Although not significant sources of CO_2, both CO and CH_4 eventually oxidize to CO_2. CO and CH_4 are the most abundant reactive carbon compounds in the atmosphere. A significant amount of atmospheric CO (roughly 25 percent) originates through the oxidation of methane. Removal of both atmospheric CO and CH_4 occurs primarily through reaction with the hydroxyl radical, OH; however, since CO and CH_4 compete for OH radicals and because the CO reacts more rapidly with CO than does CH_4, the availability of CO molecules can influence the atmospheric concentrations of OH and, thus, CH_4. Because the trend in atmospheric concentrations of OH is unknown, it is difficult to determine what fraction of the observed increase in methane concentration may be due to increased carbon monoxide emissions.

The Role of the Carbon Cycle and Atmospheric Chemistry

It is not the purpose of this chapter to address issues in either carbon-cycle or atmospheric chemistry, though it is essential that some recognition be given to the importance of both. On average, the ratio of annual atmospheric accumulation of CO_2 to fossil fuel emissions of CO_2 has been 0.58 since 1958. To understand how emissions affect the stock of carbon in the atmosphere, it is important to identify the removal mechanisms. Some CO_2 is known to be absorbed by the oceans, and some carbon cannot be accounted for, although it may be taken up by the biosphere as a result of "CO_2 fertilization" (see chapter 2 in this volume). The carbon cycle must be understood if we are to explain the rate of accumulation of CO_2 in the atmosphere.

Many of the other RIGs are also tied up in the carbon cycle. All of the carbon molecules are eventually oxidized to CO_2. The nitrogen molecules are part of the related nitrogen cycle. Thus, understanding atmospheric chemical

Table 3-1. Annual Emissions Budget Estimates for Seven Radiatively Important Gases

Source	CO_2[a] (MtC)	CO[b] (MtC)	CH_4[c] (MtC)	N_2O[b] (MtN)	NO_x[b] (MtN)	$CFCl_3$ (Mt)	CF_2Cl_2 (Mt)
ENERGY	4,846	240	50	4.0	26.0	—	—
Production							
Gas	96	—	20	—	—	—	—
Coal	—	—	10	—	—	—	—
Storage	—	40	—	—	—	—	—
End use							
Residential/Commercial		100	20	0.4	1.1	—	—
Industrial	3,400	10	—	1.2	4.1	—	—
Transport		90	—	—	11.2	—	—
Utilities	1,350	—	—	2.4	9.6	—	—
LAND-USE CHANGES	1,300	160	20	0.5	—	—	
AGRICULTURE	—	110	175	1.3	6.0	—	—
Savanna burning	—	110	30	0.4	6.0	—	—
Rice	—	—	70	—	—	—	—
Fertilizer	—	—	—	0.8	—	—	—
Cultivated soils	—	—	—	1.5	—	—	—
Cattle	—	—	75	—	—	—	—
CHEMICAL MANUFACTURE	—	—	—	—	—	0.33	0.44
Refrigeration	—	—	—	—	—	0.03	0.22
Foam-blowing uses	—	—	—	—	—	0.18	0.06
Aerosol spray uses	—	—	—	—	—	0.10	0.14
Miscellaneous uses	—	—	—	—	—	0.02	0.03

Notes: Confidence intervals are given in Wuebbles and Edmonds (1988). MtC = million tons carbon; MtN = million tons nitrogen; Mt = million tons; dashes denote zero or not significant.

Source: Wuebbles and Edmonds (1988).

[a]Carbon dioxide emissions figures are based on the total carbon content of the fuels and biomass stocks oxidized, without reference to the initial form that the carbon takes (that is, whether the carbon appears initially as CO, CH_4, CO_2, or some other carbon compound). This convention is adopted on the grounds that all carbon compounds eventually oxidize to CO_2. In principle, the emissions figures for CO and CH_4 refer to gross releases of carbon in those forms over the course of a year and make no reference to transformations of carbon from other states into CO or CH_4 or transformations of CO or CH_4 into other compounds. There is, therefore, an inconsistency between the accounting conventions used for CO_2 and those for CO and CH_4.

[b]Energy data are disaggregated on the basis of information from Kavanaugh (1987) but are scaled to match totals given in Wuebbles and Edmonds (1988).

[c]Energy data are disaggregated on the basis of information from Edmonds and Marland (1986) but are scaled to match totals given in Wuebbles and Edmonds (1988).

processes, which are important for many of the gases, is basic to understanding the composition of the atmosphere at any point in time.

ANALYSIS OF CARBON DIOXIDE EMISSIONS FROM FOSSIL FUELS: A REFERENCE PROJECTION AND UNCERTAINTY

As Schneider and Rosenberg explain (chapter 2 in this volume), the time-dependent rate of climate change depends in large part on the composition of the atmosphere. That composition depends on the rates at which a variety of greenhouse gases are emitted into the atmosphere as a result of a variety of natural and anthropogenic activities and on the rate of atmospheric removal of these gases. Constructing a radiative time profile of the atmosphere is beyond the scope of this chapter, which focuses principally on the issue of emissions.

We have chosen to direct our attention to releases from fossil fuel combustion for two principal reasons. First, fossil fuels are the dominant current global source of CO_2 emissions (approximately 80 percent). In addition, CO_2 is generally expected to account for at least half of whatever global warming has already occurred and may occur in the future. For example, Mintzer (1987) predicts that in the year 2050 carbon dioxide will be responsible for 50 percent of the "radiative forcing" attributable to the principal greenhouse gases. If CFCs are excluded from this calculation, the CO_2 contribution will be two-thirds of the total forcing. For estimates of present and past contributions to radiative forcing in this range, see Ramanathan (1988) and Ramanathan et al. (1985). For other estimates of future contributions, see Ramanathan et al. (1985) and Edmonds et al. (1987).

Second, by focusing exclusively on fossil fuel CO_2, we sidestep the need to develop a means of intercomparison to apply to multiple RIG emissions over time. The analysis of multiple RIGs is impossible without such an intercomparison scheme for assessing the trade-offs and leverages associated with different emission control strategies. By limiting our attention to CO_2 emissions, we simplify the analysis and concentrate on an important subset of human activities. In any case, the tools do not yet exist for an adequate simultaneous assessment of future emissions of all RIGs. The state of the art in forecasting future rates of RIG emissions consists of running a set of independent models and combining the results. While such a procedure is an important first step, better models must be designed and built. Without an integrated modeling regime, there is no guarantee of consistency between runs. There is no chance for regions and sectors to "compete" for scarce resources and no chance for feedback effects to be propagated through the model. In short, there is no integrated way to assess the comparative and cumulative costs of policy options to reduce emissions of the many gases released by alternative scales of human activities.

We are aware that by choosing to focus our analysis exclusively on fossil fuel CO_2 emissions we preclude an examination of some of these methodological questions and other interesting and important issues. For example, we do not investigate the interplay between differential stages of economic development and future emissions of CO_2 and other RIGs. Neither do we examine the international tradeoffs and multiple benefits that might emerge from policies fostering alternatives to current human activities that might also alter the patterns of emissions. While these issues are important and eventually must be addressed within a comprehensive analytical framework, we feel that a useful start can be made by trying to understand the complexities of energy and CO_2 emissions.

Energy and CO_2 Release

Emissions of CO_2 occur whenever fossil fuel is oxidized. Nevertheless, the rate of emission varies among fuels. The variation is primarily dependent on the relative abundance in fuels of carbon and hydrogen. Average emission coefficients for oil, gas, coal, and shale are given in table 3-2. Note that, by convention, the measure of CO_2 emissions is expressed in terms of the weight of the carbon atom only, not the weight of the entire CO_2 molecule.

In general, nonfossil energy sources—for example, nuclear, solar, hydro, geothermal—do not release CO_2 into the atmosphere. Biomass energy is a special case. Biomass contains carbon withdrawn from the atmosphere. When burned or otherwise oxidized, it releases that CO_2 back into the atmosphere. Therefore, biomass releases no net CO_2 to the atmosphere during its growth, harvest, and use cycle. When land use is changed, say from forest to pasture or vice versa, a net addition or a net reduction in atmospheric CO_2 can occur. (See Sedjo and Solomon, chapter 8 in this volume, for further discussion of this issue.)

For example, deforestation can, in effect, lead to the release of CO_2 that was sequestered in biomass at a much earlier date. Whenever forest regrowth fails to keep pace

Table 3-2. Average CO_2 Emission Coefficients by Fuel

Fuel	gC/Mj[a]	gC/kBtu[a]
Oil	19.2	20.256
Gas	13.8	14.4535
Coal	23.8	25.109
Shale[b]	27.9	29.4345

Source: Reprinted, with permission, from Edmonds and Reilly (1985), p. 266.

[a]The first column shows grams of carbon per megajoule, or million joules; the second column shows grams of carbon per thousand British thermal units (Btu). One Btu equals 1,055 joules.

[b]Shale refers to the mining of oil shale found in carbonate rock formations.

with deforestation, a net release of carbon to the atmosphere results. Similarly, a growing commercial biomass industry which systematically plants trees or other flora leads to a net removal of carbon from the atmosphere.

The transformation of primary energy (for example, coal) to a secondary or final energy carrier (electricity, for example) increases the rate of CO_2 emission per unit of final energy carrier. The CO_2 emission rate varies directly with the efficiency of transformation. For example, if coal is used to generate electricity with a transformation efficiency of 33 percent, then the rate of CO_2 emission per unit of electricity would be 71.4 grams of carbon per megajoule of electricity, or three times the rate of emission from direct combustion of coal. Similarly, the transformation of coal to synthetic oil at 50 percent efficiency would result in a CO_2 release rate per unit of "synoil" produced of 47.6 grams of carbon per megajoule, or twice the emission rate of direct combustion of coal. A shift from conventional coal combustion to the production of liquids and gaseous fuels from coal will increase CO_2 emissions unless these fuels are applied to end uses with sufficiently improved efficiencies to compensate for the carbon released in the transformation process. On the other hand, if the energy transformation allows a greater improvement in end-use efficiency than would be possible if the primary energy carrier were used directly, then CO_2 emissions could actually be reduced. This has been the case with many industrial electrical applications.

Past Rates of Fossil Fuel CO_2 Emissions

Fossil fuel CO_2 emissions have grown since World War II as shown in figures 3-1 and 3-2. Global emissions grew every year between 1950 and 1974. The average annual rate of growth over this period was 4.4 percent. Emissions declined between 1974 and 1975, apparently in response to the steep rise in the price of fossil fuels, but they resumed their upward trend in 1976, this time at an average annual rate of 3.8 percent. Global fossil fuel CO_2 emissions peaked in 1979 and then declined for four successive years. Again, the decline corresponded to a sharp increase in energy

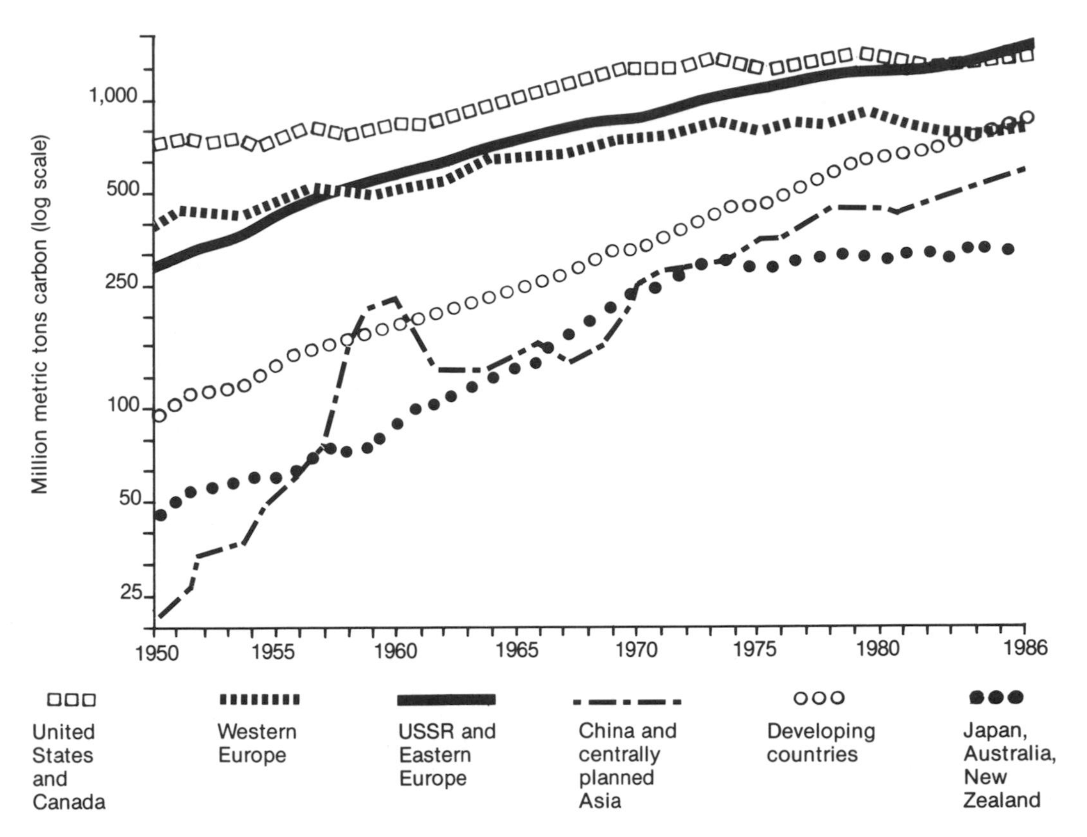

Figure 3-1. CO_2 emissions from fossil fuel combustion by major world regions, 1950–1986. *Note:* "Developing countries" represent the difference between the worldwide total (figure 3-2) and the sum of the other regions shown. They thus include several countries—e.g., South Africa—not normally classified as "developing." (*Source:* Data prepared by and obtained from Carbon Dioxide Information Analysis Center, Oak Ridge National Laboratory, Oak Ridge, Tenn., 1988.)

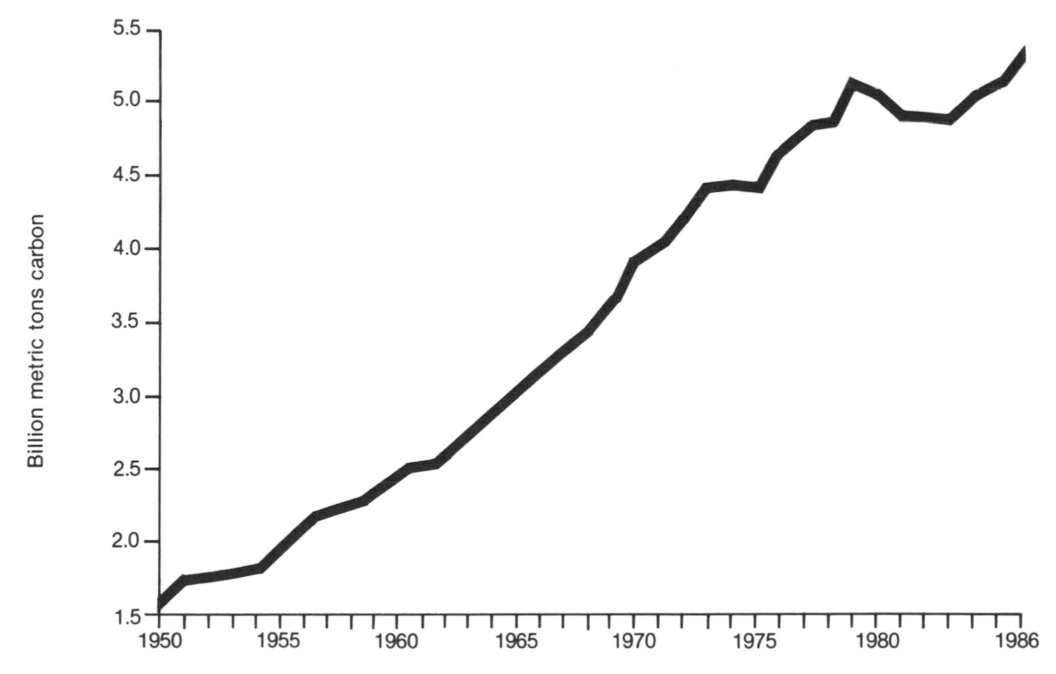

Figure 3-2. Total global CO_2 emissions from fossil fuel combustion, 1950–1986. (*Source:* Data prepared by and obtained from Carbon Dioxide Information Analysis Center, Oak Ridge National Laboratory, Oak Ridge, Tenn., 1988.)

prices. Since 1983 global emissions have returned to a pattern of annual increases averaging 3.2 percent. Between 1985 and 1986 emissions increased by 4.1 percent.

The pattern of regional distribution of emissions is also of interest. In 1950 the United States and Canada accounted for 45 percent of world fossil fuel CO_2 emissions. By 1986 this share had declined to less than 25 percent. Western Europe experienced a similar decline from 24 percent to 15 percent between 1950 and 1986. By contrast, the emissions share of the People's Republic of China (PRC)—along with that of Centrally Planned Asia—rose from 1.4 percent in 1950 to 11 percent in 1986. Similarly, the developing world increased its share of fossil fuel CO_2 emissions from 6 percent in 1950 to 15 percent by 1986. The USSR and Eastern Europe have increased their share less spectacularly, from 18 percent in 1950 to 26 percent in 1986. Since 1981 the USSR and Eastern Europe have emitted greater amounts of fossil fuel CO_2 than has any other regional grouping, surpassing the United States and Canada in 1982.

World emissions have declined in five of the thirteen years since 1973: 1975, 1980, 1981, 1982, and 1983. Western European emissions declined in eight of the thirteen years, and emissions from the United States and Canada declined in six of those years. Similarly, emissions declined in seven of these years in Japan, Australia, and New Zealand. In Western Europe and in the United States and Canada, 1986 emissions were lower than 1973 emissions. A very different pattern can be seen in the USSR and Eastern Europe, the PRC and Centrally Planned Asia, and the developing world. In these three regions emissions were higher in 1986 than in 1973 and increased every year, with the exceptions of 1980 and 1981 in the PRC and Centrally Planned Asia and 1981 in the USSR and Eastern Europe.

Projecting Future Fossil Fuel CO_2 Emissions

Future fossil fuel CO_2 emissions are uncertain and depend on the evolution of the global energy system. To explore the bounds of that uncertainty in the absence of direct policy intervention to change CO_2 emissions, we have used the results of an uncertainty (that is, Monte Carlo) analysis conducted by Edmonds et al. (1986). That analysis used the Edmonds–Reilly model (ERM) and advanced Monte Carlo techniques to explore the implications of uncertainty in seventy-nine model parameters for future global fossil fuel CO_2 emissions.

The ERM, the core of the analysis, is a long-term, global energy system model. It has four constituent parts: supply, demand, energy balance, and CO_2 emissions. The first two modules determine the supply and demand for each of eight primary energy categories in each of nine global regions. The energy supply and demand modules trace the path of energy from extraction, harvesting (biomass), or capture (solar), through transformation processes (synthetic fuels or electric power), to energy end use, with three end-use sectors (residential/commercial, industry, and transport) for the twenty-four countries comprising the Organisation for Economic Co-operation and Development (OECD) and aggregate end-use specifications elsewhere. The energy balance module ensures model equilibrium in three global fuel markets: oil, gas, and solids. The CO_2 emissions module then calculates the emission of fossil fuel CO_2 at the point of release. The model proceeds from a base year of 1975 and provides forecasts for the years 2000, 2025, 2050, 2075, and 2100.

Uncertainties in seventy-nine parameters were explored. These parameters governed population; economic growth; energy conservation; the resource base for fossil fuels, uranium, and biomass; technology descriptions for electric power generation, synfuel conversion, and solar power; environmental costs; and the effects of energy prices on overall economic activity. Monte Carlo techniques allow an analysis of the relationship between uncertainty in future values for the seventy-nine parameters and CO_2 emissions. The range of values that each of the seventy-nine variables might take on was specified. The Monte Carlo simulation systematically selected possible values for each model parameter, ran the model, and saved the results. This procedure was repeated 400 times to provide a statistically significant description of the variance of CO_2 emissions. Analysis could then be performed to assess variation in CO_2 emission forecasts and the relative importance of each of the seventy-nine input variables.

A CO_2 Emissions Reference Case

We have developed a reference case using the uncertainty analysis described above. This reference case assumes no policy intervention to reduce the rate of CO_2 emissions, but does allow for anticipated changes in important societal activities including demographic, economic, and technological factors and environmental policy interventions not directly related to CO_2. Median values for four key assumption categories are displayed in table 3-3. The four (out of seventy-nine) variables which proved to be the dominant determinants of variation in CO_2 emission forecasts are: the rate of labor productivity growth in developing countries, the rate of labor productivity growth in developed countries, the rate of improvement in exogenous end-use energy efficiency, and a parameter which implicitly reflects the relationship of energy to economic development, that is, the income elasticity of energy demand. In addition, we present our population forecast assumptions.

The rate of economic growth is determined by the growth rates of population and labor productivity. Both these factors are assumed to be larger in the developing world than

Table 3-3. Assumptions Underlying Reference Case Emissions Determination

Factor	Assumption	Period and notes
Population growth		
OECD & USSR	0.5%/yr	1975–2025
OECD & USSR	0.1%/yr	2025–2075
PRC	1.0%/yr	1975–2025
PRC	0.2%/yr	2025–2075
ROW[a]	1.7%/yr	1975–2025
ROW[a]	0.4%/yr	2025–2075
Labor productivity growth		
OECD & USSR	1.7%/yr[b]	
ROW[a]	2.9%/yr[b]	
Income elasticity		
OECD	1.00	% change in the demand for energy per % change in income
USSR	1.25	
ROW[a]	1.40[b]	
Rate of technological change		
All regions	1.0%/yr[b]	Rate of energy-efficiency improvement

Source: Edmonds et al. (1986).
[a]ROW = rest of the world.
[b]One of the four principal determinants of forecast variance.

in the developed world. The income elasticity of demand for energy is assumed initially to be 1.0 in the OECD, but 1.4 in the developing world. A value of one indicates that if the scale of economic activity doubled, the demand for energy services would double as well. The initial income elasticity of 1.4 in developing countries reflects the assumption that increases in economic activity are likely to be concurrent with sectoral shifts away from the less energy-intensive rural sector, with its reliance on traditional biomass fuels, toward the more energy-intensive industrial sector, with its reliance on modern marketed fuels. The 1 percent annual rate of energy-efficiency improvement shown in table 3-3 means that, independent of other factors, a dollar's worth of economic output by the middle of the next century will be produced with half the energy required today.

Our reference case forecast is given in table 3-4. Again we have used median values from the uncertainty analysis described in Edmonds et al. (1986). We note that the annual rate of growth of global fossil fuel CO_2 emissions averages 0.6 percent annually between 1975 and 2075. This rate is substantially lower than the post–World War II experience.

The rate of growth of global energy is, however, twice as high as that of CO_2 emissions. Thus, the ratio of CO_2 to energy declines over the period 1975 to 2075. This is due entirely to an increasing share in overall energy used in this period attributable to nuclear and renewable energy. True, fossil fuel production becomes more CO_2 intensive over this period as coal, with its high CO_2 emission coefficient, represents an increasing share of fossil fuel production. But fossil fuel use, rising at 0.6 percent annually, grows at only around half the rate of overall energy, while nuclear power and renewable energy production grow much more rapidly than overall energy in this scenario.

In our reference case the trend toward increased energy transformation continues. This trend helps to explain the fact that, from a 1975 starting point, demand for end-use energy grows only 1 percent annually, while primary energy grows 1.3 percent per year. (Rates of change from the mid-1980s are a bit lower.) Electrification grows 1.7 percent per year. Conventional oil and gas production peaks by the year 2050 and declines thereafter. The decline in oil production is particularly acute. The increasing demand for liquids and gases is met by synfuels production using both coal and biomass. Little use is made of shale or tar sand feedstocks.

Also note that the gross national product (GNP), which is calculated by the model rather than being an assumed input, grows globally 3 percent per year. Thus, substantial conservation is occurring: end-use energy demand is growing only one-third as rapidly as GNP. Half of that conservation can be explained as a consequence of the technological assumption that, independent of energy price trends, energy end-use efficiency will increase at 1 percent annually. The other half must be explained as the consequence of the higher real energy prices calculated by the model. However, while the world price of oil is higher in every period than it is in the one before, the price escalation is not even. The price of oil rises rapidly between the years 2000 and 2025, when it finally exceeds the peak oil price reached in 1981.

Table 3-4. Reference Case Forecast

Category	1975	2000	2025	2050	2075
Fossil fuel CO_2[a]	4.5	5.8	6.9	7.7	8.5
Total primary energy[b]	235	380	525	672	821
Conventional oil	100	148	148	152	106
Unconventional oil	0	0	0	0	1
Natural gas	39	42	85	110	79
Coal	75	107	124	131	185
Biomass[c]	0	14	24	41	63
Nuclear[d]	4	7	20	29	61
Solar[d]	0	0	1	1	3
End-use energy[b]	178	278	366	416	465
Electricity[b]	24	41	69	97	133
GNP[e]	6	14	30	60	112
World price of oil[f]	1.84	2.29	3.52	3.95	4.89

Source: Median values taken from Edmonds et al. (1986).

[a]Expressed in gigatons of carbon per year (GtC/yr). A gigaton = 1 billion metric tons.

[b]All primary energy categories, end-use energy, and electricity are shown in exajoules (1 EJ = 10^{18} joules). Individual energy subcategories do not sum to total primary energy due to the exclusion of hydroelectric power generation and the use of median values of output from 400 runs in the uncertainty analysis.

[c]Includes only biomass from waste and biomass plantations. Excludes traditional biomass.

[d]Primary energy equivalent.

[e]10^{12} 1975 U.S. dollars.

[f]1975 U.S. dollars per gigajoule. (A dollar per gigajoule—that is, per billion joules—is the equivalent of about $5.50 per barrel of oil in 1975 prices. Thus, $3.95 per gigajoule in the year 2050 is the equivalent of about $22 per barrel of oil in 1975 prices or $44 per barrel in 1987 prices.)

Uncertainties and Gaps in the Analysis

The reference emissions scenario involves large uncertainty bounds. Uncertainty bounds on fossil fuel CO_2 emissions, taken from Edmonds et al. (1986), are given in table 3-5.

Two results of this analysis are worth noting. First, overall uncertainty of future emissions is considerable. A range of 3 percent to -1.4 percent per year is needed to bracket 90 percent of the CO_2 emission cases. Second, roughly 25 percent of the cases result in constant or declining emissions, even though policies to control CO_2 emissions were not explicitly considered at this stage of the analysis.

ALTERNATIVE EMISSIONS SCENARIOS

Calculating statistical probability distributions of future CO_2 releases, as we have done in the foregoing section, provides a useful, but limited, perspective from which to explore the energy-policy, economic, and technological implications of attempts to control these releases in the years ahead. For that purpose, we supplement the reference case projection with two alternative scenarios describing the necessary conditions, and the feasibility of their realization, for both long-term constancy and substantial reduction in global CO_2 emissions.

Within a purely *statistical* framework, we see from table 3-5 that recent levels of CO_2 emissions (of about 5 gigatons) are maintained by the year 2050 in approximately one-quarter of the cases, while emissions decline to half that level in only about one out of twenty runs. Such low-probability outcomes might argue for dispensing with the scenario task as pointless. But it is in exploring precisely how such outcomes can occur that scenario exercises can contribute to illuminating the challenge of dealing with the CO_2 problem.

Table 3-5. Uncertainty Range in Global Fossil Fuel CO_2 Emissions, 2000–2075
(gigatons of carbon per year [GtC/yr])

Year	5%[a]	25%[a]	50%[a]	75%[a]	95%[a]
2000	3.2	4.6	5.8	7.7	14.2
2025	2.4	4.5	6.9	13.0	29.8
2050	2.3	4.3	7.7	18.7	58.1
2075	1.8	3.9	8.5	27.1	86.9

Source: Edmonds et al. (1986). See accompanying text for comments on the uncertainty analysis.

[a]The percentage of 400 runs for which CO_2 emissions were less than the values shown. For example, the 25% column shows that 25% of 400 forecast runs had CO_2 emissions lower than 4.6 billion tons per year (Gt/yr) in the year 2000 and lower than 3.9 Gt/yr in the year 2075.

Suppose, then, that a continuing upward trend in fossil fuel CO_2 emissions—if only at the relatively modest average growth rate of 0.6 percent annually implied by the 2050 reference case median level of 7.7 gigatons—is judged to be potentially too disruptive of social and economic life and of natural systems to be willingly tolerated. (Emissions of this magnitude, when combined with emissions from nonfuel sources of CO_2 and other trace gases, could mean accepting an increase of several degrees Celsius in global mean temperature.) How formidable a task would it be to achieve long-term stabilization in the volume of CO_2 emissions? Even that outcome would not preclude increases in global mean temperature. The cumulative buildup of greenhouse gas concentrations, compounded by the fact that there are bound to be at least intermediate-term increases in the volume of emissions prior to attainment of long-term constancy, implies such increases.

Three alternative scenarios for the year 2050 are outlined in table 3-6. These include our reference case, based on table 3-4, and two additional scenarios of global development: a constant emissions case and a declining emissions case—the latter taken from a World Resources Institute study by Mintzer (1987), who labels it a "slow buildup" scenario. We will discuss each of these two alternative projections.

In the constant emissions scenario, as in the reference case, we have postulated worldwide income growth of 3 percent per annum, whereas the declining emissions scenario assumes a growth rate of 2½ percent per annum. (But note: the more moderate GNP growth in the declining emissions projection is not linked to the targeted goal of a roughly 50 percent reduction in carbon emissions, for in Mintzer's multiple scenario analysis, which is not shown here, future magnitudes of energy use are wholly invariant with respect to levels of GNP; thus, for a *given level of GNP* in the year 2050, the highest energy estimate is nearly three times the lowest—or declining emissions—scenario. More on that below.)

We have purposely not varied the demographic and economic assumptions in moving from the reference to the constant emissions scenario—not because lower population and economic growth are inconceivable, or even because a concerted effort to limit greenhouse gas emissions may not suppress income growth as the necessary price of achieving that objective. (Such a policy course, and others, would be the stuff of yet more scenario constructions.) The point,

Table 3-6. Demographic, Economic, Energy, and CO_2 Data for 1985 and for Alternative Scenarios in 2050

		2050		
	1985[a]	Reference case[b]	Constant emissions[c]	Declining emissions[d]
Population (billion)	4.8	9.2	9.2	9.7
GNP (index)				
Total	100	683	683	495
Per capita	100	359	359	247
Total primary energy consumption (EJ)	323	672	583–672	354
Fossil fuels	262	393	304	179
(Coal)	(89)	(131)	(42)	(23)
(Oil and Gas)	(172)	(262)	(262)	(156)
Nonfossil fuels	61	279	279–368	175
CO_2 emissions (GtC/yr)[c]	5.2	7.7	5.2	2.5

Note: See text for further discussion of this table.

[a]Population and energy data from United Nations statistical sources. Worldwide GNP assumed to lie on trend line shown in table 3-3. CO_2 emissions based on table 3-2.

[b]Reference case taken from median values given in table 3-4.

[c]Constant emissions case derived by assuming the level of oil and gas availability given in table 3-4. Associated CO_2 emissions are derived by assuming that on average 7 percent of all oil is diverted to nonfuel uses that delay oxidation and by applying the CO_2 emissions coefficients given in table 3-2. Carbon emissions are 2.7 GtC/yr for oil and 1.5 GtC/yr for natural gas. The restriction of total emissions to 5.2 GtC/yr implies that coal emissions be constrained to 1.0 GtC/yr, or coal use cannot exceed 42 EJ/yr. In the low-range total energy and nonfossil fuel figures, we assume that the same contribution is available from nonfossil energy sources as in the reference scenario. This implies that total energy use would be 583 EJ and that the rate of reduction in energy intensity must increase from an average rate of 1.8 percent per year to 2.0 percent per year to maintain the same population and economic growth as in the reference scenario. In the high-range total energy and nonfossil fuel figures, we assume that overall energy intensity lacks this additional flexibility and that, therefore, the CO_2-constrained limit on fossil fuel use must be fully offset by recourse to nonfossil fuels.

[d]Declining emissions case taken from Mintzer (1987).

rather, is to draw out the burden on the energy sector and to depict its atmospheric profile if demographic realities and economic aspirations are to be accommodated even while emissions are stabilized. At the same time, it is important to keep in mind that the reference case itself embodies substantial deceleration from historical population and economic growth experience, particularly in the case of the population projection, which envisages the demographic transition witnessed by western societies over the past 100 years or so beginning to characterize developing nations in the decades ahead. To be specific, the projected average annual growth rates in the reference and constant emission cases are: worldwide population growth, 1 percent; per capita economic growth, 2 percent; total economic growth, 3 percent. For comparison, global population growth during the last several decades has averaged 2 percent yearly; per capita income growth, 2 to 3 percent; and total economic growth, 4 to 5 percent.

What are the energy implications of moving from the unconstrained CO_2 releases of the reference case to the constant emission scenario? Table 3-6 contemplates two possible offsets to the drastic curtailment of coal combustion necessitated by the CO_2 curb. Conceivably, that curtailment might be entirely absorbed by a corresponding reduction in total energy use, brought about by an even sharper annual reduction in energy intensity than assumed in the reference case—that is, 2 percent rather than 1.8 percent. (An additional decline in energy intensity could, in turn, come about from a somewhat enhanced annual rate of technical improvement in the use of energy.) The lower total energy consumption figure in the constant emission column of table 3-6 illustrates that circumstance. For much lower levels of energy use to prevail, past evidence suggests that much sharper price increases than those assumed here would be needed. These, in turn, could easily undermine the economic growth imperatives that we have postulated.

On the other hand, curtailed coal use could force a compensating recourse to nonfossil fuels rather than dampened overall energy use. The required eighty-nine additional exajoules of nonfossil energy (a 32 percent rise above the reference case nonfossil component) imply a greater dependence on some combination of solar, nuclear, and some other noncarbon energy source than is otherwise indicated. Neither of these wrinkles around the constant emission case strikes us as an especially formidable incremental burden. It is the proposition that holding CO_2 emissions constant through the middle of the next century compels a wholesale, absolute reduction in the world's use of coal (to less than half of what is used today and a two-thirds reduction from the 2050 reference case) that we must recognize as the key issue.

However formidable a task it would be to design a long-term energy strategy which, without penalizing human welfare, would nonetheless prevent CO_2 releases from exceeding present annual emissions, it is clearly much more challenging to define and pursue a course intended to limit releases to half current levels, the maximum specified in the declining emissions scenario of table 3-6. Yet it is vital to construct and examine such a scenario, for CO_2 reductions of at least this magnitude are believed necessary to ensure a slowdown and, eventually, constancy in the atmospheric CO_2 concentration. This would probably permit global mean temperature to stabilize, in some time period still further removed, at levels substantially below those resulting from a doubling in CO_2 concentrations (compared to pre-industrial concentrations). Our reference case implies such a doubling by the end of the next century; our constant emissions scenario implies the ability to defer doubling well beyond that point.

The principal route to achieving a 50 percent reduction in CO_2 releases is through energy-consumption practices substantially below those dictated by normal market, or income and price, factors. Changes in consumer behavior would need to occur, as well as more dramatic technological improvements in energy conversion and utilization than are assumed in the reference case and the constant emissions projection. Such an outcome would represent an unprecedented decoupling of energy and economic growth—a development whose feasibility has been argued by numerous writers, including, most recently, Goldemberg et al. (1987), to whom Mintzer turns to support the relationships reflected in the declining emissions column of table 3-6. The numbers shown in that column deserve a close look.

Note first that although Mintzer relies on a somewhat higher population estimate for the year 2050 than we do in the other two scenarios, his per capita economic growth projection is much lower than that assumed in the other two scenarios—1.4 percent, compared to 2 percent. Whether or not that low a rate of improvement in living standard adequately captures the growth imperatives among the many developing population groups of the world is debatable, but the result is a lower GNP projection in the declining emissions case (2½ percent per annum) than in the other two scenarios (3 percent). Other things being equal, one would expect lower economic growth to generate less demand for energy. In the declining emissions scenario, energy consumption in the year 2050 is barely above the level in 1985, the average annual increase over the sixty-five years being 0.1 percent per year. That is, the level of energy consumption is substantially less than would be expected by the reduced economic growth.

We thus come to a crucial difference among the three alternative projections in table 3-6—the future path in aggregate energy intensity, or energy relative to GNP. The ratio reflecting that intensity factor declines by 1.8 percent annually in the reference case, by 2 percent in the lower of the two constant emission variants, and by 2½ percent in the declining emissions scenario, whose 147 percent increase in economic output is assumed to be achievable with virtually no change in energy use.

Now, energy–GNP ratios are admittedly highly aggregative constructs which mask an enormous diversity of detailed underlying trends and phenomena. Indeed, a considerable amount of analytical dissection during the past several decades has illuminated a good deal about these underlying specifics and crosscurrents. And yet, even at the high level of abstraction with which we are dealing here, it is worth asking whether energy and economic growth, flexible as their relationship has been shown to be, are quite as independent as the third column of table 3-6 suggests. Indeed, as noted earlier, Mintzer examines a wide range of possible energy futures not reproduced here, each associated with the same economic growth rate.

In this respect, it is insufficient to point to, say, the experience of the United States during the past fifteen years, when energy consumption remained just about constant amidst a 39 percent rise in GNP. That was a period of sharp energy price increases that induced energy conservation but, in the judgment of numerous economists, also retarded economic growth. If one invites energy price increases to stem energy use, one must count as well on at least some negative impacts on economic growth as investments in energy efficiency occur at the expense of investments in other productive sectors.

The analysis by Goldemberg et al., from which the energy–economy relationship of the declining emissions scenario of table 3-6 is derived, provides many instructive examples of innovative energy-conserving technologies, products, and processes, which are described as economically justified today, let alone at the gradually increasing real energy prices that most analysts (including ourselves) project for the long term. To substantiate that assertion more firmly, it would be useful to probe three points:

1. Goldemberg et al. illustrate numerous promising energy efficiency potentials—for example, the manufacture of a new generation of fuel-efficient vehicles—without estimating the capital investment required to bring the new technologies, products, or processes on line. This makes it hard to judge economic viability.

2. If the normative goal of a zero energy growth path turns out, on further reflection or with experience, to be unachievable, what are the preferred lines of retreat? Conscious limitation of economic growth? Intensified efforts to augment the availability of noncarbon energy sources and to ensure their environmentally sound utilization?

3. As Goldemberg et al. observe, there are at any given time obvious "market imperfections" and other inhibitions impeding economically optimal energy purchase and investment decisions. But are these barriers so formidable as to hold down energy efficiency practices to the enormous extent stated? What policy initiatives would significantly alter the situation?

The last point leads us to the general recognition that constructive approaches to dealing with the greenhouse problem are not likely to succeed without including a major role for new and imaginative policy departures. We consider that issue below.

POLICIES AND INSTITUTIONS FOR MANAGING CHANGE

In exploring the challenge of managing fossil fuel use and CO_2 emissions in the interest of atmospheric integrity, one must begin with the recognition that a large part of the world appears to be committed to energy policies that assign a major long-term role to coal in particular and fossil fuels in general. But conventional liquid and gaseous fuels, more desirable than coal because of their lower carbon content, are supply limited, while shale, which exists in vast quantities, is probably the most environmentally incriminated fossil fuel of all. Hence the attraction of coal—abundant and relatively cheap to produce over a large range of its long-term supply schedule.

In terms of policies and institutions, what would a departure from that commitment to coal entail? It would require a willingness on the part of the world as a whole—or of enough individual countries to make a difference—to forswear sufficient combustion of coal or other carbon-rich fuels to head off the greenhouse danger. Consider briefly two key impediments to so managing CO_2 releases. First, at this stage of our knowledge, the potential damage from global warming lacks credible estimation, even though—as other chapters of this volume make clear—the prospect of at least some increase in the global mean temperature is in little doubt. Such an assessment requires an ability to determine how threatening *global* climate change is to *specific* regions. To judge the extent to which crop yields in one country may be subject to damaging changes in precipitation and temperature, while benign or beneficial effects occur in another, will require general circulation models (GCMs) reliable enough to tell us what climatic changes to expect where. Schneider and Rosenberg (chapter 2 in this volume) tell us that such models will not be available very soon.

Second, even if the regional consequences of increased CO_2 concentrations in the atmosphere could be calculated and found to be menacing, the implementation of defensive measures faces formidable political and institutional obstacles. Keep in mind the motivational differences arising from disparities in income, political milieu, and cultural tradition, as well as from the extent to which different countries or regions are dependent on coal-based energy systems. China is preeminent in its large dependence on coal—heightened, to be sure, by grossly inefficient mining, processing, and utilization practices. Based on enormous coal resources, China's modernization is tied acutely to the exploitation of this energy source (Smil, 1988). Other countries where coal promises to continue playing a vital role include India (45 percent of its energy consumption in recent years) and Poland (80 percent). And, of course, neither the Soviet Union nor the United States show present inclinations to keep their vast resources in the ground.

It is worth pointing out, incidentally, that while the United States, the Soviet Union, and China *each* has enough coal of its own to push long-term atmospheric concentrations of CO_2 substantially higher, there is also enough carbon sequestered in coal fields elsewhere (principally in the Federal Republic of Germany, the United Kingdom, India, and Australia) to raise CO_2 levels greatly, even if oil shale is never tapped and land-use change adds no net CO_2 to the atmosphere.

In trying to deal with this dilemma, one must count on a long leap from the readiness of a community or nation, linked by the kinship of common laws and values, to fashion internal arrangements for managing its common national resources—say by willingly bearing the cost of preserving water quality—to the readiness of the world as a whole to construct an incentive or cooperative framework within which various countries will participate—in effect, to tax itself. And there is as yet no dependable institution for knocking heads. (As we argue below, the compromises and avoidance of potentially divisive issues that achieved the 1987 Montreal Protocol to limit CFC emissions point up the possibility of greater obstacles in attempts to frame an agreement with respect to other greenhouse gases.) Notwithstanding instances where a single nation may be willing to bear the cost of a collective benefit when just its share of that benefit is perceived as justifying the price it is paying—say, U.S. nuclear export constraints not practiced by other countries—"free rider" problems always loom to frustrate such benevolence. Thus, the effect of a steep, purely U.S. tax on U.S. coal consumption and the curtailment of U.S. coal exports—unrealistic as it may be to contemplate such initiatives in the first place—could be negated by unrestrained coal combustion elsewhere.

On that question, Edmonds and Reilly (1983) studied several policy control measures, including a purely U.S. CO_2 tax accompanied by U.S. coal export curtailment. Even such a severe restraint measure did not sufficiently reduce global CO_2 emissions to forestall a doubling of atmospheric CO_2 concentrations by around 2080. The sensitivity of CO_2 concentrations to national and international taxes is explored by a number of other writers (Nordhaus and Yohe, 1983; Kosobud and Daly, 1984) with predictably similar conclusions. And since, as noted, the perceived economic sacrifice associated with such environmental discipline may be more than many countries would probably tolerate, the prospects for successful collective action face major uncertainties. In short, returning to table 3-6, one is well advised to recognize the difficulty of getting countries to move aggressively to avert the anticipated outcomes depicted in the constant emissions, much less in the declining emissions, scenario.

But what about the the Montreal Protocol and its applicability to the control of CO_2? If commercial interests can join with governments to significantly curtail future CFC releases, might not a CO_2 emission protocol be equally probable? To argue that it is requires confronting what appear to be four significant differences between the two classes of atmospheric threat, as follows.

1. *Ease of substitution.* In principle, one could argue that interfuel shifts toward noncarbon or low-carbon energy sources, along with enhanced conservation, could, like CFC controls, ensure benign atmospheric outcomes. In practice, however, the cost is apt to be orders of magnitude higher in the commitment to a policy of CO_2 stabilization, let alone of CO_2 reduction, than in the control of chlorofluorocarbons.

2. *Incidence of impacts.* There are parts of the world whose inhabitants are relatively immune to the effects of ultraviolet (UV) radiation. But from what we know today, the damaging effects of stratospheric ozone depletion are far more ubiquitous than appears to be the case for greenhouse warming, given present meteorological and climatological understanding. Hence, there exists a more tenuous basis for consensus building.

3. *Differential economic motivations.* CFC production and, to a large extent, use are centered predominantly in economically advanced countries. The cost of CFC restrictions, when measured against corresponding per capita GNPs, is incontestably lower than the burden that, say, the Chinese, Indians, Poles, and even Russians are likely to perceive as the economic impact of severely restricted coal combustion. Particularly if climatological disruption is expected to be a distant rather than a near-term occurrence, even modestly discounted future benefits, when compared with present-day costs, may militate against effective initiatives.

4. *Adaptability.* To the extent that countries are inclined, rightly or wrongly, to think in terms of moderate and gradual climatic change, optimism about the ease of adjustment—for example, through technological advance and mobility in agriculture—may inhibit willingness to submit to restrictions limiting CO_2 emissions. By contrast, and notwithstanding the desirability of broad-brimmed hats and strong levels of sunscreen, adapting to increased doses of UV radiation is not a practicable

option. Of course, evidence that points to more precipitous global warming with acute consequences would weaken the adaptability argument.

As against these four points, which make the prospect of initiatives to control CO_2 more problematic than was the CFC protocol, there is one consideration which may strengthen the push for action on the CO_2 front: policies restraining the use of carbon-based fuels would simultaneously attack multiple environmental problems aside from greenhouse warming, notably acidification, smog, corrosion, and visibility. Although that fact might not win over coal proponents who look to low-sulfur fuels and various "clean coal" technologies for attacking these local and regional pollution problems, the possibility of multiple payoffs at least increases the benefit-cost ratio associated with a regime to control CO_2.

The foregoing discussion does not justify asserting that energy strategies designed to contain the greenhouse threat are hopelessly impracticable. Indeed, the very recognition that there is a threat should provide a strong impetus to social and physical scientists and to policymakers to pursue promising and imaginative research and long-term strategies with which a society—its consciousness raised—can identify. Economically optimal use of energy in production and consumption activities deserves continued emphasis. Governmentally supported research into nonfossil, environmentally acceptable energy sources should have a high place on the public policy agenda. Even approaches to the CO_2 problem that depend on new technologies that today seem highly problematic should not be totally dismissed. Note, for example, the Brookhaven National Laboratory analysis that cautiously cites the technical possibilities and the estimated cost of scrubbing CO_2 from utility stacks and pumping it into the ocean (Steinberg and Cheng, 1987).

Without foreclosing any number of possible ways to manage the long-term CO_2 problem, during any given time span there are limits on what is feasible. Whether, for example, the proposal emerging from the recent international conference in Toronto to "reduce CO_2 emissions by approximately 20 percent of 1988 levels by the year 2005 as an initial global goal" represents an attainable target is a question well worth pursuing (Conference Statement, 1988). Since, as the Toronto declaration states, "the industrialized nations have a responsibility to lead the way," it is worth noting that, soon after the Toronto meeting, legislation introduced in the U.S. Senate set the same overall percentage reduction objective for the United States (*Congressional Record*, 1988), while another proposed bill calls for more than a 50 percent reduction in CO_2 releases from U.S. power plants by 2010 (U.S. Senate, 1988).

Both of these formulations of objectives recognize the multipronged demand and supply strategies—including, for example, a fresh look at nuclear power—that would need to be explored to achieve progress. An indication of the formidable nature of the task is this preliminary finding of a recent study prepared for the Department of Energy: the achievement of even a 10 percent reduction of U.S. CO_2 emissions by the year 2010 presupposes a significant restructuring of anticipated U.S. energy production (Edmonds et al., 1988). In that light, one cannot ignore the likelihood that significant emissions of greenhouse gases into the atmosphere are apt to continue at least well into the twenty-first century and that prudence requires a mix of mitigation and adaptation strategies. This theme recurs throughout this volume.

For some writers—Lave (1981) and Schelling (1983), for example—that conclusion appears to flow not just from a recognition of the formidable barriers that would need to be instituted to alter the prospect of continued significant emissions, but from a more general view of the forces making for social inertia. But these writers do not by any means regard the task of adaptation simply with gloom. Lave looks to such factors as income levels, mobility, education, and investment patterns—and the flexibility provided by the political and institutional milieu in which these factors operate—as clues to the outlook for human adaptability. Schelling, similarly, points to long-term economic growth as a positive cushion against the much smaller welfare losses likely to be caused by unavoidable climatological impacts. Schelling reminds us of how the Dutch, more than half of whose population resides below sea level, have protected themselves with dikes for centuries. Whatever its consequences elsewhere, he suggests that a 5-meter sea-level rise in Boston could arguably be defended against by dikes costing less than the value preserved. But it is the potentially catastrophic effects of such a change in other areas of the world that need to concern us: just consider the enormously costly burden of such protective measures for low-lying, poor countries as Bangladesh. Moreover, neither Lave nor Schelling (nor, for that matter, most other writers) wrestle with a significant issue. What are the implications of CO_2 concentrations rising substantially *above* the doubling level that serves as a standard reference? Crosson (chapter 5 in this volume) discusses that important question as it bears on society's long-term future and on near-term policies to shape that future.

Thus, it is important to record the views of persons taking a less relaxed view of these matters than Lave or Schelling—for example, Professor Wilfrid Bach, a geographer at the University of Münster, Federal Republic of Germany, and Gjerrit P. Hekstra, a contributor to this volume. Bach (1988) has urged the adoption of an international convention which would ensure that greenhouse gas emissions stay below the 2°C or so increase in mean global warming (over pre-industrial levels) which would materialize in the absence of policy intervention. Bach, along with certain ecologists, believes that the 2°C increment by the end of the next century is close to a maximum tolerable ceiling in global warming.

Hekstra, in remarks made at the workshop on which this volume is based, echoed Bach's concerns. He singled out the degree of uncertainty dominating every aspect of the greenhouse problem, the all-too-common tendency of analysts to dwell on "surprise-free" rather than "surprise-rich" futures, and the limited capacity of conventional discounting procedures and capital budgeting techniques to address long-term, and possibly highly discontinuous, phenomena in the global environment. Thinking about unthought-of policy options and resiliency to the unexpected in charting development paths were high on the list of concerns expressed by Hekstra.

Just because conceptualizing ways of dealing with these concerns is difficult should not blind us to flaws in conventional tools of analysis. It is easy to appreciate the caution expressed by Ausubel (1983) that, over the interval of a century during which many things besides climate are apt to change, standard economic models, rooted within contemporary cultural norms and employing standard economic constructs and relationships, may provide insufficient perspective. The situation, Ausubel believes, justifies multiple and divergent perspectives on long-term change. In that sense, the range of futures spanned by the scenarios presented above and the surrounding discussion is just one among numerous ways of reflecting on a major global dilemma.

ACKNOWLEDGMENT

This chapter is based in part on work conducted for the U.S. Department of Energy, Office of Basic Energy Sciences, Carbon Dioxide Research Division, under Contract No. DE-AC06-76RLO. The views expressed are those of the authors and not the U.S. government.

REFERENCES

Ausubel, J. H. 1983. "Can We Assess the Impacts of Climatic Changes?" *Climate Change* vol. 5, pp. 7–14.

Bach, W. 1988. "Modelling the Climatic Effects of Trace Gases: Reduction Strategy and Options for a Low Risk Policy," remarks prepared for conference on Climate and Development, Hamburg, Federal Republic of Germany, November 7–10.

Conference Statement. 1988. *The Changing Atmosphere: Implications for Global Security*. Report presented at the International Conference on CO_2 Emissions, Toronto, Canada, June 27–30.

Congressional Record. 1988. "National Energy Policy Act," S10282–S10306, July 28.

Edmonds, J. A., and G. Marland. 1986. *The Energy Connection to Climate Change: Gaseous Emissions* (Oak Ridge, Tenn, Institute of Energy Analysis).

Edmonds, J. A., and J. M. Reilly. 1983. "Global Energy and CO_2 to the Year 2050," *The Energy Journal* (July), pp. 21–47.

———. 1985. *Global Energy: Assessing the Future* (New York: Oxford University Press).

Edmonds, J. A., J. M. Reilly, R. H. Gardner, and A. Brenkert. 1986. *Uncertainty in Future Global Energy Use and Fossil Fuel CO_2 Emissions 1975 to 2075* Report TR036, DO3/NBB-0081 Dist. Category UC-11 (Available from National Technical Information Service, U.S. Department of Commerce, Springfield, Va. 22161).

Edmonds, J. A., D. L. Wuebbles, and M. J. Scott. 1987. "Energy and Radiative Precursor Emissions." Paper presented at the Eighth Miami International Conference on Alternative Energy Sources, Miami, Fla., December 14.

Edmonds, J. A., W. B. Ashton, H. C. Cheng, and M. Steinberg. 1988. *An Analysis of U.S. CO_2 Emissions Reduction Potential in the Period to 2010*. Draft report prepared for the U.S. Department of Energy, September 30.

Goldemberg, J., T. B. Johansson, A. K. N. Reddy, and R. H. Williams. 1987. *Energy for a Sustainable World* (Washington, D.C., World Resources Institute).

Kavanaugh, M. 1987. "Estimates of Future Trace Gas Emissions from Energy Combustion," *Atmospheric Environment* vol. 21, no. 3, pp. 463–468.

Kosobud, R. F., and T. A. Daly. 1984. "Global Conflict or Cooperation Over the CO_2 Climate Impact?" *Kyklos* vol. 37, no. 4, pp. 638–659.

Lave, L. B. 1981. "The Carbon-Dioxide Problem: A More Feasible Response," *Technology Review* (November–December), pp. 23–31.

Mintzer, I. 1987. *A Matter of Degrees: The Potential for Controlling the Greenhouse Effect*. Research Report 5 (Washington, D.C., World Resources Institute).

Nordhaus, W. D., and G. W. Yohe. 1983. "Future Paths of Energy and Carbon Dioxide Emissions," pp. 87–153 in *Changing Climate*, Report of the Carbon Dioxide Assessment Committee, National Research Council (Washington, D.C., National Academy Press).

Ramanathan, V. 1988. "The Greenhouse Theory of Climate Change: A Test by an Inadvertent Global Experiment," *Science* vol. 242, pp. 293–299.

Ramanathan, V., R. J. Cicerone, H. B. Singh, and J. T. Kiehl. 1985. "Trace Gas Trends and Their Potential Role in Climate Change," *Journal of Geophysical Research* vol. 90, pp. 5547–5566.

Schelling, T. C. 1983. "Climate Change: Implications for Welfare and Policy," in *Changing Climate*, Report of the Carbon Dioxide Assessment Committee, National Research Council (Washington, D.C., National Academy Press).

Smil, V. 1988. *Energy in China's Modernization: Advances and Limitations* (Armonk, N.Y., M. E. Sharpe).

Steinberg, M., and H. C. Cheng. 1987. *Advanced Technologies for Reducing Emissions*. Document 40730 (Brookhaven, N.Y., Brookhaven National Laboratory).

Trabalka, J., ed. 1985. *Atmospheric Carbon Dioxide and the Global Carbon Cycle* Report DOE/ER-0239 prepared for the U.S. Department of Energy (Available from National Technical Information Service, U.S. Department of Commerce, Springfield, Va. 22161).

U.S. Senate. 1988. "Global Environmental Protection Act of 1988." Bill S. 2663, July 27 (legislative day, July 26).

Wuebbles, D., and J. A. Edmonds. 1988. *A Primer on Greenhouse Gases* Report DOE/NBB-0083 prepared for U.S. Department of Energy (Available from National Technical Information Service, U.S. Department of Commerce, Springfield, Va. 22161).

Table 3-A-1. Sources, Sinks, and Atmospheric Parameters for Seven Radiatively Important Gases

Gas	Sources	Sinks	Atmospheric trends/characteristics
CO_2	*Natural sources:* Gross annual ocean release: 104–106 GtC/yr Gross annual terrestrial release: 87–120 GtC/yr *Human-related sources:* Fossil fuel use: 4.5–5.5 GtC/yr Land-use change: 0–3.0 GtC/yr	*Oceans:* Gross annual uptake: 106–108 GtC/yr Net annual uptake: 1.5–3.3 GtC/yr *Land:* Gross annual plant CO_2 uptake: 100–140 GtC/yr Net primary plant production: 60–70 GtC/yr *Atmosphere:* None	*Concentration:* 345 ppmv (1985) *Rate of increase:* 0.4%/yr *Atmospheric lifetime:* 500 years for combined lifetime for atmosphere, biosphere, plus upper ocean. Biogenic seasonal cycle causes annual variations in surface concentrations.
CH_4	*Natural sources:* Enteric fermentation (wild animals): 2–8 MtC/yr Wetlands (swamps, etc.): 60–160 MtC/yr Lakes: 2–6 MtC/yr Tundra: 1–5 MtC/yr Oceans: 7–13 MtC/yr Termites and other insects: 5–45 MtC/yr Other: 0–80 MtC/yr *Human-related sources:* Enteric fermentation (domesticated animals): 40–110 MtC/yr Rice paddies: 40–100 MtC/yr Biomass burning: 30–110 MtC/yr Natural gas and coal mining: 25–75 MtC/yr Solid waste: 0–60 MtC/yr	*Oceans:* Small *Land:* 5–15 MtC/yr (microorganism uptake by soils) *Atmosphere:* 250–450 MtC/yr (reaction with tropospheric OH) 30–70 MtC/yr (transport to and reaction with OH, Cl, or O in stratosphere) 45–75 MtC/yr (accumulation)	*Atmospheric concentration:* 1.7 ppmv (Northern Hemisphere) 1.6 ppmv (Southern Hemisphere) *Rate of change:* 1.0–1.2%/yr *Atmospheric lifetime:* 7–10 years
N_2O	*Natural sources:* Oceans and estuaries: 1–3 MtN/yr Natural soils: 3–10 MtN/yr *Human-related sources:* Fossil fuel combustion: 3–5 MtN/yr Biomass burning: 0.0–0.9 MtN/yr Fertilized soils: 0.6–1.0 MtN/yr Cultivated natural soils: 0.5–2.0 MtN/yr	*Oceans:* Small *Land:* Small *Atmosphere:* 7.5–13.5 MtN/yr (photolysis and reactions with oxygen) 3–4 MtN/yr (accumulation)	*Atmospheric concentration:* 0.31 ppmv *Rate of change:* 0.2–0.3%/yr *Atmospheric lifetime:* 150 years
$CFCl_3$	*Natural sources:* None *Human-related sources:* 330,000 tons/yr chemical industry production Uses: 8% refrigerators, air conditioners 36% closed-cell foams 19% open-cell foams 31% aerosol propellant 6% other	*Oceans:* None *Land:* None *Atmosphere:* None of significance in the troposphere Dissociated in stratosphere (photolysis and reaction with $O(^1D)$)	*Atmospheric concentration:* 0.0002 ppmv (1983) *Rate of change:* 5%/yr *Atmospheric lifetime:* 75 years

CF_2Cl_2	*Natural sources:* None *Human-related sources:* 440,000 tons/yr chemical industry production Uses: 49% refrigeration, air conditioners (including automobile air conditioners) 8% closed-cell foams 5% open-cell foams 32% aerosol propellant 6% other uses	*Oceans:* None *Land:* None *Atmosphere:* None of significance in the troposphere Dissociated in stratosphere (photolysis and reaction with $O(^1D)$)	*Atmospheric concentration:* 0.00032 ppmv (1983) *Rate of change:* 5%/yr *Atmospheric lifetime:* 110 years
CO	*Natural sources:* Plant emissions: 20–90 MtC/yr Oxidation of natural hydrocarbon: 50–500 MtC/yr Forest wildfires: 5–20 MtC/yr Oceans: 10–40 MtC/yr Methane oxidation: 75–450 MtC/yr *Human-related sources:* Fossil fuel use: 150–250 MtC/yr Agriculture: 40–170 MtC/yr Forest clearing: 80–350 MtC/yr Oxidation of anthropogenic hydrocarbons: 0–80 MtC/yr	*Oceans:* Small *Land:* Soil uptake: 110 MtC/yr *Atmosphere:* Reaction with OH: 520–1,120 MtC/yr Accumulation: 10 MtC/yr	*Concentration:* 0.20 ppmv (Northern Hemisphere) 0.05 ppmv (Southern Hemisphere) *Rate of increase:* 1–5% yr *Atmospheric lifetime:* 0.4 years
NO_x	*Natural sources:* Stratospheric oxidation of N_2O: 0.5–1.5 MtN/yr Lightning: 1–10 MtN/yr Soil emissions: 1–15 MtN/yr Oceans: < 1 MtN/yr	*Land:* Wet and dry deposition: 25–85 MtN/yr	*Atmospheric concentration:* Remote areas: 10–30 pptv Populated areas (mid-troposphere): 0–200 pptv Populated areas (lower atmosphere): >1,000 pptv Stratosphere: up to 0.02 ppmv total reactive nitrogen in upper stratosphere

Notes: The gases considered here are carbon dioxide (CO_2), methane (CH_4), nitrous oxide (N_2O), chlorofluorocarbons ($CFC1_3$ and CF_2C1_2), carbon monoxide (CO), and nitrogen oxide (NO_x).
GtC/yr = gigatons (= 10^9 tons) carbon per year.
MtC/yr = million tons (= 10^6 tons) carbon per year.
MtN/yr = million tons (= 10^6 tons) nitrogen per year.
ppmv = parts per million volume.
pptv = parts per trillion volume.
Source: Wuebbles and Edmonds (1988).

4

Sea-Level Rise: Regional Consequences and Responses

Gjerrit P. Hekstra

Climatic change and its possible impact on sea level, hydrology, agriculture, and other dimensions of human development have concerned physical and social scientists for some time. But the emergence of this issue on the world's *political* agenda is of fairly recent origin, spurred in particular by the first World Climate Conference, organized by the World Meteorological Organization in 1979. Global warming noted during the last century was interrupted prior to the mid-1960s, but since then global mean temperatures have increased by about 0.3 degrees Celsius (°C) (Wigley, 1987). This most recent warming episode and other physical insights into possible causes of climatic change permit us, cautiously and selectively, to extrapolate trends into the future (Bolin, 1987). The World Climate Conference seems to have provided the springboard for the joining of public policy discussions to ongoing scientific research efforts.

That the Netherlands is one of the countries paying increased attention to climatic change and its consequences (see Hekstra, 1986) is not surprising. Public authorities in nations with coastal lowlands have shown heightened worry about the foreseeable long-term effects of accelerated sea-level rise (SLR). By contrast with the last century, during which SLR amounted to approximately 0.12 to 0.15 meters (m), assessments of prospective change over the next 100 years range from increases of around 0.6 m to increases of 4.0 m, though scenarios can be constructed that point to the possibility of a far greater rise than the 4.0 m. Such a circumstance could result from the disintegration and rapid melting of the West Antarctic ice sheet. For practical purposes, however, low and high SLR scenarios of 0.5 m and 1.0 m respectively over the next century seem most useful. But, as will be clear from the discussion that follows, such global averages may mask both substantial variability among different coastal regions and the storm surges and other intensified perturbations which, in given localities, can greatly magnify SLR damage to human settlements, croplands, and groundwater quality.

The manner in which nations react to SLR (and other climatological dangers) depends on the values that are threatened and on the available technical means and financial resources for remediation. (A conceptual treatment of that trade-off is provided in chapter 5 in this volume, by Crosson.) Consequently, the weight given to strategies for controlling the rate of greenhouse gas emissions relative to the weight accorded to defensive and adaptive strategies to deal with SLR is likely to vary among the governments of the world, the more so if recognition of greenhouse warming as a global commons problem (a concept discussed by Crosson in chapter 5 and by Jodha in chapter 11 in this volume) fails to take hold.

This chapter focuses largely on analyzing in a detailed fashion region-specific SLR effects and on discussing adaptive and protective measures to counter them. We shall, for example, describe the development of computer models helpful to local decision making and their role in dealing with the SLR challenge. But an emphasis on defensive and reactive strategies is not intended to subordinate the critical importance of a concerted effort to lower the rate of increase in greenhouse gas emissions in order to diminish the prospect of a major SLR threat in the first place.

TRENDS AND SCENARIOS OF SEA-LEVEL RISE

Causes, Rates, and Trends

A reconstruction of the North Sea coastline since the early Holocene—about 10,000 years before the present (B.P.)—shows the sea surface rising relatively rapidly (figure 4-1) causing the shoreline to advance inland to its present position. Between 10,000 and 6500 B.P. the rate of sea-level rise was as much as 1 m per century. Between 6500 B.P. and 5000 B.P. the rate diminished to about 0.4 m per century. Thereafter, the rate diminished to 0.2 m per century until 4500 B.P., and then to 0.1 m per century to 2000

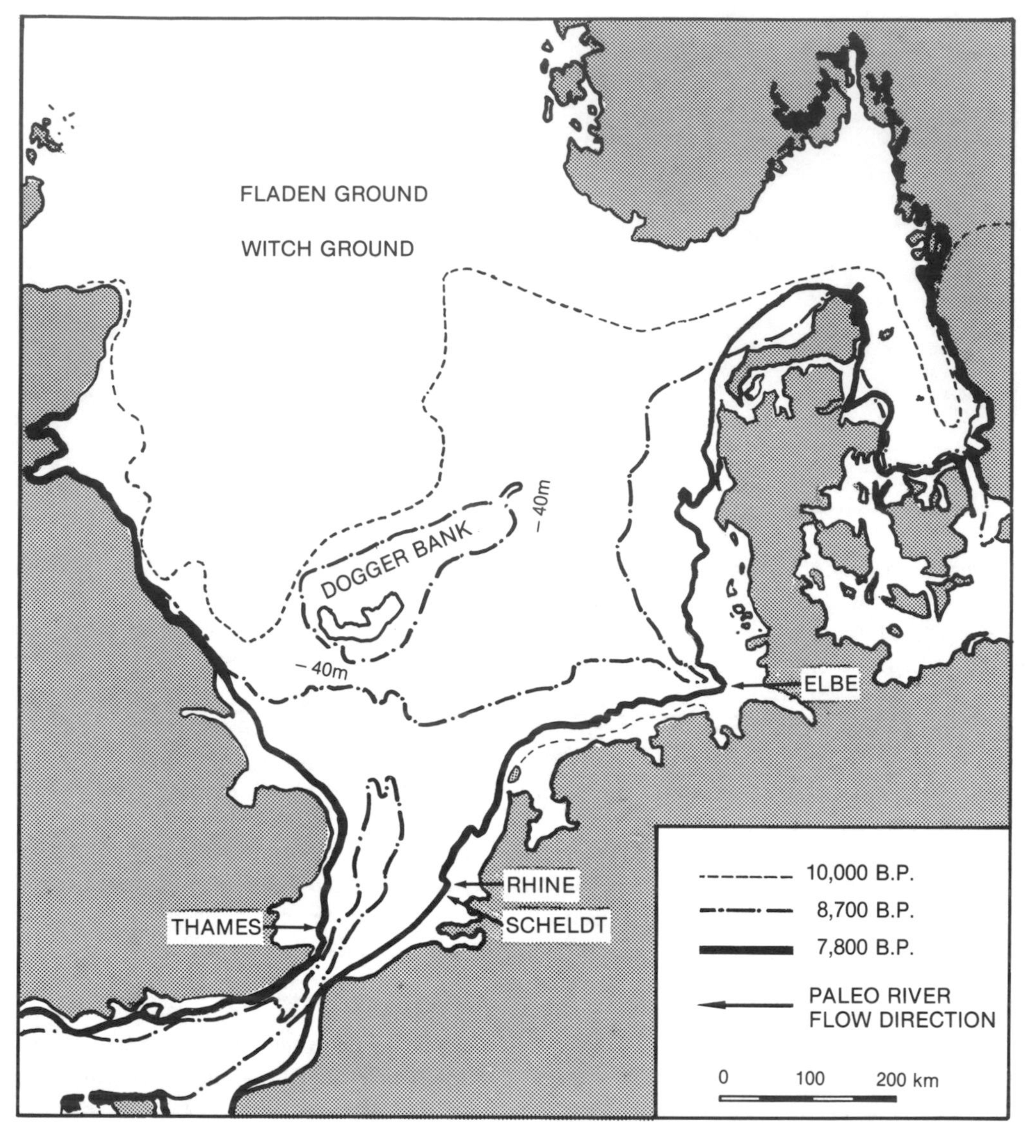

Figure 4-1. Holocene shoreline changes in the North Sea region, showing approximate (hypothetical) shoreline positions for (1) 10,000 B.P. to 10,300 B.P., (2) 8700 B.P., and (3) 7800 B.P., with sea level at 65, 36, and 20 meters, respectively, below present mean sea level. (*Source:* Reprinted, with permission, from Jelgersma, 1979.)

B.P.—close to the rate that occurred from preindustrial times up to about 1900 (Brouns, 1988).

The present rate of global SLR is attributed to the warming that has melted glaciers and some Greenland ice, as well as to thermal expansion of the oceans. Changes in the mass balance of antarctic ice will probably have little impact on SLR in the next few centuries (de Q. Robin, 1987). Indeed, it is possible that an increase of precipitation over Antarctica due to climatic change could act to reduce the rate of SLR (Oerlemans, forthcoming). Melting of arctic sea ice does not contribute to sea-level rise.

Estimates of the actual global SLR that has occurred in the past century range from 0.12 to 0.15 m (Gornitz et al., 1982; Titus, 1987). When these values are corrected for the effects of glacio-isostatic rebound—the springing up of the land mass under its lightening load of ice—the actual SLR could vary from 0.8 to 3.0 millimeters (mm) per year, depending on location. This implies that other poorly defined tectonic and orographic effects are important in determining the actual local SLR (Titus, 1986). These effects are illustrated in figure 4-2, which shows regional average annual sea-level anomalies at a number of sites around the world.

Apparently the most pronounced changes have been experienced at the Atlantic coasts of North America, with smaller changes on the east coasts of Asia. The great variability in sea level found at the west coast of Central and South America is probably correlated with the El Niño–

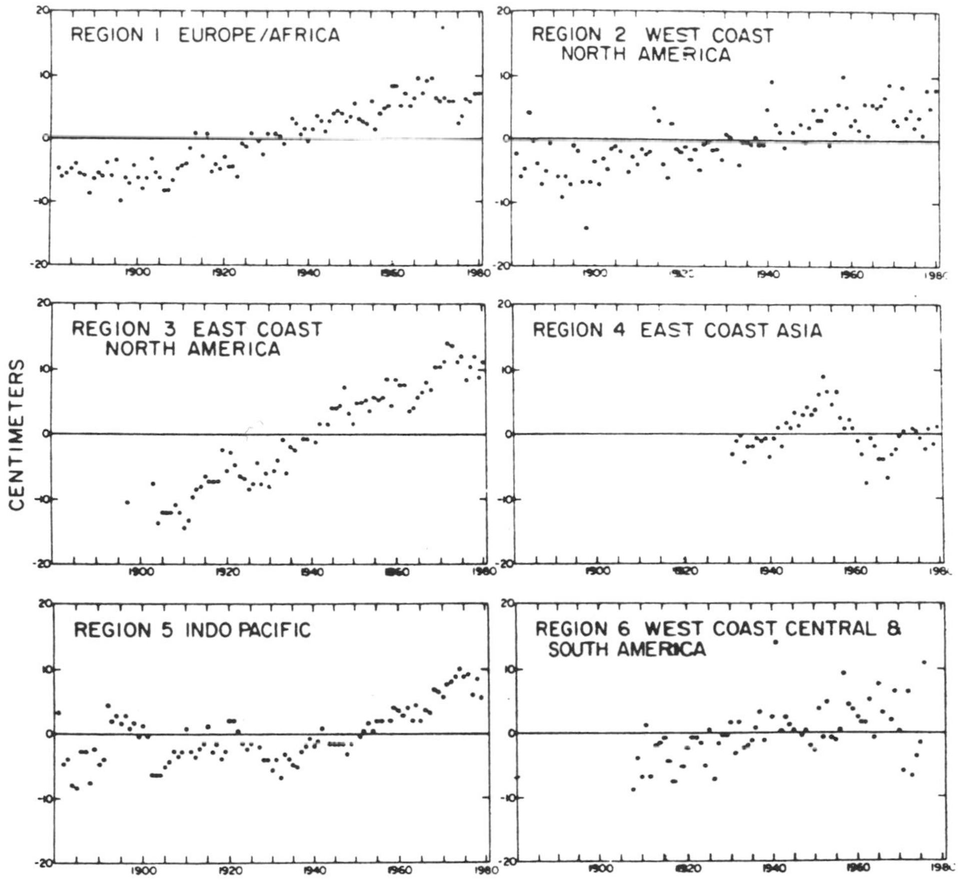

Figure 4-2. Regional averages of annual sea-level anomaly (in centimeters) at sites in six regions around the world, 1880–1980. (*Source:* Reprinted, with permission, from Barnett, 1984. © 1984 by the American Geophysical Union.)

Southern Oscillation phenomenon. A fairly constant rate of SLR occurs in the Indo-Pacific region. Barnett calculated that SLR almost doubled in rate over the period from 1920 to 1980 compared to the prior fifty years. This acceleration is likely to continue.

The time required for the ocean's surface to reach equilibrium with a warming atmosphere is considerable (Hansen, 1987). Most of the atmospheric and oceanic warming that should have occurred because of trace gases already added to the atmosphere by man is not yet observable. The great thermal capacity of the oceans (thermal inertia) is probably responsible for the delay in manifestation of the greenhouse warming to which the atmosphere and ocean may already be committed. Projections of the potential SLR to about the year 2100 range from a few decimeters to a few meters. For example, Hoffman et al. (1986) suggest a minimum rise of 0.57 m and a maximum rise of 3.68 m.

Were the atmosphere to retain more of the carbon dioxide (CO_2) emitted each year, along with a very high retention of other trace gases, and were this to coincide with an increase of solar luminosity of 0.5 percent, an SLR of about 4.3 m is possible (Titus, 1987). One worst-case scenario assumes disintegration and rapid melting of the West Antarctic ice shelf. This could cause an SLR of perhaps 6 m. Oerlemans and Van der Veen (1984) argue, however, that this extreme case is not likely to occur in the next century. A widely accepted analysis of de Q. Robin (1987) argues for an 0.8 m SLR by the year 2100 with a possible range of 0.2 to 1.65 m. This wide range in possible SLR futures stems primarily from the delays inherent in ocean warming (Hansen et al., 1985) and from the possibility of an intensified snowfall over Antarctica (Oerlemans, forthcoming).

Thus it appears that the future contribution from the melting of glaciers and Greenland ice will probably be the most important determinant of global SLR over the next century (Kuhn, forthcoming). On the local and regional scales, however, geostatic adjustments, tectonic movements, and sedimentation or subsidence due to mining activities may become significant factors determining relative change in sea level (Aubrey, forthcoming).

For planning purposes it appears that a useful and probable range of SLR over the next century would be from 0.5

to 1.0 m. A 5-m SLR can be thought of either as the worst case that might occur during the next century or as an effect that might occur within three to five centuries. The fact that sea level has risen by 65 m since about 10,000 B.P. (figure 4-1) makes these numbers less fanciful than they would otherwise appear.

Short-Term Meteorologic Events

Superimposed on the secular trends discussed above are changes in sea level determined by meteorologic factors. Physical laws hold that a fall in pressure of 1 millibar (mbar) will cause a 10-mm sea-level rise. (Mean atmospheric pressure at sea level is about 1,013 mbar.) During a series of rapid, sequential barometric lows (depressions) or a longer-lasting deep depression, an SLR of up to 0.3 m is not uncommon. Christiansen et al. (1985) in Denmark noted intra-annual variations in sea level of 0.13 m, monthly variations of 0.55 m, and diurnal variations of as much as 5 m. Along Chinese coasts, too, high sea level has occurred during years with strong barometric depressions and heavy rainfall. Sea levels may be 0.3 m higher than normal during very rainy seasons, and during typhoons surges up to 4.7 m occur regularly. In one case at Shanghai, sea level was 5.74 m above normal (Qin et al., forthcoming).

Wind direction and wind speed can greatly influence sea level, particularly in embayments and estuaries. Under gale force winds at Wilhelmshaven, West Germany, sea level has varied by as much as 3.5 m above normal astronomical tide level depending on wind direction (figure 4-3). The distribution of wind directions is not constant. Danish statistics of the periods 1876 to 1925 and 1931 to 1960 show decreased frequency of northwest winds and increased frequency of winds from the west and southwest. More recently, northwest winds as well as the number of storms appear to be increasing (Christiansen et al., 1985). Greenhouse warming is likely to lead to a decrease in the temperature gradient between the equator and poles (see Schneider and Rosenberg, chapter 2 in this volume). This could result in fewer and weaker barometric depressions, but how this global effect would distribute regionally and locally is far from being understood at this time.

Along the North Sea coast from France to Denmark, safety requirements are such that the top of a seawall or dike must be at least 16 m, and in funnel-shaped estuaries even 17 to 19 m, above mean sea level in order to reduce the probability of exceedance to less than 1 in 10,000 years. This is the so-called "Delta Norm" (Goemans, 1986). For a typical dike the calculation is done using the following allowances: storm surge level, 5.00 m above mean sea level (msl); wave run-up, 9.90 m; seiches (natural oscillations of the surface of landlocked seas or lakes amplified by the effects of storms) and gust bumps, 0.35 m; dike settlement, 0.25 m; and, finally, provision for SLR, 0.25 to 0.50 m.

PROBABLE EFFECTS AND LOCATIONS OF SEA-LEVEL RISE

Tidal Shifts and Their Effects on Surges and Currents

Changes in tidal systems that could occur as the result of SLR have been investigated for the North Sea by means of a model of the continental shelf developed by Rijkswaterstaat Netherlands (Verboom et al., 1987). Tide system calculations were made for the present situation and for sea levels 2.5 and 5.0 m higher and 2.5 m lower. At higher levels, and hence greater water depths, the tidal waves will propagate faster, and tidal amplitudes (levels and currents) will increase. The position of the nodes around which tidal waves turn (referred to as amphidromic points) will change, and the tidal amplitude along the coast will also change by several percent upward or downward (de Ronde, forthcoming). Even at a 1-m SLR with relatively small amplitudinal changes, there will still be important changes in residual currents in such places as the Wadden Sea behind the existing Barrier Island (de Ronde, forthcoming).

With changes in currents come changes in sand transport, erosion, and sedimentation. If sedimentation keeps pace with SLR, assuming the existence of available material, a shallow bottom will break the waves that roll onto the coast. Model calculations show, however, that for a 5-m SLR over an unchanged seabed, wave height at a sandy dune coast might increase from 0.9 m at present to 2.1 m (de Ronde, forthcoming). Amphidromic changes may influence the height of the storm surge itself, especially in conjunction with wave transformation by the specific shape of a basin or river mouth. For example, if a surge enters an estuary from a slightly different angle, the pattern of wave modes and of wave peaks may change. Interference of wave modes and peaks can lead to higher surges. The amphidromic changes, however, do not affect the frequency and severity of storms.

The impact of tidal shifts on major oceanic currents like the Gulf Stream and on the pattern and frequency of the Southern Oscillation is probably negligible, but the opposite effect—that is, the impact of changing currents, such as a displacement of the Gulf Stream, on tidal patterns and local SLR—could be important. This author knows of no models that take account of such major events as shifts in direction of the Gulf Stream or the Labrador Current.

Coastline Regression

The sea at rocky coasts is usually bounded by steep cliffs, with only narrow beaches of boulders and gravel. At such places as the Atlantic coast of the British Isles, Scandinavia, and Chile, SLR would have much less impact than at lowland coasts of more erodible sands and mud such as

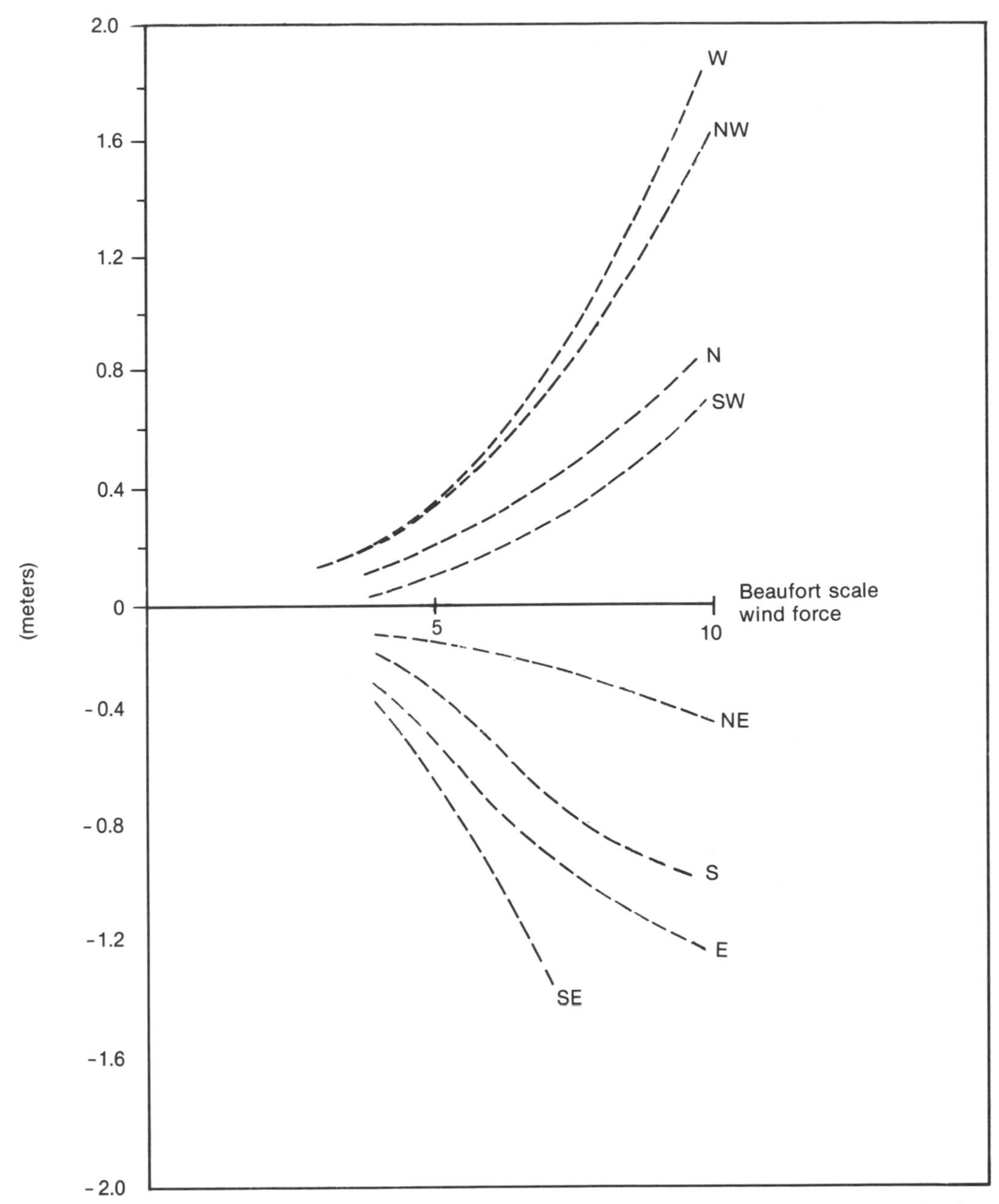

Figure 4-3. The effect of variations in wind speed and wind direction on the high tide level at Wilhelmshaven, West Germany, in meters above or below the astronomical tide level. (*Source:* Redrawn, with permission, from Brouns, 1988.)

those found along the southern North Sea, the Baltic Sea, many deltaic lowlands of the Mediterranean, and many more coastal swamps, marshes, and wetlands around the world. A beach that has attained equilibrium with coastal erosion and sedimentation processes will respond to SLR by losing material from the upper part of a profile and gaining it in the near shore area until a new equilibrium is established (Bruun, 1962). The alterations of the beach profile usually start with storms, followed by finer adjustments at each high tide. Bird (1986) lists the factors that favor the initiation or acceleration of beach erosion, reduction of sand supplied to the shore due to the construction of breakwaters, seawalls, and other protective works.

Those coastlines with sand dunes from Belgium to Denmark have beach slopes from 1:60 to 1:80. Thus, coastline regression there with a 1-m SLR will be 60 to 80 m.

Barrier island coasts such as those found at the Waddenzee in the Netherlands, off the coasts of Texas and Florida, along the Pamlico Sound of North Carolina, and in many places off the west coast of Africa will respond in a variety of ways to sea-level rise, depending on bottom and sediment availability. Some barrier islands are naturally rolling on shore as sand is eroded from the side exposed to the sea but is replenished by sediments deposited on the quieter side. When the supply of new sediments is not great enough to keep pace with erosion, then barrier islands diminish in size and can be completely submerged, or "drowned."

Coastal lagoons—embayments partly or wholly sealed off from the sea by sediment deposition and other spills and with salinity gradients from fresh river inflow through brackish water to seawater—will tend to be trimmed back or to become more saline, or both with SLR.

In river deltas and estuaries, sediment transport from inland is the main factor determining the impact of SLR. If sediments are trapped by man-made dams in upstream reservoirs, deltas will change into estuaries, and SLR will help the process of eroding away deltaic deposits. The tidal volume of the river mouth or estuary will increase until a new equilibrium is established between the widened mouth or inlet and the sedimentation-erosion balance. The Mississippi Delta, with its accelerating land loss of 100 square kilometers (km^2) per annum, is an example of simultaneous sedimentation decrease, land subsidence, and SLR (Day, 1987).

Since the construction of the Aswan High Dam, the Nile Delta has shrunk because of a lack of sediment supplementation (El-Fishawi, 1987). Similar processes have occurred in the Mediterranean region for the Ebro (Marino, 1987), the Gulf of Cadiz (Zazo et al., 1986), the Venetian and Po deltas (Sestini, 1987), and the northern Aegean (Perissoratis and Zimianitis, 1987). Such effects are no doubt to be found in the deltas of hundreds of rivers where dams have been constructed. With SLR the deltas will disappear more quickly unless protected by coastal works.

The still-growing deltas of some of the large Asian rivers such as the Ganges-Brahmaputra, the Irrawaddi, and the Mekong behave differently. Almost all tropical coasts are naturally protected by mangroves (salt-tolerant forest swamps). But mangroves fringing coastlines are disappearing rapidly now throughout Asia, Africa, and Australia because of land reclamation, fish-pond construction, mining, and waste disposal. Where mangroves persist, they typically stand in front of zones of salt marshes and freshwater vegetation and as a natural, living barrier to the sea, absorb surges and trap sediment, thereby protecting the hinterlands.

In many cases the advance of mangroves has caused coastal accretion, with salt marsh and saltwater communities following in succession as sedimentation builds up the substrate to appropriate levels. Thus, intertidal deposition terraces are formed (Bird, 1986). This is particularly true of the Bay of Bengal coastline in Bangladesh, where sediment supplementation from the Himalayas is the source of land accretion. However, ongoing destruction of the mangroves remains appalling, and when sediment is no longer "trapped," the coast becomes more and more exposed to surges.

Although mangroves regenerate quickly in areas that have been abandoned and would likely spread back to suitable habitats as the sea level rises, present land-use practices will prevent this. On the Asian coasts, land immediately behind the mangroves is commonly used intensively for fish ponds and rice paddies. It is this land that would, barring human intervention, be colonized by retreating mangroves as sea level rises and sand deposition is diminished. In drier areas, as in Northern Australia, mangroves are backed by bare, hypersaline plains, and in these areas SLR will result in colonization by mangroves of presently unvegetated tracts as the coastline retreats (Bird, 1986).

Within large developing countries such as Argentina, Brazil, and India, certain land areas threatened by inundation will likely be protected and others abandoned. Not all coastal stretches will be defended at any price in the highly industrialized countries either. Massive expenditures for coastal protection schemes are more likely to be considered near urban and industrial centers and densely populated agricultural coastal plains. The Vilan Plain in northeast Taiwan is such an area of intensive agriculture, and it is already artificially protected against the sea at high costs (Hsu, 1985).

Coastline recession of several hundred meters could occur in the Gulf of Thailand, specifically the bight of Bangkok. Coastal villages would be displaced and many people deprived of land and resources. The mangrove fringe has been largely cleared in this region, while the landward canals, dug to bring fresh water to the rice fields, are now becoming channels for the intrusion of salt water and storm surges from the sea. Many of the existing rice fields here can be converted into brackish fish ponds, but there seems little chance that funds will be available for constructing seawalls in Thailand, as is now done in the Netherlands, for example (Bird, 1986).

At certain points along the eastern coast of South America similar but less dramatic processes have been reported. Most alarming is the situation in the state of Rio de Janeiro, where the destruction of mangroves is virtually uncontrolled (Leatherman, 1986).

Indonesia is worthy of special discussion. It possesses 15 percent of all the world's coastline, and at least 40 percent of its land surface is vulnerable to a sea-level rise of 1 m, either directly by inundation or indirectly by increased coastal erosion and saltwater intrusion. Additionally, it is the richest of countries in terms of wetland ecosystems, both in quantity and diversity (Silvius et al., 1986). At this time, however, millions of inhabitants from overpopulated Java and Bali are being transferred to the tidally influenced swamps of Sumatra and Kalimantan (Borneo). There the immigrants are faced with increasingly serious problems that limit their prospects for agricultural adjustment (for example, rapidly declining soil fertility, soil compaction, biochemical oxidation and toxification, and saltwater penetration). Sustaining agriculture is already problematic

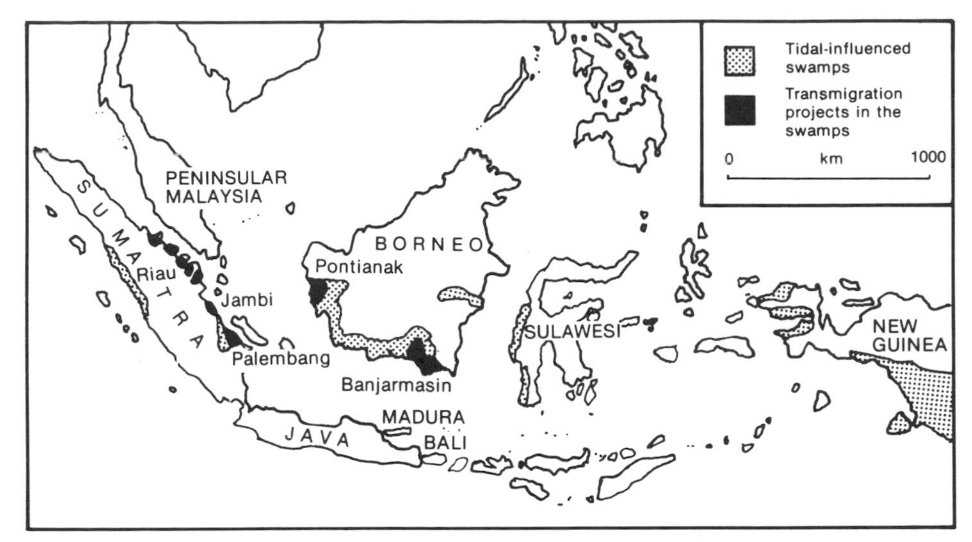

Figure 4-4. Location of official transmigration projects in the tidal wetlands of Indonesia. (*Source:* Reprinted, with permission, from fig. 5.IV.1, p. 179, in M. G. Wolman and F. G. A. Fournier, eds., *SCOPE 32: Land Transformations in Agriculture.* © 1987 by John Wiley & Sons, Chichester.)

(Ruddle, 1987), and it will become even more so due to SLR. The areas of transmigration in Indonesia are shown in figure 4-4.

How coastal nations will respond to a relatively rapid SLR will vary considerably with economic and political factors, in particular population pressure. Industrialized countries including Japan, Korea, Taiwan, Singapore, South Africa, Australia, and New Zealand will likely respond differently from most of the Asian and African coastal nations. Where beaches are valued for recreation and tourism and are close to cities and seaside towns and resorts, as in Australia, Asia, Africa, and certain Pacific islands, they will likely be protected against erosion and maintained by expensive beach nourishment programs of sand supplementation (Bird, 1986).

Saltwater Penetration

Intrusion of salt water is a well-known phenomenon in all coastal lowlands and has been particularly well studied in the Netherlands (de Ronde, forthcoming). Intrusion has occurred at least 50 kilometers (km) upstream of the mouth of the Rhine River. Groundwater of the surrounding polders (low-lying land reclaimed from a body of water and protected by dikes) has been affected. It is also possible for inland brackish water to seep underground into parts of the country that are below or close to sea level. The presence of a sandy dune coast with a freshwater pocket or lens overlying deeper salt water does not prevent seepage into lowlands behind the coastal dune (Volker, 1987).

The best way for the Netherlands and for countries with similar tidally influenced, agricultural coastal lowlands to counteract saltwater penetration is through the conservation of freshwater from the rainy season into the drier season. This can be accomplished by raising the levels of lake and polder bossoms (the complex of canals and ditches) by several decimeters. This kind of action, however, raises the possibility of conflict with other agricultural interests. Farmers usually prefer that surplus water in spring be pumped out as rapidly as possible to allow farm operations—plowing, mowing, grazing—to begin. Surplus water in springtime is relatively unpolluted. At times of water shortage during the drier summer season, farmers want river water to replenish polder bossoms. However, in summer the rivers contain less water and higher concentrations of pollutants. In Western Europe, where there is overproduction in agriculture, the possibility of abandoning farming in some of the polders that are far below sea level and turning them into freshwater lakes for fisheries and recreation should be seriously considered.

In developing countries there may be insufficient means for manipulating freshwater reservoirs to prevent the loss of agricultural land to seawater intrusion. This unavoidable loss may be partially compensated by turning lost paddy fields into fish or shrimp lagoons. These can be equally productive and economical and entail lower maintenance costs than do rice paddy fields. This tactic could help developing countries cope with SLR while improving national food security. In the meantime, plant breeding is proceeding to select for more salt-tolerant crops. Such a tactic can postpone awkward decisions about abandoning croplands, at least for some decades.

Loss of Coastal Lowland Ecosystems

Marine intertidal zones are of great significance as hatcheries for fish, fry, shrimp, oysters, and other biota that serve as food for geese, ducks, and wader birds. This water-locked wealth is now threatened almost everywhere, but in combination with SLR, human impact will become still more dramatic in the future. Ecosystems, however, cannot easily be valued in financial terms and hence do not often appear in economic balances.

How coastal ecosystems will react to SLR can be deduced partially from slow subsidence processes such as those occurring at glaciers on the Black Sea coast and in the Wadden Islands (in particular, Ameland) because of the mining of natural gas and oil. At Ameland 0.26 m of subsidence has occurred over the past twenty years. Coastal erosion is greater at the North Sea side. At the Waddenzee side the intertidal flats now become dry for much shorter periods at low tide. With SLR the area where birds can forage on the marshes and slicks will decrease progressively. The lens of fresh groundwater in the dunes rises to the surface and is less deep, while salt water intrudes from underneath. The vegetation of the dunes, in particular of the valleys, will shift to more xerophytic and salt tolerant types (Dankers et al., 1987).

THE IMPACT OF SEA-LEVEL RISE ON SOCIETY

The total length of coastline in the world is somewhere between one-half million and 1 million kilometers. If maximum storm surges and the upstream effects of flooding and saltwater intrusion are taken into account, then all land up to 5 m above mean sea level is, within one to three centuries, potentially vulnerable in one way or another to SLR. Vulnerable areas throughout the world are shown in figure 4-5, and the details of vulnerable areas in Europe are shown in figure 4-6. The land area that would be subject to inundation or indirectly influenced, deteriorated, or made vulnerable by saltwater intrusion is on the order of 5 million km^2—about 3 percent of the land area, but one-third of the total area of cropland in the world (Wind, 1987). Much of the threatened land is densely populated and holds many large cities. As many as 1 billion people may live in these areas. Characteristic values of arable land are \$300,000 per km^2 (equal to \$1,200 per acre) in Bangladesh and ten times this much in the Netherlands (Wind, 1987).

Sea-level rise can cause not only losses of land and property but also damage to engineering structures and to the entire socioeconomic system of a country. In 1987 the Delft Hydraulics Laboratory with the support of the United Nations Environmental Programme (UNEP) and the Netherlands Government developed the Impact of Sea Level Rise on Society (ISOS) project in order to enhance national and regional awareness of the impacts of SLR and to develop strategies for responding to these impacts. The ISOS project aims to build a global network of experts in national agencies who can develop and use software and data bases with which to model the policy options appropriate for each case study area. Results are to be presented through workshops for policymakers in each area. The first full case study, now in progress, is for the Netherlands and is financed by its Ministry of Public Works. Apart from its contribution to developing the model, the network, and the data base, UNEP will also support two urgent case studies—for Bangladesh and the Maldive Islands in the Indian

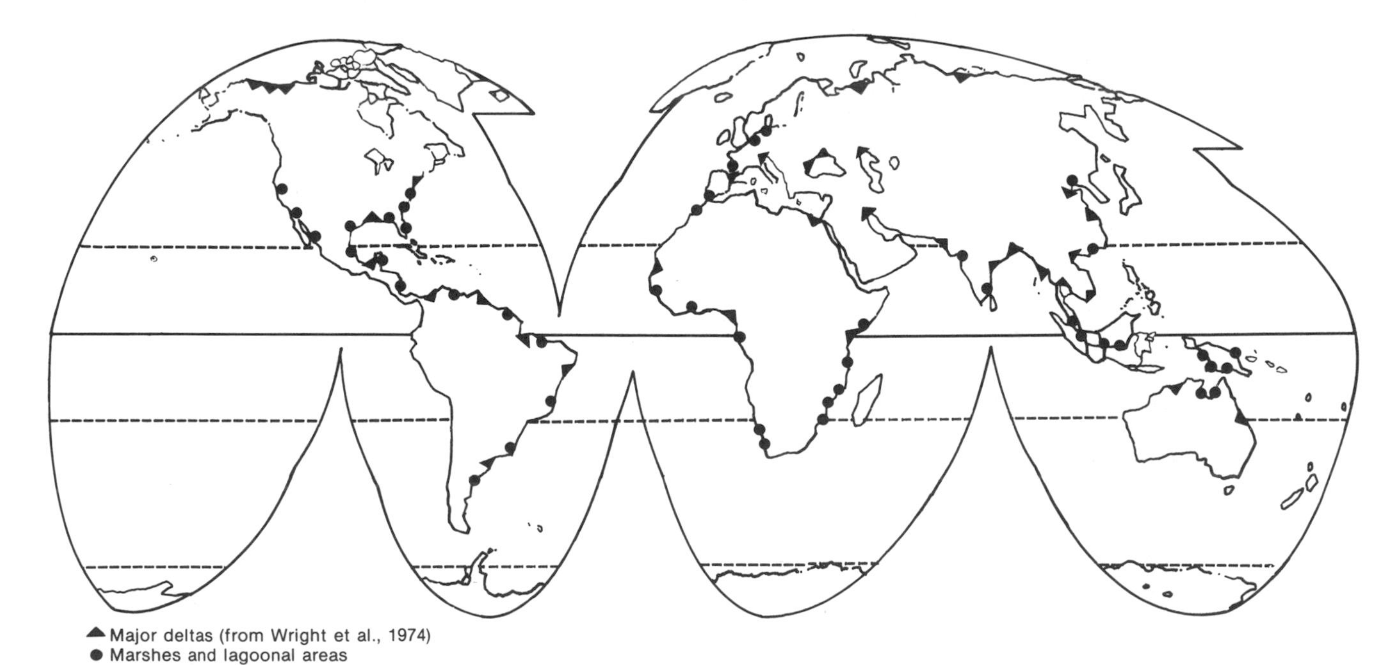

Figure 4-5. Areas of the world vulnerable for one reason or another to a rising sea level. (*Source:* Redrawn, with permission, from Jelgersma, 1986.)

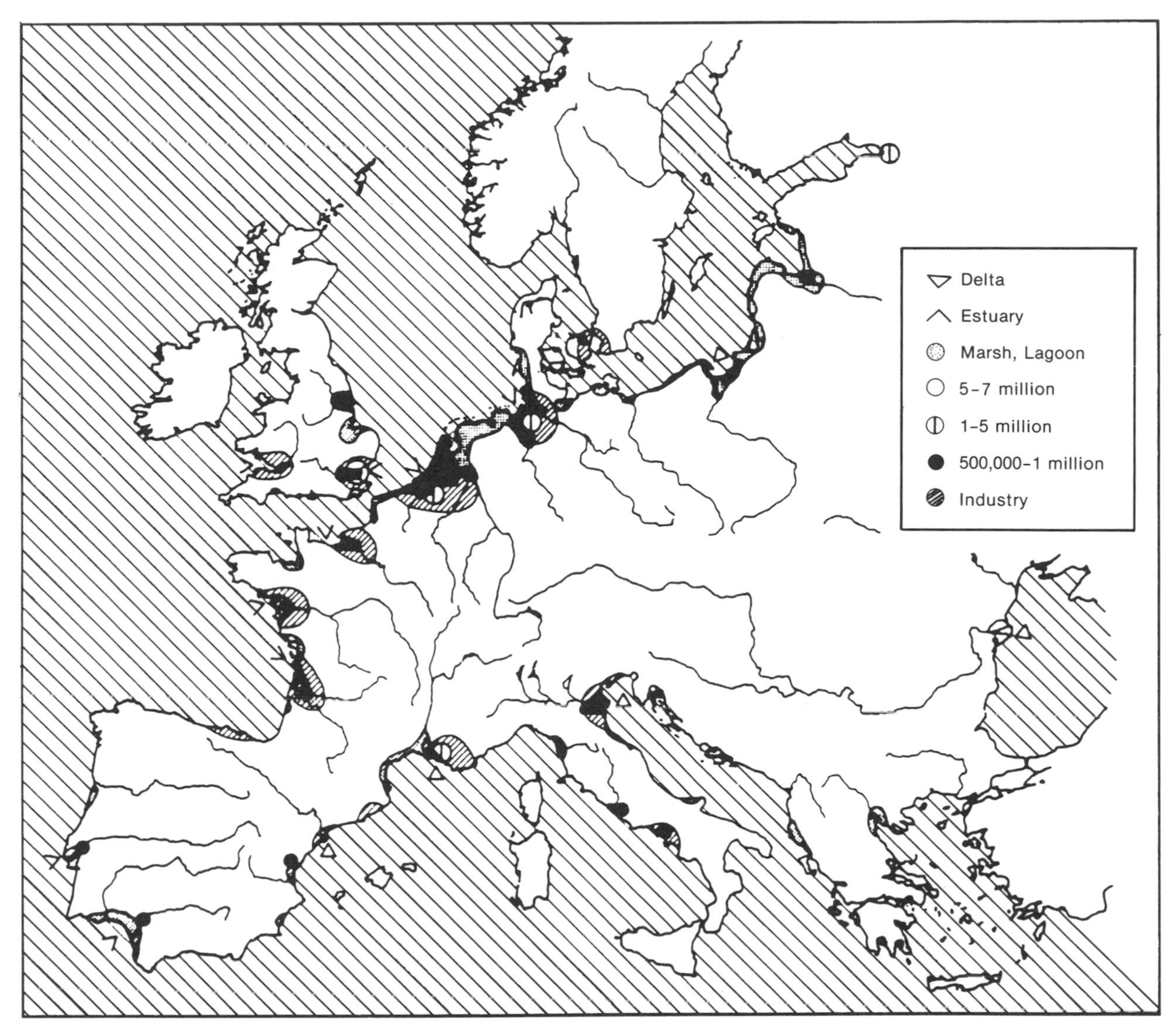

Figure 4-6. European areas vulnerable to a rising sea level. (*Source:* Reprinted, with permission, from Jelgersma, 1986.)

Ocean. A full description of these studies is given by Wind (1987).

Figure 4-7 diagrams the analytical framework used by Delft Hydraulics Laboratory as part of the ISOS project to describe the effects of SLR on society (Wind, 1987). First, various scenarios of SLR provide input to the model. For the selected country or region, social, demographic, and economic scenarios have to be developed as well. Furthermore, realistic measures that can be taken with regard to port and shipping systems, water resources, and flood protection must be defined on the basis of on-site studies. The result of the ISOS analysis is an assessment of impacts and effects in a form that can be presented to decisionmakers.

The ISOS plan was developed with an awareness of political realities. The manner in which societies respond to SLR and its effects are known to be determined by many factors. At any one time many problems compete for the attention of decision makers. Experience indicates that only sudden events or disasters trigger sufficient awareness to cause countermeasures to be considered, even though these events or disasters are foreseeable. Consider, for example, the case of the major flooding in 1953 in the southern Netherlands. The response to that disaster was the so-called Delta Plan. Construction work actually began in the early 1960s. Completion of all coastal defenses to the level of safety now demanded will occur in the early 1990s. The total costs over that thirty-year period amount to about U.S. $7.5 billion. Over the entire next century the Netherlands may need to spend additional billions to cope with a 1-m sea-level rise.

Other major costs and related factors that influence policy decisions include the interests that are endangered, the availability of finances, the employment opportunities that a response strategy provides, the political responsibilities

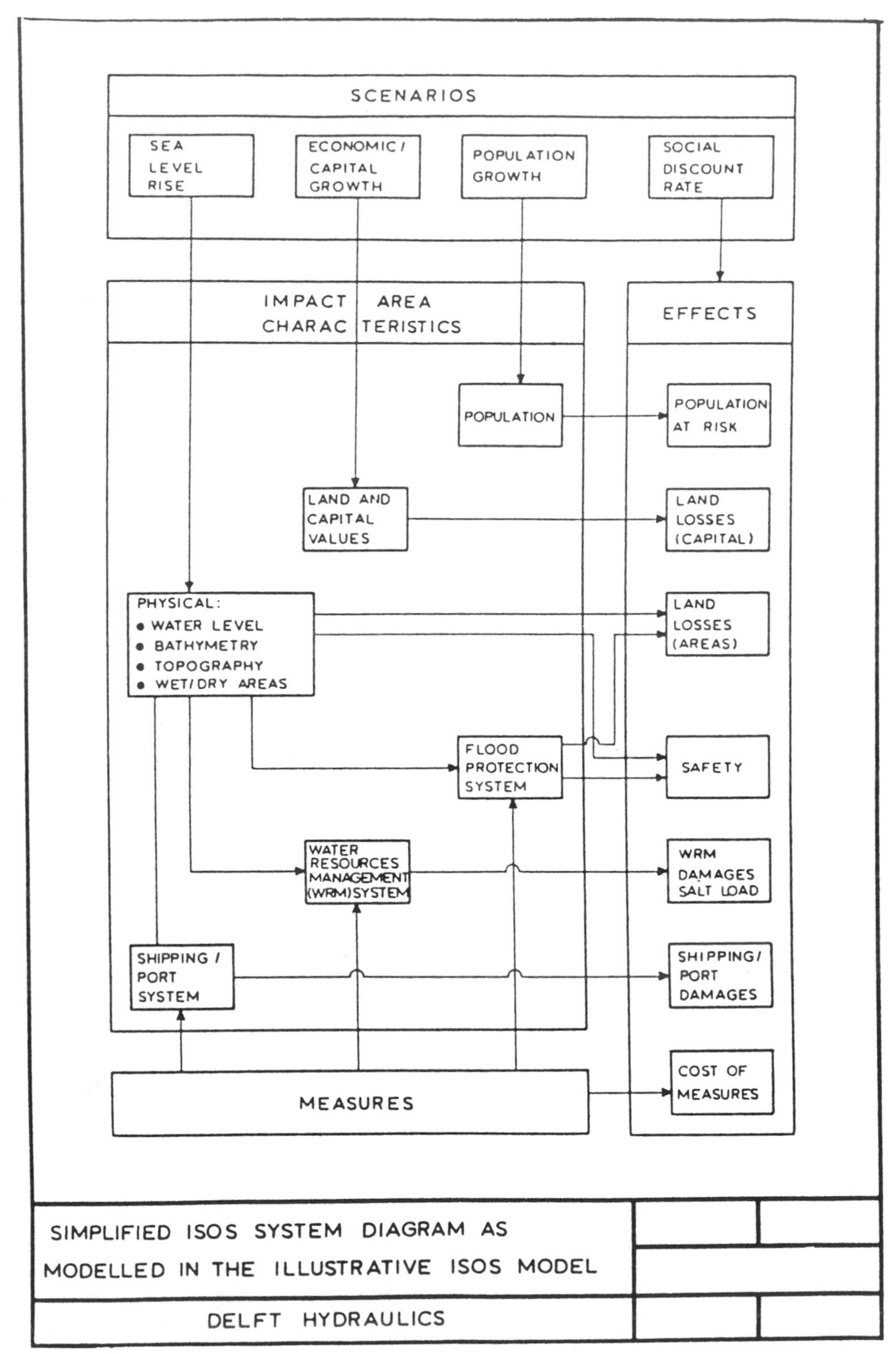

Figure 4-7. The analytical framework used by Delft Hydraulics Laboratory to describe the effects of SLR on society. (*Source:* Reprinted, with permission, from H. G. Wind, ed., *Impact of Sea Level Rise on Society: Report of a Project-Planning Session,* Delft, Aug. 27–29, 1986. Rotterdam, Netherlands–Brookfield, Vt., A. A. Balkema, 1987.)

involved, and the national prestige. An estimate of the lead time has to be made for each case study. Whatever the choice of the decision makers—defensive action, planned retreat, abandonment—each can be elaborated in various forms for its technical, environmental, financial, social, and, perhaps, political consequences.

The ISOS model is rather technocratic in approach; that is, it deals with values most highly regarded by policymakers. The most important impacts considered in the model relate to losses of land (both in area and in capital terms), cost of providing security against flooding, damages related to water resources management and to the shipping and port systems, and the inland impact of saltwater intrusion on agriculture and on drinking water supplies. Notoriously lacking in the model is a means of evaluating the risks to natural ecosystems.

The model is programmed on microcomputer software. The model in its present version has two different possible uses: (1) exploration of involved problems, mechanisms, and tradeoffs and (2) analysis of actual strategies to find the "best" one.

In order to determine whether one strategy is better than another, the model provides results in concrete terms (for example, monetary losses, number of people and estimated capital value at risk, and safety of the flood protection system). Monetary quantities should be interpreted critically, however, in relation to the effect of the social discount rate—that is, how great a sacrifice society is willing to make now for the benefit of future generations. Thus, the lead time required for a decision to take effect is another important factor to consider.

Preliminary Results for the Netherlands

Rijkswaterstaat Netherlands has already calculated the cost of a 1-m SLR, assuming that defense is the policy option chosen, rather than planned retreat or abandonment. The cost of new defense measures will be about twice the current cost of coastal defense maintenance. Strengthening dikes, dunes, beaches (sand shorelines), and shore faces (nonsand shorelines such as mudflats) over the next century will cost about U.S. $3 billion; the current cost of maintenance is about U.S. $30 million per year. An interesting detail is that a storm surge barrier in the Rotterdam waterway currently under study as an alternative to the urgent and costly strengthening of the dikes of the lower rivers becomes all the more attractive with the prospect of SLR. Inland adaptations to a 1-m SLR are estimated to cost U.S. $1.5 billion, primarily for pumps and other needed changes in infrastructure. Operation costs will increase by about U.S. $5 million per year. Adaptation of harbors, locks, bridges, and the like will cost about U.S. $0.5 billion. The total cost of adaptation to a 1-m SLR can be estimated at about U.S. $5 billion to $6 billion spread over about 50 to 100 years (de Ronde, forthcoming). By comparison, the Delta works cost about U.S. $7.5 billion over 30 years.

It appears, then, that the Netherlands can cope technically and financially with a projected sea-level rise of 1 m occurring over 50 to 100 years. But it is obvious that many developing countries that lack the Netherlands' means and have much longer coastlines to protect are in a less enviable position. The ISOS model can help them determine which areas to defend or how best to minimize the costs and losses of a planned retreat.

DEVELOPING INTERNATIONAL POLICY RESPONSES TO SLR

International Solidarity Regarding Effects of Sea-Level Rise

Throughout history coastal nations have muddled through inundations by the sea. Lands have been lost, as have lives, and property has been damaged repeatedly as nations have struggled to defend against storm surges. International relief for the peoples most affected after dramatic disasters is always heartening, but it is usually limited to humanitarian aid. When it comes to the long term, only the affected countries themselves generally, but not always, are willing to consider prevention.

But sometimes donor aid to developing countries takes long-term environmental protection into consideration. Mainly this occurs when the donors' own interests are served in delivering technology, equipment, and expertise. As yet no system exists for developing international commitments to deal with the effects of globally induced changes on specific areas. The ISOS model, for example, deals only with national and local decision making.

Countries in North America, Europe, and other regions with heavily industrialized coastlines have begun to make preliminary, but still very crude, cost estimates of the possible effects of SLR (see, for example, Titus, 1986). It is still extremely difficult to total the potential global costs of SLR in a manner similar to that being done for the Netherlands. One way to express global costs might be as a percentage of the present cost of protection and maintenance. This crude statistic can then be refined in each country or region of a country with the help of the ISOS model or some similar approach. Vellinga (1987) made a first attempt at doing so at a meeting sponsored by the United Nations Environmental Programme (UNEP), World Meteorological Organization (WMO), and International Council of Scientific Unions (ICSU) held in Villach, Austria, in September 1987. The findings of this meeting and a follow-up meeting held at Bellagio, Italy, November 1987 are reported by Jäger (1988) and have been used by the Advisory Group on Greenhouse Gases. Results of Vellinga's assessment are presented in figure 4-8, which shows

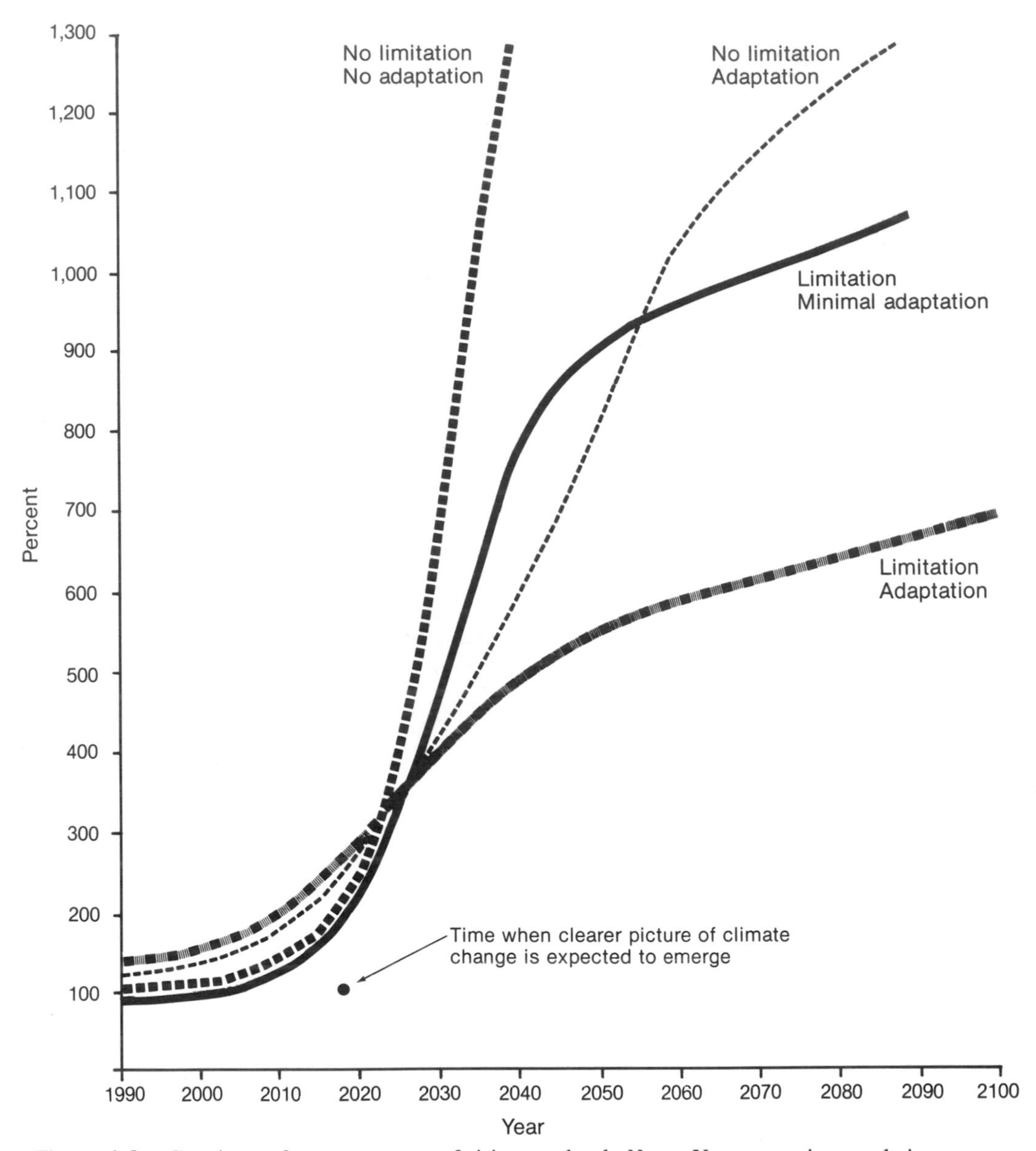

Figure 4-8. Cost issues for management of rising sea level. *Notes:* Very approximate relative costs are shown as percentages. No limitation: no additional actions on greenhouse gas emission. Limitation: reduce emissions by 2 percent per year. No adaptation: retreat from coastal damage. Minimal adaptation: ad hoc measures after disasters. Adaptation: selective coastal engineering measures. (*Source:* Redrawn, with permission, from Vellinga, 1987.)

four scenarios. The lower curve in the figure represents the ideal ("Limitation/Adaptation") scenario, which assumes that all of the world is engaged in a strenuous effort to limit the emissions of greenhouse gases by at least 2 percent per year and that appropriate measures of defense against SLR begin immediately. The upper curve stems from a scenario which assumes that no attempts are made at limitation or adaptation (selective coastal engineering measures). Within forty to fifty years the losses incurred would be overwhelmingly expensive. In Indonesia even an 0.5-m SLR could cause losses three or four times greater than that nation's current investment in defense. The intermediate scenarios assume either that unlimited emissions of greenhouse gases are accompanied by a positive adaptation program or that greenhouse gases are limited to some extent and only minimal adaptation (ad hoc measures after a disaster) is attempted. It seems most likely that some countries will emphasize limitation and others adaptation and that the world at large will do some of both, but probably not as much as might be needed.

Figure 4-8 is based on two assumptions: (1) that the costs of climatic change and of SLR in particular are already clear enough to justify action and (2) that the measures needed to abate or adapt to climate change and SLR must be accomplished within the next three to four decades—the lead time needed for major engineering works—if they are to be effective in time. Figure 4-8 implies that, regardless of scenario, the relative costs and impacts will be, by and large, the same until about 2025 to 2030 but will diverge widely thereafter. However, the lack of divergence until 2025 to 2030 should not be considered an argument for postponing decisions until then, because the lead time required for response will still be on the order of three to four decades.

Costs are postponed in the two scenarios that involve little or no adaptation over the next two decades. However, irreparable damage may occur later. The scenarios that involve immediate attempts at adaptation risk the possibility that SLR will be less dramatic than is now projected or that it may not occur at all. In such a case abatement or adaptation will have been a waste of capital.

One disadvantage of early engagement in defensive strategies is that decisions will likely be made on the basis of present-day technology. In the case of coastal defense structures, today's technology is heavy and clumsy, involving sand, clay, boulders, asphalt, and concrete. Technical innovations such as the application of super fiber tissues, new alloys, and other synthetic products have not yet penetrated into such works to any large scale. Properly, such technical innovations, after testing in the industrialized countries, should be made easily available to the developing countries. Experience in other sectors has shown, however, that technologies transferred to the developing countries, often as "aid," are mostly obsolete. Transfer of state-of-the-art technology should be a feature of worldwide solidarity for dealing with the SLR question.

SLR and Strategies for Mitigating Climatic Change

Thus far in describing figure 4-8, measures for limiting the causes of climatic change have not been discussed in detail. It is obvious, however, that reduction in the emissions of carbon dioxide and other radiatively active trace gases must be paramount. Yet SLR is only one of the components of global change induced by the prolific growth of human numbers and activities that has led to the increase in greenhouse gas emissions. This chapter is not the appropriate place for a lengthy discussion of the various policy proposals for mitigating and adapting to climatic change, but the reader is referred to other chapters in this volume and to Bach (1987 and 1988), Mintzer, (1987), and Jäger, (1988), for discussions of these approaches.

SLR as a Lever to Reduce Greenhouse Warming

Climatic change may in some cases affect primary photosynthetic production and hydrology negatively and dramatically (for example, by causing desertification, crop failure, and forest dieback). Nevertheless, SLR is the most clearly foreseeable direct effect of climatic change and is likely to be the most visible direct effect and to have the greatest impact on society, particularly in many of the most prominent and dominant nations. The Advisory Group on Greenhouse Gases (AGGG) has proposed policies for coping with climate change (Jäger, 1988). Among their priorities for action is limitation of the impact of SLR by the identification of vulnerable areas and the prevention of further development of such vulnerable areas.

The AGGG, in consideration of both coastal and river valley risk management, recommends priority attention be given to identification of vulnerable areas and prevention of further development of such areas as ways to limit the impacts of SLR. The more detailed recommendations (unpublished) of the Bellagio meeting of November 1987 read about as follows:

- International unions of geographic and geodetic, coastal, hydrologic, and soil sciences should develop—through their respective national societies—maps to identify coastal areas vulnerable to the effects of climatic and sea-level changes, river regulation, and intensification of land use, or any combination thereof. Mapping of areas at risk should preferably be completed in five years.
- In the longer term the United Nations Disaster Relief Organization (UNDRO), in cooperation with WMO and UNEP, should coordinate national efforts to assess the vulnerability of developments in coastal regions and large river valleys to climatic and sea-level changes and should stimulate governments to develop policies to accomplish the assessments.
- The Global Sea Level Observing System should give urgent attention to strengthening the monitoring of sea-level changes. The WMO, together with the ICSU, should foster improved modeling techniques for predicting frequencies and intensities of storms which, in combination with sea-level rise, will increase the risk of flooding of coastal areas and river valleys.
- In its studies of the impacts of climate change on tropical islands and coasts, the UN University Study Program should consider the effects of pollution, coral harvesting, and mangrove destruction on the natural flood protection in tropical coastal areas.
- Governments should see to it that the planning for large new industrial, tourist, and urban facilities near the sea and in river valleys should involve consideration of

the possible risks of sea-level rise, changes in river discharges and erosional patterns, and reduced natural protection by coastal (mangrove) forests, sand banks, and coral reefs.

Mintzer (1987) has provided some other detailed recommendations. Despite the best possible control of greenhouse gas emissions by combining the use of technical innovations and societal adaptations, there will be, he says, a slow greenhouse warming of between 1°C and 3°C by 2075. With modest policies (best present technologies only), there could be by 2075 an increase of between 2°C and 6.5°C and if present wasteful trends continue between 3°C and 8.5°C. For long-term planning purposes the contingency of an SLR up to 1 m should be considered, even with the slowest of the warming scenarios.

A FINAL WORD

Sea-level rise caused by climate warming is certain to cause problems for many of the world's coastal countries. The actions that we take, or fail to take, in the next few decades will greatly determine our effectiveness in dealing with these problems. Farsighted and stringent measures to reduce emissions of gases that can influence climate and the stratospheric ozone layer are imperative. Ministers of the environment, when they assemble at international conferences, of which there will be several on the issue of climate change in coming years, should see to it that the rising levels of concern about the future of our climate and the biosphere are translated into appropriate political action.

REFERENCES

Aubrey, D. Forthcoming. "Regional Variations Due to Isostatic Adjustment, Tectonic Movements, Sedimentation and Other Local Processes," in T. M. L. Wigley and R. A. Warrick, eds., *United Nations Environmental Programme/CEP/USEPA International Workshop on Climatic Change, Sea Level, Severe Tropical Storms and Associated Impacts*.

Bach, W. 1987. "Reducing Emissions of Greenhouse Gases." Discussion paper presented at the European Workshop. Noordwijkerhout, Netherlands. October 16–21.

———. 1988. "The Endangered Climate," in F. Krause and W. Bach, eds., *Energy and Climate Change: What Can Western Europe Do?* Report for the Netherlands' Ministry of Housing, Physical Planning and the Environment, no. 15.

Barnett, T. P. 1984. "The Estimation of 'Global' Sea Level Change: A Problem of Uniqueness," *Journal of Geophysical Research* vol. 89 (C5), pp. 7980–7988.

Bird, E. C. F. 1986. "Potential Effects of Sea Level Rise on the Coasts of Australia, Africa and Asia," in J. G. Titus, ed., *Effects of Changes in Stratospheric Ozone and Global Climate, Proceedings of the International Conference on Health and Environmental Effects of Ozone Modification and Climate Change*, vol. 4, *Sea Level Rise* (Washington, D.C., U.S. Environmental Protection Agency).

Bolin, B. 1987. "The Role of Carbon Dioxide and of Other Greenhouse Gases for Climatic Variations and Associated Impacts." Paper presented at the European Workshop, Noordwijkerhout, the Netherlands, October 16–21.

———, B. R. Döös, J. Jäger, and R. A. Warrick. 1986. *The Greenhouse Effect, Climatic Change and Ecosystems, SCOPE* vol. 29 (New York, Wiley).

Brouns, J. J. W. M. 1988. "The Impact of Sea Level Rise on the Dutch Coastal Ecosystems." Draft discussion paper for the workshop Climate Change and Marine Ecosystems, Rijksinstituut voor Natuurbeheer, Texel, Netherlands, November 12–17.

Bruun, P. 1962. "Sea Level Rise as a Cause of Shore Erosion," *Journal of Waterways and Harbours Division*, American Society of Civil Engineering, vol. 1, pp. 116–130.

Christiansen, C., J. T. Moller, and J. Nielsen. 1985. Fluctuation in Sea Level and Associated Morphological Response: Examples from Denmark. *Eiszeitalter und Gegenwart* vol. 35, pp. 89–108.

Dankers, N., K. S. Dijkema, G. Londo, and P. A. Slim. 1987. *De ecologische Effecten van bodemdaling op Ameland*. Report 87/17, RIN, Texel, Netherlands.

Day, J. W., Jr. 1987. "Consequences of Sea Level Rise: Implications for the Mississippi Delta," in H. G. Wind, ed., *Impact of Sea Level Rise on Society: Report of a Project-Planning Session* (Rotterdam, Netherlands–Brookfield, Vt., A. A. Balkema).

de Q. Robin, G. 1987. "Projecting the Rise in Sea Level Caused by Warming of the Atmosphere," in Bolin, et al., *The Greenhouse Effect, Climatic Change and Ecosystems, SCOPE* vol. 29 (New York, Wiley).

de Ronde, J. G. Forthcoming. "What Will Happen to the Netherlands if Sea Level Accelerates?" in T. M. L. Wigley and R. A. Warrick, eds., *United Nations Environmental Programme/CEP/USEPA International Workshop on Climatic Change, Sea Level, Severe Tropical Storms and Associated Impacts*.

El-Fishawi, N. M. 1987. "Sediment Movement at the Retreating Coasts of the Nile Delta, Egypt." Paper presented at the European Workshop, Noordwijkerhout, the Netherlands, October 16–21.

Goemans, T. 1986. "The Sea Also Rises: The Ongoing Dialogue of the Dutch with the Sea," in J. G. Titus, ed., *Effects of Changes in Stratospheric Ozone and Global Climate, Proceedings of the International Conference on Health and Environmental Effects of Ozone Modification and Climate Change*, vol. 4, *Sea Level Rise* (Washington, D.C., U.S. Environmental Protection Agency).

Gornitz, V., S. Lebedeff, and J. Hansen. 1982. "Global Sea Level Trend in the Past Century," *Science* vol. 215, pp. 1612–1614.

Hansen, J. E. 1987. "Prediction of Near-Term Climatic Evolution: What Can We Tell Decision-makers Now?" Statement presented before the U.S. Congress, Senate Committee on Energy and Natural Resources, November 9.

Hansen, J., G. Russell, A. Lacis, D. Rind, and P. Stone. 1985. "Climate Response Times: Dependence on Climate Sensitivity and Ocean Mixing," *Science* vol. 229, pp. 857–859.

Hekstra, G. P. 1986. "Will Climatic Change Flood the Netherlands? Effects on Agriculture, Land Use and Well-Being." *Ambio* vol. 15, no. 6, pp. 316–326.

Hoffman, J. S., J. Wells, and J. G. Titus. 1986. "Future Global Warming and Sea Level Rise," in Sigbjarnarson, ed., Iceland Coastal and River Symposium (Reykjavik, Iceland, National Energy Authority).

Hsu, T. L. 1985. "Coastal Plains of Taiwan," in E. C. F. Bird and M. L. Schwartz, eds., *The World's Coastline* (New York, Van Nostrand Reinhold).

Jäger, J. 1988. "Anticipating Climatic Change: Priorities for Action." *Environment* vol. 30, no. 7, pp. 12–15, 30–33.

Jelgersma, S. 1979. "Sea Level Changes in the North Sea Basin," in E. Oele, R. T. E. Schuttenhelm, and A. J. Wiggers, eds., *The Quaternary History of the North Sea*. Acta Univ. Uppsala. Symp. Univ. Uppsala: 233–248.

———. 1986. "A Future Sea-Level Rise: Its Impacts on Coastal Lowlands." Paper presented at UN Economic and Social Commission for Asia and the Pacific, Seminar on Geological Mapping in the Urban Environment, Oct. 29.

Kuhn, M. Forthcoming. "Past and Possible Future Contributions from Glaciers to Sea Level Rise," in T. M. L. Wigley and R. A. Warrick, eds., *United Nations Environmental Programme/CEP/USEPA International Workshop on Climatic Change, Sea Level, Severe Tropical Storms and Associated Impacts*.

Leatherman, S. P. 1986. "Coastal Geomorphic Impacts of Sea Level Rise on Coasts of South America," in J. G. Titus, ed., *Effects of Changes in Stratospheric Ozone and Global Climate, Proceedings of the International Conference on Health and Environmental Effects of Ozone Modification and Climate Change*, vol. 4, *Sea Level Rise* (Washington, D.C., U.S. Environmental Protection Agency).

Marino, M. G. 1987. "Sea Level Rise on the Western Mediterranean: The Ebro Delta Case." Paper presented at the European Workshop, Noordwijkerhout, the Netherlands, October 16–21.

Mintzer, I. M. 1987. *A Matter of Degrees: The Potential for Controlling the Greenhouse Effect* Research Report 5 (Washington, D.C., World Resources Institute).

Oerlemans, J. Forthcoming. "Possible Changes in the Mass Balance of Greenland and Antarctica Ice Sheets and Their Effects on Sea Level," in T. M. L. Wigley and R. A. Warrick, eds., *United Nations Environmental Programme/CEP/USEPA International Workshop on Climatic Change, Sea Level, Severe Tropical Storms and Associated Impacts*.

———, and C. J. van der Veen. 1984. *Ice Sheets and Climate* (Dordrecht, The Netherlands, Reidel).

Perissoratis, C., and V. Zimianitis. 1987. "Land-Sea Relationship at the Northern Aegean: A Case Study and Impacts of a Three-Meter Rise of Sea Level." Paper presented at the European Workshop, Noordwijkerhout, the Netherlands, October 16–21.

Qin, Y. S., S. L. Zhao, and S. X. Cang. Forthcoming. "Sea Level Change in the Eastern Coastal Region of China in the Last 100 Years and Its Possible Impacts in the Future," in T. M. L. Wigley and R. A. Warrick, eds., *UNEP/CEP/USEPA International Workshop on Climatic Change, Sea Level, Severe Tropical Storms and Associated Impacts*.

Ruddle, K. 1987. "The Impact of Wetland Reclamation," pp. 171–203 in M. G. Wolman and F. G. A. Fournier, eds., *SCOPE 32: Land Transformations in Agriculture* (New York, Wiley).

Sestini, G. 1987. "The Impact of Sea Level Rise and Temperature Increase on the Deltaic Lowlands of the Eastern Mediterranean." Paper presented at the European Workshop, Noordwijkerhout, the Netherlands, October 16–21.

Silvius, M. J., E. Djuharsa, A. W. Taufik, A. P. J. M. Steeman, and E. T. Berczy. 1986. *The Indonesian Wetland Inventory: A Preliminary Compilation of Information on Wetlands of Indonesia*. A joint study by the governments of Indonesia and the Netherlands. PHPA, AWB/INTERWADER, EDWIN, Bogor, Indonesia.

Titus, J. G. 1987. "The Causes and Effects of Sea Level Rise," in H. G. Wind, ed., *Impact of Sea Level Rise on Society: Report of a Project-Planning Session* (Rotterdam, Netherlands–Brookfield, Vt., A. A. Balkema).

———, ed. 1986. *Effects of Changes in Stratospheric Ozone and Global Climate, Proceedings of the International Conference on Health and Environmental Effects of Ozone Modification and Climate Change*, vol. 4, *Sea Level Rise* (Washington, D.C., U.S. Environmental Protection Agency).

Vellinga, P. 1987. "Sea Level Rise, Consequences and Policies." Paper presented at the Villach Workshop, Beijer Institute, Stockholm, September 28–October 2.

Verboom, G. K., R. P. van Dijk, and J. G. de Ronde. 1987. *Detailed Tidal Flow and Storm Surge Modelling of the Northwest European Continental Shelf.* (The Hague, Rÿks Water Staat (State Water Service)).

Volker, A. 1987. "Impacts of the Rapid Rise of the Sea Level on Flood Protection and Water Management of Low Lying Coastal Areas," in H. G. Wind, ed., *Impact of Sea Level Rise on Society: Report of a Project-Planning Session* (Rotterdam, Netherlands–Brookfield, Vt., A. A. Balkema).

Wigley, T. M. L. 1987. "Climatic Scenarios." Paper presented at the European Workshop, Noordwijkerhout, the Netherlands, October 16–21.

Wind, H. G., ed. 1987. *Impact of Sea Level Rise on Society: Report of a Project-Planning Session* (Rotterdam, Netherlands–Brookfield, Vt., A. A. Balkema).

Zazo, C., J. L. Goy, and J. C. Dabrio. 1986. "Late Quaternary and Recent Evolution of Coastal Morphology of the Gulf of Cadiz, Southwestern Spain." Paper presented at the International Symposium on Harbours, Port Cities and Coastal Topography, Haifa, Israel, September 22–29.

5

Climate Change: Problems of Limits and Policy Responses

Pierre R. Crosson

Present emission rates of carbon dioxide (CO_2) and the other principal greenhouse (radiatively important) gases (RIGs)—methane (CH_4), nitrous oxide (N_2O), and chlorofluorocarbons (CFCs)—exceed the capacity of the oceanic, terrestrial, and tropospheric sinks to absorb them. Consequently, their concentrations in the troposphere are increasing and will continue to increase so long as emissions exceed sink capacities.

Do indefinitely rising atmospheric concentrations of these gases imply indefinitely rising global temperatures and attendant changes in regional climates? I have found no discussion of this question in the climate change literature. That literature focuses almost exclusively on the climate changes resulting from a doubling of atmospheric CO_2, or its warming equivalent in CO_2 and other RIGs. Yet the focus on equivalent CO_2 doubling (written henceforth as $2 \times CO_2$) is simply a scientific convention. No one disputes that RIG concentrations will in time exceed $2 \times CO_2$ if emissions continue to exceed sink capacity, nor, so far as I know, does anyone argue that a continuing gap between emissions and sinks would eventually set in motion natural processes to increase sink capacity enough to close the gap. Finally, so far as I know, no one disputes that indefinitely rising concentrations of RIGs would trap an indefinitely increasing amount of infrared radiation from the earth's surface and atmosphere, implying indefinite global warming.

Absent arguments to the contrary, I assume that an indefinitely persistent gap between emissions and sinks of RIGs implies indefinite global warming and related changes in regional climates.

So what? Why should we care? We care because we believe that indefinite global warming would in time impose unacceptably high costs on some parts of the global community and probably, eventually, on all parts. The costs would take many forms: more expensive production of some goods and services; losses of unpriced environmental values, such as those of wetland and forest habitats, or the costs of measures to avoid these losses; damage to perceived national security interests, for example, a loss of comparative advantage in world agricultural markets; and political problems arising from shifts in the distribution of income and other measures of welfare, both intranationally and internationally. The unpriced costs could easily outweigh the priced costs and would be especially difficult to deal with because they are so hard to quantify.

Most of these costs will fall on future generations, not on us. But the way we manage the activities which emit CO_2 and other RIGs can make a significant difference in the magnitude of the costs. And most of us accept the ethical notion that we should not impose higher costs willy-nilly on future generations if there are things we can do to avoid it. Most of us, that is to say, accept the precept of intergenerational equity in resource management.

Acting on the precept is easy if we can do it at little cost to ourselves. And some analysts, including Lovins et al. (1981), Keepin et al. (1986), and Goldemberg et al. (1987) have suggested that in fact we can do this. They argue that the world community could shift to patterns of energy use which would eliminate the CO_2 emissions-sink gap without sacrificing per capita income growth. The grounds for this argument are considered below, and at greater length in Darmstadter and Edmonds (chapter 3 in this volume), but clearly if the argument is correct, intergenerational equity requires that we begin to move immediately to adopt the energy-use patterns that would do the job, at least so far as CO_2 is concerned. Moreover, in deciding to act now to slow or avert climate change we would not need to contend with the high present uncertainty about the prospective amount of global warming and its consequences. The uncertainty is of little importance if the costs of controlling climate change are low to negligible. It would be enough to be reasonably sure that some amount of climate change is probable and that it would impose some, not necessarily high, costs on future generations.

The intergenerational issue becomes complicated, however, if the cost to our generation of halting climate change is not low. (For convenience of expression, "our generation" means people now living.) That cost inevitably would show up as a lower level of per capita income and other measures of welfare for our generation. Yet there are members of our generation already suffering from inadequate income and welfare. If we accept a lower level generally to avoid imposing higher costs on future generations, we make it more difficult to improve the lot of those disadvantaged members of our own generation.

There is an additional complication. The higher the cost to us of not imposing costs of climate change on future generations, the lower our income; the lower our income, the less our capacity to transfer capital, tangible and intangible, to those generations. Thus, in seeking to meet our intergenerational obligation in one dimension—by controlling climate change—we may violate it in others, perhaps enough to make future generations worse off than they would be if we did less, perhaps nothing, to avoid climate change.

The nub of the intergenerational issue as it is affected by climate change is to find an equitable sharing of the costs of change within and across generations, understanding that the costs are broadly defined to include both priced and unpriced losses and that they include costs of controlling change as well as costs of not controlling it. If the costs of control to our generation are low, then the issue is easily resolved. We begin immediately to bring emissions of the RIGs into balance with their respective sinks. But if the costs to us are not low, the issue is not easily resolved. If costs of climate change are to be equitably shared within and across generations, then we must have at least rough estimates of what they are. Determining costs of control would be difficult enough. But discovering the costs of not controlling would be even more difficult because of the great uncertainty about the prospective amount of climate change and its cost consequences and because many of the costs would be difficult, if not impossible, to quantify.

I argue below that the cost to our generation of dealing with climate change in a way consistent with our obligation to future generations is not likely to be low. The question of how we determine a fair intergenerational sharing of the costs of change, therefore, must be faced. We can do this most usefully by considering the conditions under which the global community might come to consensus about a maximum acceptable amount of warming.

ARE THERE MAXIMUM ACCEPTABLE LIMITS?

Just as the question of whether an enduring gap between RIG emissions and sinks implies indefinite global warming

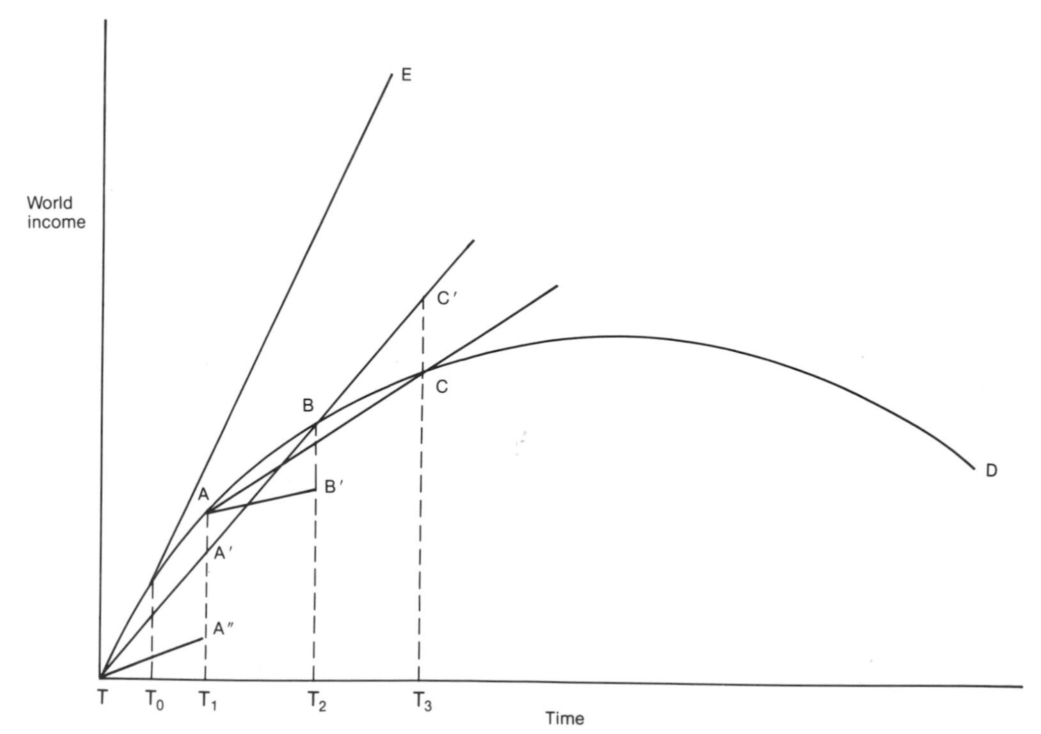

Figure 5-1. Hypothetical paths of world income with and without global warming and related climate change, case 1.

has not been systematically addressed in the climate change literature, neither has the related question of what constitutes acceptable limits to the amount of change. Keepin, Mintzer, and Kristoferson (1986, p. 86) suggest that the amount of global warming resulting from 2 × CO_2 possibly "is the upper limit of 'acceptable' climatic changes," but they do not develop the argument. And the Jäger report (1988, p. 28) implies the existence of a maximum, asserting, but without discussion, that "unlimited warming would sooner or later become intolerable."

Is there some amount of global warming about which the world community would agree, more than that is too much? There is no clear answer to the question. Everyone would agree that warming well short of that sufficient to make the oceans bubble would be too much, but that does not define a generally acceptable upper limit.

The Income Loss Criterion

I suggest that the question of maximum acceptable global warming can best be addressed by considering the losses of world income likely to result from unfettered warming in relation to the losses of income incurred by halting warming. To facilitate discussion I have constructed figure 5-1. The line TE traces the hypothetical path of world income in the absence of global warming and related climate change. TABCD is the path of world income if nothing is done to halt global warming. This curve takes into account the costs of the adjustments to climate change people will make on their own, the costs of whatever policies governments might pursue to promote adjustments beyond those that people will make anyway, and the costs which remain after all these adjustments are made.

Lines TA″ and TA′BC′ trace, respectively, the growth of world income if efforts begin "now" (time T) to halt warming by T_1 (line TA″) or T_2 (line TA′BC′). Curve AB′ represents the income growth path if control efforts do not begin until T_1 and aim to halt warming by T_2. Line AC traces income growth if control efforts begin at T_1 and aim to halt warming by T_3.

The income growth paths in figure 5-1 are strictly hypothetical, but their patterns reflect certain assumptions I believe to be defensible: (1) after a certain time (represented by T_0 in figure 5-1), and allowing for all least-cost adjustments, unfettered global warming will slow the growth of world income at an increasing rate and eventually cause it to decline; (2) all alternative ways of halting warming require a commitment of resources and the adoption of practices, such as reduced use of fossil fuels, that entail some sacrifice of near-term world income; (3) at least one alternative would eventually yield greater income after warming is halted than if it were not halted (for example, TA′BC′ after Time T_2); (4) the cost in lost income of halting warming diminishes as the time over which halting is achieved increases (for example, if halting occurs by T_1, income grows along TA″; if halting occurs at T_2, income grows along TA′BC′).

I suspect the assumption that unfettered warming would eventually reduce world income will appear plausible enough. The assumption that halting warming would impose slower near-term income growth, however, contradicts the arguments of several analysts (see Mintzer, 1987; Goldemberg et al., 1987). I consider the issue briefly below, and Darmstadter and Edmonds (chapter 3 in this volume) discuss it in more detail. The assumption that the cost of halting warming declines as the time available for halting increases is plausible a priori—that is, it is plausible that we could close the gap between CO_2 emissions and sinks with less sacrifice of income if we had 100 years in which to do it rather than only 10—and plausibility is the only claim I make for it here. However, the assumption is of key importance for thinking about policy responses to climate change, and research on the issue should be high on the agenda of scholars and policymakers concerned about such changes.

In principle there may be an income growth path for each future date at which warming would be halted, however, only one of the paths, that which bisects the unfettered warming growth curve at the earliest date, would be a candidate for consideration in policies to halt warming. If efforts to halt warming begin "now," then TA′BC′ in figure 5-1 is the path of interest for policy. Of all the possible alternatives for halting warming starting "now," it imposes the least cost in lost income. Other alternatives—path TA″, for example—could halt warming sooner than time T_2, but always at a higher cost than the alternative reflected in TA′BC′. For example, if halting occurred at T_1, world income then would be represented by the vertical distance T_1A″. If alternative TA′BC′ were followed, halting would not occur until T_2, but world income in T_1 would be higher than if alternative TA″ were adopted. The difference between the two growth paths after T_1 probably would not diminish much, and might increase. The relatively large sacrifice of income involved in halting warming by T_1 might make it difficult for growth along TA″ ever to catch up with growth along TA′BC′ even though the TA″ path would halt warming sooner.

Growth along TA′BC′ also involves a sacrifice of income (measured by the space TABA′T). However, it offers compensation in higher world income growth after T_2. Consensus in the world community that efforts should begin "now" to halt warming would imply that the warming occurring between T and T_2 would be the maximum acceptable amount. Such a consensus would also imply a willingness by our, and near-term, generations to accept the loss of income represented by TABA′T so that later generations (those living after T_2) could enjoy higher income than if we did not begin to halt warming "now."

The discussion indicates that the maximum acceptable amount of warming would not be set at some arbitrary number, say 10°C above the present global average level. Nor would it be set at that amount of additional warming occurring by some arbitrary date, say 2050 or 2100. Rather, the maximum acceptable amount of warming would be set at that point where world income with unfettered warming equals world income with warming halted. If more warming than this occurs, then subsequent world income would be less than if warming were halted; so such additional warming would be "too much." If less warming than this occurs, subsequent world income would be greater with warming than if it were halted; so warming would be halted "too soon." In effect, this way of stating the issue says that the maximum acceptable amount of warming is that at which the cost of additional warming equals the cost of halting additional warming, with costs measured by sacrifices of world income broadly defined to include both priced and unpriced values.

So far the discussion has focused on the consequences for world income growth of a global consensus that efforts to halt warming should begin "now," following the least-cost alternative represented by TA′BC′ in figure 5-1. As noted, the consensus would imply a general willingness to accept the prehalting loss of income represented by TABA′T so that people living after T_2 could enjoy more income than they would if warming were not halted by that date.

But forming a global consensus that people living between "now" and T_2 should bear the cost of halting warming by T_2 may prove difficult, for reasons discussed below. Poor countries in particular may argue that present and near-term generations should bear less of the costs of warming and more distant generations should bear more. In terms of figure 5-1, these countries could note that if efforts to halt warming were deferred until time T_1, people living between "now" and T_1 would have more income (TAA′T in figure 5-1) than they would if efforts to halt begin "now." To be sure, if efforts to halt are deferred until T_1, and halting is not achieved until T_3, the long-term growth of world income would be less than if efforts to halt begin "now" (growth along AC is less than it is along TA′BC′). But, the poor countries might argue, that outcome would still be a fairer intergenerational distribution of the costs of warming than if efforts to halt begin "now." And deferring efforts to control warming until T_1 still would be consistent with the precept of intergenerational equity that global warming must *eventually* be halted.

The argument for deferring efforts to halt warming generally would be stronger the more successful are policies for adjusting to climate change, that is, the more these policies succeed in reducing the impact of climate change on world income growth. To explore this, consider figure 5-2. As in figure 5-1, the line TE is the path of world income growth in the absence of global warming and related climate change. The curve TABCD is the path of

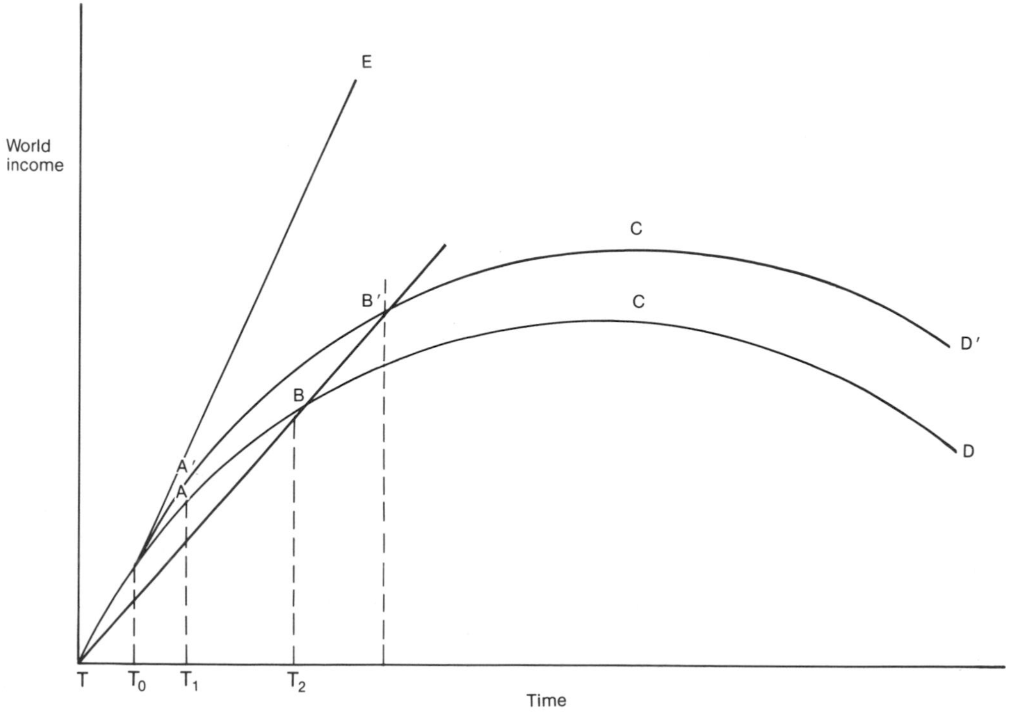

Figure 5-2. Hypothetical paths of world income with and without global warming and related climate change, case 2.

world income growth with unfettered global warming under one set of least-cost adjustments to climate change. It exactly replicates curve TABCD in figure 5-1. Line TBB′ in figure 5-2 is the least-cost path of world income growth when efforts to halt warming begin "now." It exactly replicates line TA′BC′ in figure 5-1.

Now suppose that a little investigation reveals an even less costly set of adjustment practices such that world income is less affected by unfettered warming and would grow along the path traced by curve TA′B′C′D′. In this circumstance, the date at which warming would be halted is sometime after T_2, and the maximum acceptable amount of warming is greater than before.

But there is a paradox here, at least given the assumptions underlying figure 5-2. Although the more successful adjustment policies increase the amount of world income consistent with unfettered warming (compare curves TABCD and TA′B′C′D′ in figure 5-2), they also increase the loss of prehalting income if efforts to halt warming begin "now." That is, TA′B′T, the space representing the prehalting loss of income with TA′B′C′D′, is larger than TABT, the space representing the prehalting loss of income with TABCD.

If this is generally true, it suggests that the greater the potential of adjustment policies for easing the impact of climate change on world income, the greater the difficulty of getting a global consensus for beginning efforts "now" to halt global warming. However much some countries may resist such a consensus because of unwillingness to accept the prehalting loss of income, their resistance likely would be stronger if the income loss is higher. In short, more successful adjustment policies are likely to increase the number of people who believe that a higher, rather than a lower, maximum amount of warming is consistent with intergenerational equity in the distribution of the costs of warming.

Problems of Applying the Criterion

So far the discussion has ignored all the empirical problems of identifying the various paths that world income growth might take, with and without warming and with alternative policies to adjust to, and eventually halt, warming. Of course, in the real world these problems are mammoth, indeed insurmountable, at least for some time to come. Many of the costs of unfettered warming, such as threats to national security and losses of unpriced environmental values, will not be reflected in conventional measures of world income. And some of the costs incurred to halt warming, such as the resources needed to overcome institutional barriers to constructing more energy-efficient buildings, will be hard to identify in conventional income measures.

Apart from these problems of identifying and quantifying both priced and unpriced costs, the vast uncertainty about the rate of warming, its implications for regional climates, the consequences of regional climate change for the whole range of human activity, and the future economic and technical conditions determining quantities of greenhouse gas emissions prohibit even rough estimates of the two kinds of cost anytime soon.

Because of these vast uncertainties, the global community would have difficulty agreeing on a maximum acceptable amount of warming even if the view were widely shared that in principle such a maximum exists. International differences in the distribution of warming costs also would be a major obstacle to agreement. Some nations probably will lose little if anything from $2 \times CO_2$ warming, and others may gain. The greater the differences among nations in this respect, the less likely is a global consensus on a maximum acceptable amount of warming. More generally, the less evenly distributed among nations are the costs of any given amount of unfettered warming, the less likely is an international agreement that that amount is unacceptably high.

The discussion of figure 5-1 suggested that global agreement on a warming limit also would depend on the temporal distribution of the costs of warming. Nations likely will have different views of what constitutes a fair intergenerational distribution of warming costs, depending on each nation's present level of per capita income. Per capita income differences now are large. In 1985 the average per capita gross national product (GNP) of a group of the poorest countries, which accounted for 50 percent of the world's population, was $270 per year (World Bank, 1987). The group of countries that the World Bank calls the "industrial market economies," with 15 percent of the world's population, had an average per capita GNP of $12,000 in 1985. The United States had a per capita GNP of $16,700 (World Bank, 1987).

Whatever the costs of halting warming, rich nations will find the per capita burden less than will poor nations. Rich nations, therefore, could afford to be more generous to future generations than could poor nations. This is not to accuse poor nations of selfishness. Rather, it is to recognize that their present levels of per capita income are so low that intergenerational fairness requires that the present generation in poor nations get much higher priority for income growth relative to future generations than in rich nations.

Thus, existing per capita income differences among nations likely will yield sharply different perceptions of what constitutes a fair intergenerational distribution of the costs of global warming and related climate change. The poor nations will likely argue for shifting more of the costs to future generations, the rich nations for shifting less. In this case, the poor nations would not be willing to accept as low a limit on global warming as the rich nations.[1]

Summary

The discussion points to the following propositions.

1. The nations of the world probably would agree that beyond some amount unfettered global warming would impose unfair costs on future generations. Consequently, the nations probably would accept the principle that there is some maximum acceptable amount of global warming.

2. Within the world community, nations will decide their positions on maximum acceptable warming primarily with reference to the amount and intergenerational distribution of the costs to them of unfettered warming and of halting warming. A useful guideline is to assume that each nation will define the maximum as that amount of warming at which the nation's income with unfettered warming would be the same as it would be if warming were halted.

3. Achieving a global consensus on the maximum acceptable amount of warming will be difficult because the conditions for a maximum will vary widely among nations, as follows.

a. Despite the great uncertainty about the implications of global warming for regional climates, there is little doubt that the consequences will be uneven across nations, and some likely will gain, at least with the amount of warming now generally expected, with $2 \times CO_2$. In terms of a figure 5-1 for each country, the income curves of unfettered warming (TABCD) will vary widely in position and scale, although not in general shape. For countries that would gain from warming, the earlier portions of TABCD would lie to the left of TE, the without-climate-change income growth curve. Assuming for the moment that countries have similarly shaped income curves for halting warming (although the scale of income, of course, is different), then intercountry differences in the position of curve TABCD would imply different dates for halting warming, hence different judgments about the maximum acceptable amount of warming.[2]

b. Countries will differ also in their ability to adjust to climate change. Ability to adjust no doubt is determined in complex ways. Degree of exposure to the consequences of climate change must be part of it. Low-lying countries with a large share of territory close to the sea, such as Indonesia, will have an inherently more difficult time adjusting to sea-level rise than will Switzerland, for example. Countries whose institutional structures permit a ready shift of resources among uses in response to changing conditions of scarcity and surplus should have greater adjustment ability than countries with less flexible institutions. The market economies probably would be more favored than nonmarket economies in this respect. Other things being equal, rich countries may have more adjustment ability than poor countries, although this is not certain. For example, rich countries would have more capacity to develop new crop varieties better adapted to a changing climate regime. However, the more diversified cropping patterns now followed by farmers in many poor countries might permit a low-cost response to changing climate without the need to invest in the development of new crop varieties.

As noted above, ability to adjust to global warming is reflected in the income curve of unfettered warming (TABCD in figure 5-1). Consequently, intercountry differences in adjustment ability will foster differences in the position of each country's TABCD curve, leading to different judgments about the maximum acceptable amount of warming.

c. The costs of halting warming surely will differ among countries. Any global agreement to halt warming inevitably would require constraints on the use of fossil fuels, particularly coal. Countries relatively rich in these resources would confront higher costs of halting warming than countries with a more diversified energy resource base. In terms of figure 5-1, for countries rich in fossil fuels the income growth paths for halting warming would rise less rapidly relative to the income path of unfettered warming than in countries less well endowed with fossil fuels. The less well endowed countries, therefore, likely would accept a lower maximum warming than the better endowed countries.

d. Rich and poor countries are likely to differ about what constitutes a fair intergenerational distribution of the costs of global warming. Rich countries will find the costs of halting warming less burdensome than will poor countries, so rich countries would be more inclined to accept a lower maximum amount of warming.[3] The rich countries thus would be more generous to future generations than the poor countries would be. No moral distinction between the poor and the rich is implied. The rich have always found it easier to be more generous than the poor. But the difference between the two sets of countries could be a major obstacle to a global consensus about the maximum acceptable amount of warming.

4. These various obstacles to consensus would exist even if there were complete certainty about the future of global warming and its consequences for world income, both in total and across countries. But the vast uncertainty about future warming and its income consequences compounds the difficulty of achieving global consensus about a maximum acceptable amount of warming. Because of the uncertainty, no country at present has much basis for judging what its income curve of unfettered warming or its income curves for halting warming might look like.

As just noted, reducing, or even eliminating, the uncertainty would not remove all obstacles to a global consensus. But it surely would help.

In short, all nations probably would accept the principle that some amount of global warming would be too much. But they likely will differ sharply about what that amount is. Strategies for achieving a global consensus on how and when to halt warming under these conditions are discussed below.

A Note on the Maximum Rate of Warming

Discussions of maximum acceptable warming often run in terms of the maximum acceptable rate per decade rather than the maximum total amount. The report of the World Climate Programme, based on meetings in Villach, Austria, in October 1987 and Bellagio, Italy, the following month, sets 0.1 degree Celsius (°C) per decade as the target rate of warming which the global community should aim at (Jäger, 1988). The target "was selected after considering the observed limited ability of natural ecosystems and societies to adapt successfully to faster change" (Jäger, 1988, p. 25). The target was set in the context of scenarios showing that in the absence of new control policies, the decadal rates of warming would range from 0.3°C to 0.8°C between the present and the middle of the next century. The lower rate assumes "current trends in RIG emissions, a reduction of chlorofluorocarbon emissions according to the Montreal Ozone Protocol, and a moderate climate sensitivity" (Jäger, 1988, p. iii). The higher rate assumes "a large increase of RIG emissions and a high sensitivity of the climate response" (Jäger, 1988, p. iii). These estimates are roughly consistent with those in Bolin et al. (1986) showing that the equivalent doubling of CO_2 concentrations from preindustrial levels sometime in the middle third of the next century would increase global average temperatures between 1.5°C and 5.5°C.

Evidently the 0.1°C per decade target was adopted in the Jäger report for tactical policy reasons, not as a rejection of the concept of a maximum acceptable total warming. On the contrary, the report states that the target decadal rate "would be supplemented with absolute limits on temperature . . . since unlimited warming at any rate must sooner or later become problematic" (Jäger, 1988, p. 21). A subsequent statement (p. 28) is less equivocal, replacing "problematic" with "intolerable."

In any case, whether the warming limit is set as a rate or as a total is a distinction without a difference for thinking about climate change policy. In both cases the principle for rational target setting is the equating of income losses of continued warming with the income losses of meeting the target. The 0.1°C per decade target set in the Jäger report implicitly assumes that for higher rates (for example, the 0.3°C or 0.5°C decadal rates that the report estimates in the absence of new control policies) the income losses of warming exceed the income losses of meeting the target and that the losses of bringing warming to less than 0.1°C per decade would exceed the losses of maintaining that rate, at least for the foreseeable future.

Getting a global consensus on the proposed target *rate* of warming would face exactly the same obstacles as getting consensus on the maximum *total* warming: uncertainties about the costs of continued warming above the target and about costs of achieving the target, intercountry differences in the distribution of priced and unpriced costs, and intercountry differences about a fair intergenerational sharing of the two kinds of cost. To be sure, the costs of reducing warming within a given time frame from whatever they otherwise would be to 0.1°C per decade probably would be less than the costs of halting it entirely. But this is a difference of degree, not of kind. The questions that must be asked and the policy issues that must be faced are the same whether the warming target is set as a rate or as a total.

IMPLICATIONS OF 2 × CO_2 WARMING

As indicated above, most of the discussion of the consequences of climate change has focused on global warming resulting from a doubling of atmospheric CO_2, or its warming equivalent in CO_2 and other greenhouse gases (2 × CO_2 warming). I follow this convention and consider whether the consequences of 2 × CO_2 warming for world income growth are likely to be so severe as to produce a consensus in the global community that that amount of warming would be "too much."

Is 2 × CO_2 Warming Too Much?

Schelling (1983) concludes that climate change over the 100 years from 1983 probably would not reduce global per capita income by more than a few percentage points below what it otherwise would be. Putting it somewhat differently, he states that "the living standards that might have been achieved by 2083 in the absence of climate change would be achieved instead in the late 2080s" (Schelling, 1983, p. 475). For example (mine, not Schelling's), if global per capita income in the absence of climate change were to grow 2 percent annually from 1983 to 2083, it would increase 7.2 times over that period. If that much growth does not occur until 2088, the delay suggested by Schelling, the annual growth rate from 1983 would be 1.9 percent.

In a global perspective, a climatically induced loss of 0.1 percent per year in world income growth is not likely to appear very threatening to the world community. Schelling argues that most of the loss would be in agriculture, which suggests that countries more specialized in agriculture would take the threat more seriously than those less specialized. Nevertheless, it seems unlikely that in the Schelling scenario the more exposed countries would be able to induce the less exposed to join a consensus that the costs of 2 × CO_2 warming would be higher than the costs of preventing it.

Thus, Schelling's analysis suggests that the global community would not likely agree that 2 × CO_2 warming would exceed the maximum acceptable amount. However, certain aspects of the argument merit additional attention. One concerns the rate at which 2 × CO_2 warming would occur.

The Rate of Warming

Schelling does not specifically link his conclusion about income consequences of climate change to 2 × CO_2 warming. However, the 100-year period he uses is at the far end of the period now anticipated for 2 × CO_2 warming to occur. And his discussion (p. 472) implies that 2 × CO_2 warming is in fact what he has in mind.

But the estimates of when 2 × CO_2 warming could occur range between 50 and 100 years from the mid-1980s. If the costs of warming vary directly with the rate, as seems plausible, then if 2 × CO_2 warming occurs in 50 years, the loss of world per capita income growth could be greater than the 100-year loss. If the additional loss were substantial, then governments would be more inclined to agree that the costs of 2 × CO_2 warming were unacceptably high. Whether they would in fact agree would depend on the pervasiveness of the additional costs among countries and on the costs of preventing 2 × CO_2 warming—which, I have argued, likely would be higher if prevention had to be achieved in 50 years rather than in 100. But the point here is that if the costs of climate change vary directly with the rate of change, then judgments about governments' willingness to accept 2 × CO_2 warming depend in part on the amount of time over which warming is expected to occur.

Agricultural Costs

Schelling's conclusion that the loss of world per capita income from 2 × CO_2 warming would be small rests on his belief that most of the loss would be in agriculture. Since income from agriculture is small relative to total world income and, if history is a guide, will become even smaller as world income grows, then the loss of income from climate change also would be small. As Schelling puts it (p. 475), "A rise of 10 percent or even 20 percent in the cost of producing food would be a few percent of world income at the outside."

Elsewhere, I have concluded that 2 × CO_2 warming by roughly the middle of the next century probably would have little effect on the economic and environmental costs of world agricultural production (Crosson, forthcoming). The reasoning was along the following lines. In a without-climate-change scenario, projections of world population and income growth suggest that world demand for food and fiber could rise between 1.0 and 1.4 percent annually from the 1980s to the middle of the next century, the period I assumed for 2 × CO_2 warming. This rate of increase in demand would be substantially less than the 2.5 percent annual production increases experienced in the last 40 years. Because of technological advance, real economic costs of world agricultural production declined over those 40 years. Environmental costs must have increased, but probably not as much as the decline in economic costs. Total production costs, therefore, likely were declining, or at least not rising. The prospects for new agricultural technologies, such as biotechnology, appear highly promising; there is no reason to believe that the potential for technological advance will be any less in the next 50 to 75 years than it was in the last 40.[4] The fact that in the last 40 years world agricultural production grew 2.5 percent annually with constant to declining costs suggests that with future demand growing at only 1.0 to 1.4 percent annually, we ought to be able to do at least as well in containing costs.

Imposing 2 × CO_2 warming on this scenario did not change the conclusion. Changes in regional climates likely would impose significant cost increases on some countries, or on some regions in some countries, and these changes would pose significant policy problems for those regions and countries. However, in other regions and countries agriculture would benefit from 2 × CO_2 climate change, at least during the early years and perhaps during the entire 50- to 75-year period. There were no compelling reasons to believe that on balance the losers would lose more than the winners would gain.[5]

I concluded, therefore, that 2 × CO_2 warming by the middle of the next century would not increase costs of world agricultural production, a more optimistic conclusion than Schelling's belief that 2 × CO_2 warming by 2083 could increase agricultural production costs 10 to 20 percent.

If Schelling is right that the main burden of 2 × CO_2 warming will fall on agriculture, then my conclusion would strengthen his belief that the effect of such warming on the growth of world per capita income would be small. Before accepting this conclusion, however, it is important to consider some other costs of 2 × CO_2 warming.

Other Costs

The warming and related changes in climate resulting from 2 × CO_2 doubtless would impose a variety of costs other than those on agriculture. Damages to environmental values (Batie and Shugart, chapter 9 in this volume), increased costs of water management (Frederick and Gleick, chapter 10 in this volume), and costs of sea-level rise (Vellinga, 1987; Hekstra, chapter 4 in this volume) are examples. No doubt there are others. Here I deal only with costs of sea-level rise because I suspect these costs may be more effective in forging a global consensus to halt warming than any other set of climate change costs for several reasons:

1. Many of the costs will be directly measurable in economic terms, for example, the lost value of abandoned coastal recreational facilities, the cost of relocating the facilities farther inland, or the cost of protecting the facilities by building sea walls or taking other defensive measures. The measurability of the costs should help to narrow differences about their importance.
2. The costs may be high relative to other costs of climate change.
3. The costs may be more clearly attributable to climate change than, say, agricultural costs, where rapid technological advance could more easily mask the cost effects of climate change.
4. The costs of sea-level rise will be more widely distributed among nations; that is, there will be fewer winners and more losers than, for example, in agriculture.

Much depends, of course, on how much the sea level rises with 2 × CO_2 warming. As with all aspects of climate change, there is much uncertainty about this, caused, for example, by the lag time of ocean warming behind the initial warming of the atmosphere (thermal lag) and by the possibility that warming may increase the amount of snowfall in Antarctica, tending to diminish any sea-level rise or even to lower sea levels (Hekstra, chapter 4 in this volume).

Hekstra refers to the "widely accepted analysis of de Q. Robin" (see de Q. Robin, 1986) indicating a sea-level rise of 0.8 meters (m) by 2100, with a range of 0.2 to 1.65 m. For his purposes Hekstra assumes a rise of 0.5 to 1.0 m "over the next century." Thomas (1986) estimates the rise at 1.1 m by 2100, with a range of 0.9 to 1.7 m, but notes that the rise could be substantially less if thermal lag is greater than now expected.

In all of these estimates, sea level rises because of thermal expansion of the water and the melting of mountain glaciers and Greenland ice (Hekstra, chapter 4 in this volume; Titus and Seidel, 1986). There evidently is a consensus that the disintegration and melting of the West Antarctic ice sheet could cause a 5-meter rise in addition to the increases already indicated. Hekstra cites Oerlemans and van der Veen (1984) as arguing that this is not likely to occur in the next century. De Q. Robin (1986) comes to the same conclusion. Thomas (1986) asserts, however, that the melting could begin in that period if warming continues beyond that resulting from 2 × CO_2.

Hekstra (chapter 4 in this volume) does not say explicitly that the estimates of sea-level rise he considers are those to be expected with 2 × CO_2 warming, although the period he cites—within the next century, or by 2100—is in the upper end of the range now estimated for such warming. Thomas (1986) specifically bases his estimate of sea-level rise (1.1 m by 2100 within a range of 0.9 to 1.7 m) on 2 × CO_2 warming of 3°C by 2050, *with no additional warming after that.*

Hekstra (chapter 4 in this volume) asserts that approximately 1 billion people (one-fifth of the world's present population) live within the land area that would be directly affected by a 1-meter rise in sea level. Many of these people live in large cities where the per capita market value of land and other assets must be substantially higher than in smaller cities and rural areas. I think it probable, although I have not sought the evidence, that more, perhaps substantially more, than one-fifth of the world's market-valued assets would be directly affected by a 1-meter sea-level rise, if Hekstra is right that 1 billion people would be affected.

A 1-meter sea-level rise would also affect unmarketed environmental values provided by coastal lagoons, estuaries, and wetlands (Batie and Shugart, chapter 9 in this volume; Hekstra, chapter 4 in this volume; Vellinga, 1987). Hekstra refers to Indonesia as dramatically illustrating possible damage to these values. He cites evidence indicating that Indonesia is the richest country in the world in amount and diversity of coastal wetlands ecosystems. He also notes that Indonesia has 15 percent of the world's coastline, at least 40 percent of which would be vulnerable to a 1-meter sea-level rise.

The costs of sea-level rise are of three sorts: (1) the cost of protecting property and people against the effects of sea-level rise, for example by building seawalls and putting buildings on elevated foundations; (2) the net value of the assets lost, and losses of human life, where protection against flooding from typhoons and hurricanes is not provided or is not complete; (3) the cost of relocating people and property to inland areas.

No global estimates exist of the costs resulting from a rise in sea level caused by 2 × CO_2 warming. Vellinga (1987) makes some very rough estimates of percentage increases in such costs relative to present costs of protection and maintenance. With unfettered warming and protection in the form of selective coastal engineering measures, Vellinga estimates that the costs by 2100 would be higher by more than 1,300 percent. With measures to reduce RIG emissions by 2 percent per year and the same protective measures, the cost increase by 2100 would be about 700 percent. Conceptually these estimates may be low because they apparently do not include costs in areas which are not currently protected against the sea but would be threatened by a 1-meter rise in sea level.

The Vellinga cost estimates clearly are extraordinarily rough. However, if at least one-fifth, and probably more, of the world's market-valued assets and a substantial portion of its unmarket-valued assets would be directly affected by a 1-meter sea-level rise, then the absolute cost of such a rise could be high (that is, relative to the climate-induced rise in agricultural production costs) even if the percentage increases were substantially less than those estimated by Vellinga.

On balance, I think it likely that the costs of sea-level rise caused by 2 × CO_2 warming would cause world income to grow more slowly than it would otherwise in, roughly, the second half of the next century. If Schelling (1983) is right about the effects of warming on agricultural costs, then, in my judgment, the costs of sea-level rise could be comparable. If I am right that agricultural costs would rise little, then costs of sea-level rise would be greater than agricultural costs.

Whether the cost of sea-level rise in this scenario would be enough to forge a global consensus that 2 × CO_2 warming would be "too much" is, like everything else in this area of work, highly uncertain. However, for reasons given at the beginning of this section, the probability of consensus likely would be higher per dollar of sea-level-rise costs than per dollar of agricultural costs.

BEYOND 2 × CO_2 WARMING

Estimates of the amount of 2 × CO_2 warming vary, with Bolin et al. (1986) giving a global average increase varying from 1.5°C to 5.5°C, while Schneider and Rosenberg (chapter 2 in this volume) give from 2°C to 5°C. Whatever the numbers may in time prove to be, it seems virtually certain that they will continue to rise so long as atmospheric concentrations of CO_2 and other RIGs continue to increase. It is reasonable to believe that whatever the costs of 2 × CO_2 warming may be, they will be greater, and will continue to increase, as warming rises beyond that resulting from 2 × CO_2.

The first half of this chapter is based on this plausible proposition. However, the proposition provides no insights into how much costs might rise per increment of post–2 × CO_2 warming. In the earlier discussion I assumed that the costs of warming would increase at an increasing rate, but there is no empirical support for this. So far as I know, no one has systematically addressed this issue even in a speculative way, probably because the sparse research in this area has focused mostly on the consequences of 2 × CO_2 warming.

One line of argument nevertheless seems worth considering. The literature on sea-level rise strongly suggests that with a global temperature increase well beyond the mean of 3°C–3.5°C expected with 2 × CO_2, continued thermal expansion of the ocean and melting of smaller glaciers and ice caps would result in more sea-level rise than that resulting from 2 × CO_2 warming. Thomas (1986) addresses this directly. He first assumes that an equivalent doubling of CO_2 concentrations by 2050 would increase global average temperature by 4.5°C, not the 3°C he used to derive his sea-level rise estimate of 1.1 meters. That is, to explore the sea-level consequences of post–2 × CO_2 warming, Thomas adopted a temperature increase with 2 × CO_2 in the upper part of the currently expected range. He then assumed that GHG concentrations would continue to increase and that by 2100 the global average temperature would rise another 1.5°C, to 6°C. In this "high rise" scenario, sea level rises 2.3 meters. Little if any of the additional rise is attributable to melting of the West Antarctic ice sheet.

Thomas stresses that he considers this result to be an extreme upper limit reflecting an exceptionally severe amount of warming. It is worth noting, however, that the 6°C increase that produces the result is just outside the high end of the range (5.5°C) that Bolin et al. (1986) consider likely with 2 × CO_2.

As noted, Thomas's "high rise" scenario does not involve the melting and breakup of the West Antarctic ice sheet. De Q. Robin (1986) notes the great stability of the sheet across a considerable range of global average temperatures and concludes that its collapse is not imminent. He also notes, however, that present knowledge is too limited to rule out the possibility that a global temperature rise of 3.5°C would start such a collapse by 2100. The 3.5°C increase reflects 2 × CO_2 warming. De Q. Robin does not speculate on the likelihood or timing of collapse should warming continue past 2 × CO_2 and reach the 6°C by 2100 assumed by Thomas.

I concluded above that the costs of a 1-meter rise in sea level probably would be high enough to have some small effect on world income growth after the middle of the next century. The costs of a 2.3-meter rise surely would be higher, probably substantially so. Whether they would be enough higher to begin to concentrate the minds of governments on the problem of halting global warming is unknowable. Indeed, it is not certain that a rising probability of eventual collapse of the West Antarctic ice sheet and an additional 5- to 6-meter rise in sea level would produce a global consensus to halt warming. But when one considers that, by the time the global average temperature would have risen enough to do that, other costs of warming also would have increased well above 2 × CO_2 levels, then the likelihood of consensus to halt warming should increase sharply as well.

THE TASK OF HALTING WARMING

If global warming is to be halted, the gaps between emissions of RIGs and their various sinks must be closed. The work of Darmstadter and Edmonds (chapter 3 in this volume) suggests that this would be a formidable undertaking. Assuming annual world income growth of 3 percent and per capita growth of 2 percent, their "reference case" scenario shows global carbon emissions from energy use alone rising from 5 billion tons (gigatons [Gt]) in 1985 to 7.7 Gt in 2050 to 8.5 Gt in 2075. These projections assume substantial increases in energy-use efficiency; that is, the ratio of world energy use to world income declines 70 percent from 1985 to 2050, or 2 percent annually.

According to Darmstadter and Edmonds, tropical deforestation currently adds net carbon releases of about 1.3 Gt annually, and in their reference case this rises to 1.5 Gt in 2075. Darmstadter and Edmonds do not project emissions of methane and nitrous oxide, but these also can be expected to rise over the next 100 years, reflecting continued agricultural growth (rice paddies release methane, for example, and fertilizer and newly cultivated soils release nitrous oxide). Emissions of CFCs should decline in accordance with the Montreal Protocol. However, these substances have atmospheric lifetimes of 75 years or more, and they have no tropospheric sinks. Consequently, even reduced emissions of CFCs will continue to contribute to rising global temperatures for many years.

The 2 percent annual increase in world per capita income assumed in the Darmstadter and Edmonds reference case is modest by historical standards,[6] and it implies that the present large absolute per capita income gap between the rich and poor countries not only would persist but probably would widen. The poor countries are expected to account for about 85 percent of world population growth over the next 100 years. If world per capita income growth is only 2 percent annually and, in the developed countries, is 0.5 percent, low by historical standards, then maximum per capita income growth in the developing countries would be 2.25 percent annually. At that rate, per capita income in the poorest countries, currently with 50 percent of the world's population, would rise in 100 years from the 1985 amount of $270 (World Bank, 1987) to $2,500. Per capita income in the industrial market economies would grow from $12,000 (World Bank, 1987) to $19,700. The income gap between the poor and the rich countries would widen substantially even though per capita growth in the rich countries would be low by historical standards. In short, the assumption of 2 percent annual growth in world income over the next 100 years appears modest both by historical experience and when measured against present and prospective differences between per capita income in rich and poor countries.

Darmstadter and Edmonds emphasize the great uncertainty attending all long-term projections of CO_2 and other greenhouse gas emissions. And in the many model runs made to contend with this, they found a 90 percent probability that carbon emissions from energy use in 2075 would be between 1.8 Gt and 86.9 Gt, too wide a range to be of any use for policy purposes. They found a 50 percent probability that in 2075 carbon emissions from fuel combustion would be between 3.9 Gt and 27.1 Gt. (Recall that energy-use emissions of carbon were 5 Gt in 1985 and 8.5 Gt in the Darmstadter and Edmonds reference case for 2075.) These projections do not include net carbon emissions from land-use changes, estimated by Darmstadter and Edmonds at 1.3 Gt annually at present and projected by them to rise to 1.5 Gt in 2075. Combining this projection with their declining emissions scenario (from Mintzer, 1987) gives total carbon emissions in 2075 of 4 Gt. This is approximately 2 Gt more than the amount of carbon now absorbed by the oceans, suggesting that even in the minimum Darmstadter and Edmonds scenario, the CO_2 emissions-sink gap would not be completely closed by 2075. And the gaps for other greenhouse gases still would be widening. The rate of global warming would be drastically slowed, but it would not be halted.

The "slow build-up scenario" developed by Mintzer (1987)—the declining emissions scenario of Darmstadter and Edmonds—assumes a very slow rate of increase in world per capita income and a very rapid rate of increase in energy efficiency. The assumed rate of growth in world per capita income, 1.4 percent per year (Mintzer, 1987, p. 19), would provide very little improvement in living conditions for the world's poorest people. For example, the 50 percent of the world's population with per capita income of $270 in 1985 (World Bank, 1987) would have an income in 2075 of $940 if the rate of increase were only 1.4 percent annually. Mintzer provides no argument for why world per capita income would grow no more rapidly than this, and there cannot be much question that governments in the developing countries would strive for much faster rates. If they did, and they succeeded, total world energy use could grow faster than in Mintzer's slow build-up scenario, even if the improvements in energy efficiency it depicts were achieved.

But the argument that those improvements can be readily achieved is not compelling. Fossil fuel prices rise two to three times in Mintzer's slow build-up scenario, which evidently strengthens incentives to use the fuels more efficiently. The Darmstadter and Edmonds reference case scenario incorporates comparable increases in fossil fuel prices and gets substantial improvements in energy efficiency; however, the improvement is far less than in Mintzer's slow build-up scenario. The Mintzer scenario also assumes policies to substitute solar and nuclear energy for fossil fuels, but provides neither details of these policies nor justification for why they would have the dramatic effect on energy efficiency depicted in the scenario.

I conclude that closing the gaps between emissions of CO_2 and other greenhouse gases and their respective sinks while maintaining satisfactory world income growth will be extraordinarily difficult. This is to say that so far as CO_2 is concerned, the Darmstadter and Edmonds reference case (7.7 Gt of carbon released from energy use in 2050 compared with 4.7 Gt in 1975) now looks more likely than Mintzer's slow build-up scenario. In the world of Darmstadter and Edmonds, warming is slowed, but it is not halted, and the world community will have to deal eventually with the consequences of more than $2 \times CO_2$ warming.

SUGGESTED POLICY GUIDELINES

If the analysis here of the consequences of global warming is approximately correct, the principal global threat is from warming that would occur beyond an equivalent doubling of atmospheric CO_2. To say this is to imply that the costs of warming up to $2 \times CO_2$ probably would be less than the costs of halting that warming and that the costs of warming beyond $2 \times CO_2$ probably would be greater than the costs of halting warming at approximately the $2 \times CO_2$ amount.

Given the vast uncertainties in this area of work, this is a bold statement, and I do not pretend that the analysis here provides a firm basis for it. The statement is not wholly arbitrary, however, and it may prove useful in setting a rough, first-approximation policy objective for the global community concerned about climate change.

The discussion above entitled "Are There Maximum Acceptable Limits?" identifies five main barriers to an international agreement setting and enforcing such limits (for example, holding emissions of greenhouse gases to not more than an equivalent doubling of atmospheric CO_2):

1. The great uncertainty about the likely amount of warming over the next 75 to 100 years, the likely costs of warming in lost income, and the probable distribution of the costs among countries;
2. The perception among some countries that they may benefit from warming, at least that amount up to $2 \times CO_2$;
3. Differences among countries in the costs of adjusting to climate change;
4. Differences among countries in the perceived costs to them of joining an international agreement to halt global warming; and
5. Disagreements among countries, particularly between the rich and the poor, about what constitutes a fair intergenerational sharing of the costs of warming and of halting it.

To make headway toward an international agreement to limit the atmospheric concentration of greenhouse gases to the equivalent of $2 \times CO_2$ will require progress in reducing these barriers. Because the great uncertainty about all aspects of climate change strengthens the barriers, research in both the natural and social sciences clearly has a key role to play in the campaign to reduce them. Following are some suggested areas of research which appear particularly promising.

1. Reduce scientific uncertainty about the amount and timing of global warming associated with $2 \times CO_2$. Present estimates are so wide as to be almost useless for policy. For example, the estimate that equivalent $2 \times CO_2$ will increase global average temperatures between 1.5°C and 5.5°C by some time in the middle third of the next century (Bolin et al., 1986) implies that the decadal rate of warming after 1984 would vary between 1.1°C (+5.5°C by 2035) and 0.2°C (+1.5°C by 2065). With these differences, the costs of warming and of halting it would surely be higher, probably far higher, if the target date were 2035 instead of 2065. And the policy strategies also would be far different. A clear prospect of +5.5°C by 2035 would indicate a strategy far more heavily weighted toward halting warming than toward adjusting to it than would an equally clear prospect of +1.5°C by 2065. The present variation in estimates of the amount and timing of equivalent $2 \times CO_2$ warming provides people in the policy community with little guidance in choosing between these two policy strategies.
2. Reduce scientific uncertainty about the implications for regional climates of alternative amounts of global warming, say that resulting from $2 \times CO_2$ and that resulting from $2.5 \times CO_2$.
3. Reduce scientific uncertainty about the amount and distribution among countries of the costs (broadly defined, as noted throughout this discussion) of the warming scenarios associated with $2 \times CO_2$ and, for example, $2.5 \times CO_2$. In doing this analysis, consider scenarios in which the amount and timing of warming vary widely, as noted above. Consider focusing the analysis initially on the costs of sea-level rise and of freshwater management for agricultural and other uses.
4. Reduce scientific uncertainty about the costs of alternative measures to close the emission-sink gaps for greenhouse gases by dates corresponding to estimates of when $2 \times CO_2$ warming would occur. The Mintzer (1987) argument that major reductions in fossil fuel use could be achieved over the next 50 to 75 years at no sacrifice of world income growth would deserve close attention in this area of research.
5. Reduce scientific uncertainty about the costs of adjusting to climate change under several scenarios of the amount and timing of warming associated with $2 \times CO_2$ and, for example, $2.5 \times CO_2$. It may be most productive

to consider initially the adjustment costs for sea-level rise, agriculture, and water resources.

6. Explore the different perceptions of rich and poor countries about what would constitute a fair intergenerational sharing of the costs of global warming and of halting it. If investigation supports the hypothesis here that the poor countries would opt for later generations to absorb more of the costs, while the rich countries think our generation should absorb more, then consider how the difference might be reconciled. The research described above to reduce scientific uncertainty would help. However, even if this research were highly successful in reducing the various uncertainties about climate change and its consequences, the intergenerational issue between rich and poor countries likely would remain because of the present vast income difference between them. In this case, some mode of compensation from rich to poor countries might be worth investigating. Such an approach would have to be carefully considered, among other reasons because of the likelihood that both rich and poor would engage in strategic bargaining, each angling initially to give less, or get more, than they would be willing to settle for in the end. The recent U.S. negotiations with Spain, the Philippines, and Greece about American military bases in those countries might provide some insights into this kind of bargaining process.

This listing of research areas is not exhaustive, but I believe it includes most areas with the highest promise of easing the way toward international agreements to slow, and eventually halt, global warming. The emphasis on research as a high-priority policy instrument for achieving agreements does not exclude other action to that end, such as congressional hearings and their equivalent in other countries, to alert the public and policy people to climate change as an important and enduring policy issue. The emphasis on research does reflect my conviction, however, that reduction of the many present uncertainties about climate change and its consequences is a necessary condition for achieving enforceable international agreements to hold the change within generally acceptable limits.

NOTES

1. This difference between rich and poor nations implies that the poor discount future income at a higher rate than the rich. If it is accepted that finding a fair intergenerational sharing of the costs of unfettered warming and of halting it is a legitimate policy issue, then discounting becomes not only an inevitable but also an entirely appropriate part of the policy process dealing with climate change.

2. For any country, the slope of its income growth path for halting warming would depend on the effect on the country of the international agreement to halt warming. The more successful the country in shifting the burden of halting warming to others, the more steeply its income path for halting warming would rise.

3. Rich countries also rich in fossil fuels, such as the United States, would be pulled both ways: favoring a higher maximum warming than countries poor in fossil fuels because of the higher cost of sacrificing the fossil fuel alternative; favoring a lower maximum than poor countries because of greater ability to absorb the cost. Poor countries rich in fossil fuels, such as China, would favor a higher maximum on both counts.

4. For a brief discussion of long-term prospects for technological advance in agriculture, see Easterling, Parry, and Crosson (chapter 7 in this volume).

5. Evidence cited in Easterling, Parry, and Crosson (chapter 7 in this volume) reinforces this conclusion.

6. From 1960 to 1985 world per capita income increased between 3.0 and 3.5 percent annually (Crosson, forthcoming, based on World Bank, 1987).

REFERENCES

Bolin, B., B. Döös, J. Jäger, and R. Warrick, eds. 1986. *The Greenhouse Effect, Climatic Change and Ecosystems* (New York, Wiley).

Crosson, Pierre. Forthcoming in 1989. "Climate Change and Mid-Latitudes Agriculture: Perspectives on Consequences and Policy Responses," *Climatic Change.*

De Q. Robin, G. 1986. "Changing Sea Level," in B. Bolin, B. Döös, J. Jäger, and R. Warrick, eds. *The Greenhouse Effect, Climatic Change and Ecosystems* (New York, Wiley).

Goldemberg, J., T. Johansson, A. Reddy, and R. Williams. 1987. *Energy for a Sustainable World* (Washington, D.C., World Resources Institute).

Jäger, J. 1988. *Developing Policies for Responding to Climate Change.* Impact Study of the World Climate Programme, World Meteorological Organization and United Nations Environment Programme, Geneva and Nairobi.

Keepin, W., I. Mintzer, and L. Kristoferson. 1986. "The Rate of Release of CO_2 as a Function of Future Energy Developments," in B. Bolin, B. Döös, J. Jäger, and R. Warrick, eds., *The Greenhouse Effect, Climatic Change and Ecosystems* (New York, Wiley).

Lovins, A., L. Lovins, F. Krause, and W. Back. 1981. *Least-Cost Energy: Solving the CO_2 Problem* (Andover, Mass., Brick House).

Mintzer, I. 1987. *A Matter of Degrees: The Potential for Controlling the Greenhouse Effect* (Washington, D.C., World Resources Institute).

Oerlemans, J., and C. van der Veen. 1984. *Ice Sheets and Climate* (Dordrecht, The Netherlands, Reidel).

Schelling, T. 1983. "Climate Change: Implications for Welfare and Policy," in *Changing Climate: Report of the Carbon Dioxide Assessment Committee* (Washington, D.C., National Academy of Sciences).

Thomas, R. 1986. "Future Sea Level Rise and Its Early Detection by Satellite Remote Sensing," in J. Titus, ed., *Effects of Changes in Stratospheric Ozone and Global Climate* vol. 4, *Sea Level Rise* (Washington, D.C., U.S. Environmental Protection Agency).

Titus, J., and S. Seidel. 1986. "Overview of the Effects of Changing the Atmosphere," in J. Titus, ed., *Effects of Changes in Stratospheric Ozone and Global Climate* vol. 4, *Sea Level Rise* (Washington, D.C., U.S. Environmental Protection Agency).

Vellinga, P. 1987. "Sea Level Rise, Consequences and Policies." Paper presented at the workshop on Developing Policies for Responding to Climate Change, Villach, Austria, September 29–October 2.

Warrick, R., H. Shugart, M. J. Antonovsky, with J. Tarrant and C. Tucker. 1986. "The Effects of Increased CO_2 and Climatic Change on Terrestrial Ecosystems," in B. Bolin, B. Döös, J. Jäger, and R. Warrick, eds., *The Greenhouse Effect, Climatic Change and Ecosystems* (New York, Wiley).

World Bank. 1982 and 1987. *World Development Report* (New York, Oxford University Press for the World Bank).

6

Assessing and Managing the Risks of Climate Change

Paul R. Portney

Nine years ago marked the real beginning of my interest in climate issues. At that time John W. Firor and I collaborated on a chapter on the global climate for a Resources for the Future book (Portney, 1982). Since that time I have done very little research related to climate policy, although I have tried to keep abreast of both research and policy developments as an interested layperson.

For the past four or five years I have been almost exclusively preoccupied with the management of a different class of environmental problems, namely, the health and ecological risks associated with conventional air and water pollution, the improper disposal of hazardous wastes, pesticide residues on foodstuffs, and other such problems that are normally the province of the Environmental Protection Agency. Although this group of problems is often thought of as being quite different from the problem of global warming, there are interesting similarities, too.

I have two purposes in the present chapter. First, I will point out some similarities and differences between the assessment and management of climate problems on the one hand and what I will call ordinary environmental regulatory problems on the other. Second, I wish to discuss the applicability of a familiar rule often invoked in thinking about the management of traditional environmental risks, namely, that "an ounce of prevention is worth a pound of cure." Specifically, I want to see of what use that rule is in thinking about the optimal management of climate-related risks.

COMPARING GLOBAL WARMING WITH OTHER ENVIRONMENTAL PROBLEMS

Let us first consider differences between the problem of global warming (the so-called greenhouse effect) and the problems associated with more conventional environmental risks. Again, by the latter I mean traditional or familiar air and water pollution problems, hazardous wastes, pesticide-related risks, and the like. First, greenhouse warming is what we might refer to as a broad, macroscale problem; that is, sources from all around the world contribute carbon dioxide (CO_2) and other trace gases to the atmosphere, which then trap outgoing terrestrial radiation headed back into space. Gram for gram, carbon dioxide from a small wood fire in Tanzania is no more or less benign than that emanating from a coal-fired power plant in Tokyo. And because of the rapid, turbulent mixing that occurs in the atmosphere, CO_2 emissions in either place eventually contribute to the climate alterations that will be felt by all parts of the globe. There are no lasting "pockets" or "hot spots" due to local emissions. On the other hand, more traditional environmental risks are generally much more narrowly focused. Often, in fact, they are confined to a particular location and pose no threat at all to individuals elsewhere. There are certainly exceptions, acid rain being one, but the generalization is not a bad one.

Second, and as suggested above, there are hundreds of millions of anthropogenic sources that contribute to greenhouse warming. These range from coal-fired power plants and other types of industrial facilities to individual automobiles. They include individual homes where wood, coal, or other fossil fuels are burned as cooking or heating fuels. (There are also millions of natural sources which are ignored in this discussion.) By contrast, traditional environmental problems often, although not always, arise from far fewer sources. In fact, concern often centers on a single chemical plant, smelter, power plant, petroleum refinery, sewage treatment plant, or landfill when we are dealing with traditional environmental problems.

Third, no regulatory authority exists to enable us to address the chemical causes of global warming, with the exception of the regulations now (or soon to be) in place limiting emissions of chlorofluorocarbons (CFCs), another greenhouse gas, because of concern about their role in stratospheric ozone depletion. Carbon dioxide emissions are not regulated anywhere, although they may be limited indirectly by regulations that limit the extent to which power plants may rely on coal or other fossil fuels. By contrast, there are a number of regulatory authorities in place to cope with traditional environmental health risks. These include the Clear Air and Clean Water Acts (1970 and 1972, respectively), the Toxic Substances Control Act of 1976, the Resource Conservation and Recovery Act of 1976, the Occupational Safety and Health Act (1973), the Consumer Product Safety Act (1972), and a number of other environmental, safety, and health statutes.

Fourth, international cooperation of an almost unprecedented dimension is required to deal with the problem of greenhouse warming. This follows directly from the worldwide scale of the problem and the vast number of sources that are implicated in it. On the other hand, solitary action can be quite effective when dealing with traditional environmental problems. Indeed, individual nations, and even states and local governments, all have their own environmental protection laws, their own laws governing occupational health, the safety of food and drugs, and so on.

Fifth and finally, some nations will probably benefit from greenhouse warming. These are the nations where the climate may become more hospitable, where warmer temperatures may mean enhanced agricultural or silvicultural yields, where altered rainfall may make economic activity more productive, and where change may produce other benefits. In other words, warming per se may be positive. On the other hand, there are no benefits whatsoever to having higher levels of carcinogens in the environment in and of themselves. They may enter the environment as a result of valued activities, but they do not themselves benefit anyone.

There are, of course, respects in which the problem of global warming and the traditional environmental problems mentioned here are similar. First, for both types of problems, significant delays may exist before we recognize their onset. In the case of greenhouse warming, the absorptive capacity of the ocean can delay recognition of climate modification; we may already be locked into a certain amount of temperature modification before we begin to notice its first signs. Similarly, in thinking about carcinogenesis, there may often be thirty- to forty-year latency periods between the time someone is exposed to a cancer-causing agent and the time a tumor appears. This means that clear cause-and-effect relationships are difficult to recognize for both sets of problems and, hence, policymaking is made more difficult.

A second similarity between the two kinds of problems has to do with the signal-to-noise ratio; that is, other factors besides greenhouse gases can affect climate. Indeed, there is much cyclical variability in temperatures and rainfall that has nothing to do with CO_2 or other gases. Similarly, diet, smoking, occupational exposures, genetic predisposition, and other factors can often mask the effects of a particular carcinogen. As with the latency period, this means that it is difficult in both cases to determine whether or not a problem has begun to develop and to identify measures to alleviate it.

A third similarity between the two kinds of problems has to do with the methods we use to assess those risks formally. To assess traditional risks to health or the environment, we rely to a large extent on toxicological studies using experimental animals and occasionally even on computer studies that model human or plant physiological responses to environmental stresses (National Research Council, 1983). Similarly, in understanding the problem of greenhouse warming, we rely heavily on general circulation models to make predictions of climate change. While both of these methods are partially experimental, both incorporate theoretical constructs and rely to a large extent on extrapolation from computer models and experimental data to real-world situations.

Human epidemiological studies are also used in the assessment of the health risks associated with traditional environmental contaminants. In such studies we examine situations where humans have been exposed to the contaminant in question, often in an occupational or accidental setting, and then attempt to draw inferences about the effects of exposures to the same substances in ambient settings, often at much lower levels. A similar process can be said to take place in estimating the effects of climate change. By looking at past records of climate changes or previous climate-altering events and by exploring the consequences associated with those events, we try to draw conclusions about what may happen in the future as a result of greenhouse warming. This can be thought of as a sort of epidemiological analysis in climate research.

Another similarity between the two types of problems has to do with the variability of potential responses. For instance, the adverse effects on health of exposures to traditional air or water pollutants, hazardous wastes, or other contaminants depend greatly on individual susceptibility. Some individuals may be highly vulnerable and thus may be adversely affected by low levels of various environmental contaminants, while other, hardier individuals may suffer no adverse effects whatsoever even when exposed to high levels of the substance in question. The basic point is this: even uniform exposures to a particular environmental contaminant may have very different effects on different individuals.

Similarly, different areas will be affected very differently by global climate change. Some areas may experience warmer temperatures, others cooler temperatures; some areas may experience more rainfall, others less; some areas may enjoy increased agricultural output while others suffer reductions. The important point is that different areas will respond very differently to global warming in the same way that different individuals may respond differently when exposed to environmental contaminants.

A fifth similarity between the two problems concerns the pervasiveness of uncertainty. It is amusing to read in discussions of climate change apologies by researchers concerned that their predictions of temperature change may be off by 20 percent, 50 percent, and the like. When one talks about uncertainty in the context of carcinogenic risk assessment, it is not unusual to find disclaimers that the true risk may be three, four, or even more orders of magnitude different from the predicted level. Environmental risk assessors would be thrilled to be as close to the true number as climate modelers appear to feel they will be. In spite of this difference, one cannot comprehend the problems faced in making policy in either area if one does not have a healthy appreciation for the great uncertainties involved. Surely these uncertainties complicate policymaking in both areas more than any other single factor.

Risk Management Strategies

Having discussed these similarities and differences, which have primarily to do with the assessment of both kinds of problems, I would now like to turn my attention to parallels in risk management. Of particular interest here is the appropriate role of prevention as a strategy for dealing with both greenhouse warming and the more traditional environmental regulatory problems. There is almost universal appeal, of course, in the old bromide "an ounce of prevention is worth a pound of cure." It is offered up as a justification not only for most environmental regulatory policies but, indeed, for government policies of all sorts. Virtually no one ever says, "Let's sit around and wait for the problem to happen and figure out how to deal with it then." Somehow that just doesn't sound like good public policy.

Not unexpectedly, perhaps, the economist tends to view the choice between prevention and cure somewhat differently. From an economic perspective, one needs to know the relative prices of prevention and cure (or adaptation), the effectiveness of both types of measures, and the probability distribution of possible outcomes before one can choose between the two alternatives. For surely it is the case that not every environmental risk problem merits extraordinary preventive measures. For instance, not every abandoned hazardous waste disposal site will come to pose serious risks to health and the environment. Not every ocean-going oil tanker will have an accident entailing large releases of oil into the ocean, although recent events in Alaska dramatize what terrible consequences can accompany such an accident. Not every ton of benzene emitted into the air will initiate a tumor. Prevention would, of course, be an ideal strategy if we could identify in advance which disposal sites would become serious health risks, which tankers would be involved in spills, or which sources of benzene would cause cancer. In that case, we could target our preventive measures on those sites, tankers, or sources, and thus avoid spending money where no problem would ever have occurred.

But the fact is, we cannot target this carefully in general. This means that we must take preventive measures at many, perhaps all, problem areas—be they waste disposal sites (which number in the hundreds of thousands), oil tankers, or benzene sources. This, in turn, implies that prevention can be expensive. Thus, depending on the seriousness of the consequences of an adverse outcome (drinking water contamination, an oil spill, and so on), it may make sense to wait and target resources to the areas where problems occur. Similarly, in a recent book Louise Russell (1986) points out that even in the area of health care, disease prevention is not always to be preferred to treatment. Once again, it depends on the cost and efficacy of preventive measures and the cost and efficacy of treatment measures.

This suggests, then, that prevention will not be the only (and may not even be the major) response to the problem of greenhouse warming. (See also Schelling [1984] for a discussion of this point.) When, then, will prevention be an attractive policy option? We can identify several preconditions. While these are not rigorously established here, they may be of some use in thinking about the wisdom of preventive policies in the case of greenhouse warming.

First, preventive policies are more attractive the less expensive they are. Anything that makes us better off looks better the cheaper it is. Second, preventive policies are more attractive the more efficacious they are. Even cheap goods or services tend not to be bought if they do not work at all. These points are quite obvious, but worth restating.

The third and fourth points are somewhat less obvious. Preventive measures are more attractive the more likely it is that a problem will develop. In other words, we might not take preventive measures at 100,000 different waste disposal sites if we were confident that only one would come to pose health or environmental risks (although we might if we thought that serious illness would result in that case). Similarly, we might elect to forgo expensive preventive measures if we thought significant global warming was just a remote possibility.

Fourth, preventive measures are more attractive when what is being forestalled would be very undesirable or irreversible. Put differently, some outcomes are so trivial that they do not merit any preventive measures regardless of how inexpensive. Others are so extreme that society goes to great lengths to avoid any chance they will happen (executing an innocent person, to pick a nonenvironmental example). This consideration has to be factored into thinking about climate policy.

Combining these characteristics, then, we would find attractive a policy that would prevent greenhouse warming if the policy were inexpensive and effective, if the likelihood of warming were great, and if the consequences of warming would be dire. Conversely, one need not be a rocket scientist to figure out the circumstances under which preventive policies are unattractive. This will be the case if they are particularly expensive, if they are not effective in delaying or preventing greenhouse warming, and if we are not particularly troubled by the consequences of greenhouse warming even should it occur.

How do preventive policies stack up against these criteria in the case of greenhouse warming? Consider first expense and effectiveness. Certainly some relatively inexpensive preventive options exist. For instance, the process of controlling chlorofluorocarbon emissions on a worldwide basis has already begun because of global concern about stratospheric ozone depletion. Because chlorofluorocarbons also act as greenhouse gases, actions taken to prevent ozone depletion will also help us on the global warming front. In some sense, then, these are free preventive policies to guard against greenhouse warming.

Similarly, there are some energy conservation measures that can and should be taken in the United States and elsewhere because they, too, will slow the accumulation of greenhouse gases in the atmosphere. If proponents are correct in their claims, some of these energy conservation policies should be undertaken purely on economic grounds. Certain of them, it is alleged, not only will reduce fossil fuel combustion but also will save money for those who undertake them. These would be the most desirable of all preventive policies. They would give the lie to the old economic adage that there is no such thing as a free lunch.

One must question, however, how widespread such energy conservation opportunities are. It is no doubt the case that through better insulation of homes, more fuel-efficient cars, more energy-efficient appliances, and better use of such technologies as passive solar heat, we can reduce energy use and, concomitantly, carbon releases into the atmosphere. But beyond some point, such measures surely have a positive economic cost. That cost may be worth bearing when one takes into account the external costs associated with global warming; nevertheless, we have to face the fact that such measures will not be free.

While I would like to believe otherwise, I am skeptical of claims that, for instance, a 50 percent reduction in energy consumption in industrialized countries can be had with no strain on the growth of the gross national product in these countries (e.g., Goldemberg et al., 1987). Such claims about energy conservation are, I believe, counterproductively exaggerated. Similarly, in thinking about which alternative energy sources we may substitute for fossil fuels, we have to be realistic in assessing the opportunities. Once again, some low-cost prevention may be available—perhaps through more use of natural gas, for instance—but I am skeptical about the extent to which alternative energy sources can substantially reduce carbon dioxide emissions at low cost. Of course, nuclear power is squeaky clean on CO_2 grounds but presents its own set of problems, which I will not address here. Finally, turning to such measures as pumping the CO_2 produced in electricity generation into the deep ocean, I must conclude that preventive actions like this will not be economically feasible for a very long time to come.

Let us consider the seriousness of the effects of greenhouse warming that preventive measures would be designed to forestall. Some of the possible effects could be very serious. I refer here to significant sea-level rise (which could exact a high human and economic toll), precipitation changes, temperature increases in the tropics that might have very serious ecological effects, and certain effects on forests that could greatly reduce their economic productivity as well as the recreational opportunities and aesthetic amenities they provide. These potential adverse effects are very serious indeed, and we must pay careful attention to them.

Nevertheless, at least some of the effects of global warming in some areas will be positive, and it makes little sense to ignore this fact. This is a bone in the throat of those who would warn us about global warming. They would much prefer us to believe that, sooner or later, the effects of warming would be adverse over all parts of the globe. This would make it easier for them to build support for policies to limit emissions of carbon dioxide and the other greenhouse gases. However, it seems inevitable to me—and, more importantly, to others more knowledgeable about global warming—that some parts of the world would benefit from warmer temperatures and altered rainfall patterns. The countries most prominently mentioned in this regard are the Soviet Union and Canada, both of which might see a substantial expansion in their acreage suitable for agricultural production. Other parts of the globe might also benefit to some extent from increased temperatures and the altered weather patterns that would accompany them, but it is more difficult to speculate where those areas might be.

To see why the presence of beneficiaries from greenhouse warming makes it a much more difficult problem to deal with, consider the case where *all* countries would suffer from global warming. Even in this circumstance, it would be difficult to get them to take concerted action on account of what is known as the "free-rider problem" (see Olson, 1965). Even countries that would benefit from slowing the rate of warming might find it in their interest to continue with their own emissions of carbon dioxide, chlorofluorocarbons, or other trace gases. This is because by refraining from control measures in their own country, they would avoid the associated costs, yet they would enjoy the benefits of reduced warming if all other countries restricted their CO_2, CFC, and other emissions. In this regard, then, even if everyone were to be made worse by global warming, some countries would have an incentive to cheat on a global agreement in order to avoid their share of the costs.

The global warming problem is made much more complicated when some countries would prefer to see temperature increases for whatever reason. In that case, the problem is a pure redistributional one where we must convince some countries to accept cooler temperatures than they would prefer so that other countries might avoid harmful warming.

A further complication concerns the role of the less developed countries in limiting carbon dioxide emissions. They have a fair basis for complaint. Any worldwide compact would require them to cut back carbon dioxide emissions from fossil fuel use and to limit other activities such as deforestation. Such cutbacks, they argue, would reduce economic opportunities for their citizens, and all because the wealthier, more developed countries have already filled the atmosphere with greenhouse gases during the process of increasing the economic well-being of their own citizenry.

Thinking further about the possible consequences of warming, it is interesting to speculate whether the equivalent of a doubling of carbon dioxide concentrations in the atmosphere really would be that bad. There is a tendency to fear any departure from the status quo, and this may explain at least part of the concern about global warming. Would we want to return to today's climate if we lived in a world whose climate matched that predicted for a CO_2-doubled world? In other words, if we started in a warmer, more humid world, would we find reasons to resist a return to a world like that which we enjoy today? Suppose we had all grown up in a mini-ice age: would we find reasons to oppose a shift to today's far milder climate?

I am not suggesting that we should passively accept the predicted consequences of global warming. As I have emphasized, many of those consequences appear to be unacceptable, and it is not too soon to develop appropriate policy responses. Furthermore, there is something to be said for a reluctance to tinker with natural processes on such a grand scale as that envisioned in climate scenarios. What I am suggesting is that part of our antipathy toward the prospect of a world where CO_2 concentrations are equivalently doubled may be due to the inherent biases we feel in favor of the status quo.

CONCLUSION

There are many respects in which the problem of global warming differs significantly from the problems associated with regulating traditional environmental pollutants. These differences have to do primarily with the geographic scope of the problem, the number of contributors to it, and the difficulty of taking specific control measures. On the other hand, there are respects in which these two problems are quite similar. These have to do with the methods—experimental and otherwise—that must be used to assess each type of problem, the uncertainties which attend the selection of appropriate policy responses, and, in particular, the importance of choosing between preventive and adaptive measures to deal with both problems.

For the problem of greenhouse warming, it is clear that the optimal policy will combine both preventive and adaptive strategies. On the preventive side, favored policies will include energy conservation, use of less-polluting fuels (for instance, substituting natural gas for coal or fuel oil), further reductions in chlorofluorocarbon emissions into the atmosphere, and a variety of other approaches designed to prevent the atmospheric buildup of greenhouse gases. If any of these policies also will save us money, it is past time that we should act. Even those that entail higher spending may be excellent buys if they can significantly reduce the chances for global warming.

We can also envision policies designed to make it easier to adapt to altered temperature and climate conditions when and if they should occur. I have said very little about possible adaptive measures here; this is not because they are unimportant. Given the consensus that some significant warming is almost sure to occur, these approaches deserve serious attention. These include actions now that would begin to relocate population centers and economic activities away from shorelines, research designed to make agricultural and silvicultural production more resistant to climate change, and other measures designed to make it easier to live in a world with altered temperature and climate conditions. To be sure, these adaptive measures will not be perfect substitutes for preventing climate change, but depending on their cost and efficacy, they may be the best that we can do.

How do we choose between preventive and adaptive strategies? This requires careful qualitative and, where possible, quantitative balancing of the benefits and costs associated with both sorts of policies. Making such evaluations is very difficult indeed. In doing so, we must struggle against letting quantifiable effects obscure or even drive out other consequences that do not so easily lend themselves to quantification. Very often the things that matter most will also be the most resistant to enumeration or easy valuation. But we must keep firmly in mind the fact that such valuations and trade-offs are made whenever decisions are made. There is something to be said for making such valuations explicitly and in the open. That is what policy analysis has to offer in the case of greenhouse warming, or in any other area for that matter.

ACKNOWLEDGMENTS

For their helpful comments on an earlier draft, thanks are due Adam M. Finkel, A. Myrick Freeman III, and Michael Gough.

REFERENCES

Goldemberg, J., T. B. Johansson, A. K. N. Reddy, and R. H. Williams. 1987. *Energy for a Sustainable World* (Washington, D.C., World Resources Institute).

National Research Council. 1983. *Risk Assessment in the Federal Government: Managing the Process* (Washington, D.C., National Academy Press).

Olson, Mancur. 1965. *The Logic of Collective Action* (New York, Shocken Books).

Portney, Paul R., ed. 1982. *Current Issues in Natural Resource Policy* (Washington, D.C., Resources for the Future).

Russell, Louise. 1986. *Is Prevention Better Than Cure?* (Washington, D.C., Brookings Institution).

Schelling, Thomas C. 1984. "Anticipating Climate Change," *Environment* vol. 26, pp. 6–35.

Part II

Natural Resource Sectors

7

Adapting Future Agriculture to Changes in Climate

William E. Easterling III
Martin L. Parry
Pierre R. Crosson

This chapter proceeds on the assumption that, however desirable it is to avert the greenhouse warming discussed by Schneider and Rosenberg (chapter 2 in this volume), some warming will occur in coming decades and that this warming, with changes in other climate features, will demand that agriculture make adjustments.

Following Schneider and Rosenberg, we assume that atmospheric concentrations of carbon dioxide (CO_2) and other greenhouse gases will have increased to the warming equivalent of twice the preindustrial concentration of CO_2 (henceforth $2 \times CO_2$ warming) by the middle third of the next century and that this will induce an increase between 1.5 and 5.5 degrees Celsius (°C) in the global average temperature.[1] If equivalent CO_2 doubling occurs in the mid-2030s, the maximum decadal rate of warming would be about 1.1°C. If the doubling occurs in the mid-2060s, the minimum decadal rate of warming would be about 0.2°C.

We assume that the prospects of climate change raise two broad questions for agriculture: How will future agriculture be affected? Will the effects require policy actions to achieve and maintain adequate food security and other desirable social and economic goals? Our focus is on agriculture in the midwestern United States, but we draw extensively from research throughout the world.

We begin by describing a decision-making framework for midwestern farmers. We then construct a scenario of how the key elements of that framework might look by roughly the year 2050 in the absence of climate change. Finally, we consider how that scenario might be affected by climate change.

A FRAMEWORK FOR AGRICULTURAL DECISION MAKING IN THE MIDWEST

It is useful to think of the agricultural industry as a collection of individual firms (farms) producing to meet a demand for agricultural goods. Each farm requires a combination of basic production inputs: land, labor, materials, capital, and climate. A farm's output is produced at a given level of technology and management and organized within a certain climatic, economic, social, and political environment. The economic aspects are particularly important in market economies. The market reflects demand and supply conditions, hence prices, for production inputs and farm outputs. It can be assumed that the goal of the individual farm firm is to maximize profits, that is, to maximize the difference between gross revenues and total costs.

Sonka and Lamb (1987) argue that any study of the effects of climate on agriculture must take explicit account of the economic environment in which farm firms operate, recognizing that this environment reflects not only local and regional conditions but also those at the national and global levels. That is to say, the farm enterprise is part of an interconnected system of local, regional, national, and world markets.

Figure 7-1 (adapted from Sonka and Lamb, 1987) is a schematic of an integrated agricultural production system. It emphasizes the annual and interannual planning process on the farm. Farmers' expectations of future market conditions for inputs and outputs affect current planning and production decisions. Expectations about climate, other environmental conditions such as water and soil, and government policies also are part of the decision-making framework. The role of the market as a price-setting mechanism is important as a determinant both of farm profits and of the mix of farm outputs. Any given farmer is a price taker; that is, market prices are independent of the level of his output. For farming as a whole, however, prices and output are interdependent.

This schematic emphasizes the dynamic nature of the relationship between climate and agriculture. Climate change affects expectations, which then lead to adjustments of production decisions. These adjustments are part of both

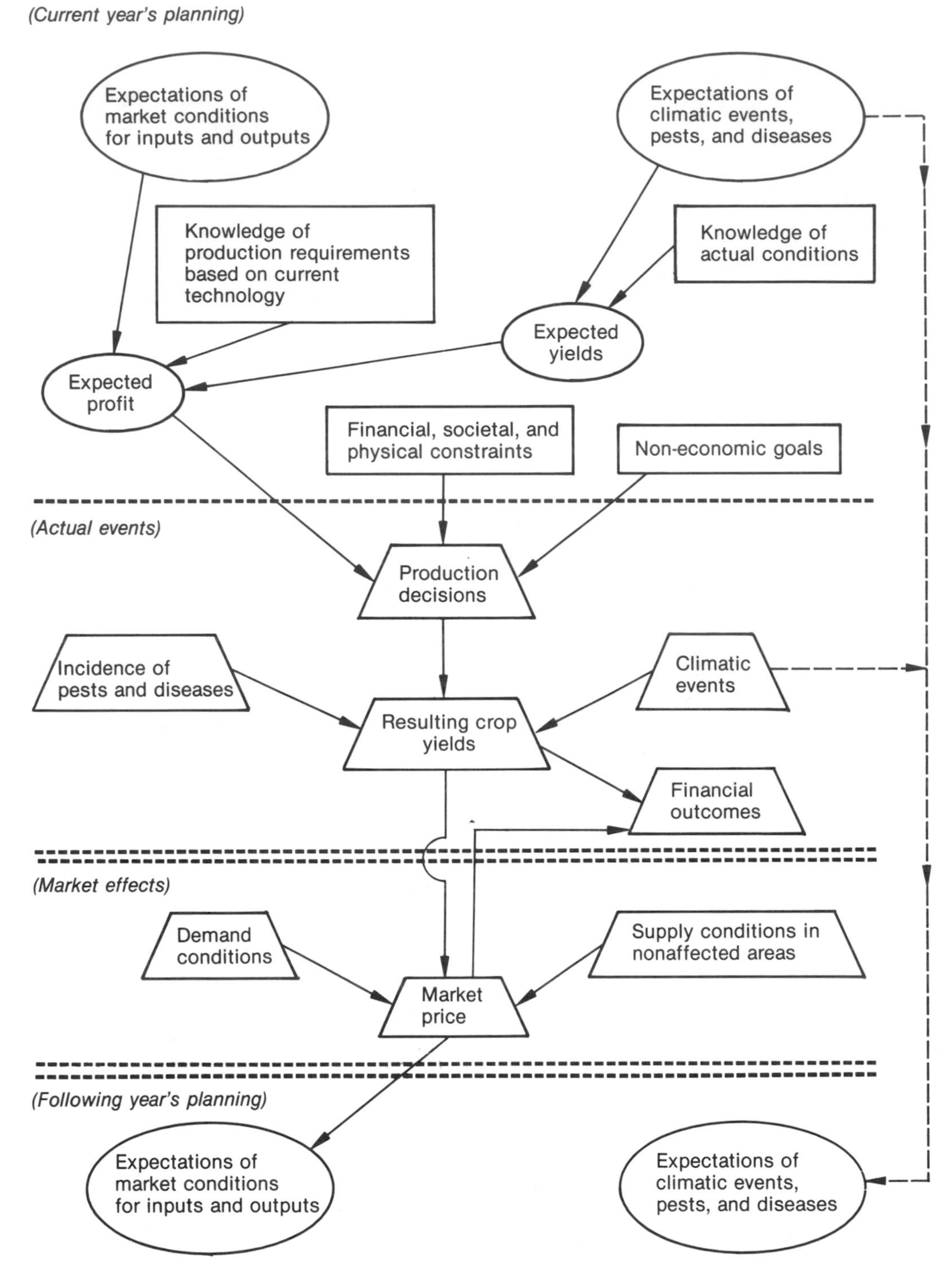

Figure 7-1. Farm-level decision framework. (*Source:* Redrawn, with permission, from Sonka and Lamb, 1987. © 1987 by Kluwer Academic Publishers.)

annual planning cycles and longer term planning decisions (say of five to ten years).

The relationships shown in the figure 7-1 schematic must be examined together to gain insights into how climate change will influence the whole farm system. We use this schematic to organize our discussion of what a typical midwestern agricultural system might look like in a future without climate change[2] and how climate change might alter that future.

A FUTURE WITHOUT CLIMATE CHANGE

What might the economic, technical, and policy environment for agriculture in the Midwest look like fifty to seventy-five years from now in the absence of climate change? A key element of that environment is the price of agricultural commodities. Those prices are set in world markets, so we begin by considering long-term trends in world demand and supply conditions for food and fiber. The pace of technical change also will critically affect the future performance of midwestern agriculture, and rising concern about environmental impacts of agricultural production will be significant as well. So we consider both of these. Finally, shifts in agricultural comparative advantage among countries around the world will have major consequences for the performance of midwestern agriculture, so this, too, is briefly discussed.

World Agricultural Demand

World population and income growth are the main long-term determinants of the growth of world demand for food and fiber. The present population of the world is about 5 billion. World Bank (1984) projections indicate roughly 10 billion by the last quarter of the twenty-first century, with most of the growth occurring by 2050. More than 80 percent of the growth will occur in the developing countries of Asia, Africa, and Latin America.

Since 1960 world per capita income has been growing at an average annual rate of 3.0 to 3.5 percent (World Bank, 1984). Nordhaus and Yohe (1983) use a projection of little more than 1 percent annually in the twenty-first century. Crosson (1986) assumes an annual rate of 2 percent in this period, noting that this is conservative by experience since 1960 and citing evidence (Goeller and Zucker, 1984) that supplies of energy and other natural resources should be adequate to support such growth. In this discussion we assume the 2 percent annual growth rate.

This quick look at the numbers suggests that population growth alone could increase world demand for food and fiber by 60 to 80 percent from the mid-1980s to the middle of the next century. Income growth in the developed countries will not add much to this because people in those countries already are well fed. In the developing countries, however, present levels of per capita income and nutrition are so low that a significant proportion (20 to 30 percent) of any additional income is spent on food. Per capita income growth in these countries, therefore, could add substantially to the growth of world demand for food and fiber. Indeed, the contribution could roughly match that of world population growth. Under this set of assumptions, world demand for food and fiber in 2050 could easily be two to two and one-half times the level of the mid-1980s.

World Agricultural Supply

This rate of increase in world demand (between 1.0 and 1.4 percent annually) is low by comparison with the roughly 2.5 percent annual rate of increase in world agricultural production experienced from the early 1950s to date. During this period world agricultural prices declined as supply increased faster than demand, reflecting the deployment of additional land, water, labor, and especially new technology. Thus, judging from the experience of the last thirty years, there is reason to believe that world agricultural capacity could easily run ahead of the relatively slow growth of world demand.[3] The prospect in this case would be for a continuation of the long-term declining trend of world prices for agricultural commodities, including the wheat, corn, and soybeans of special interest to midwestern farmers.

The Position of Midwestern Agriculture

A scenario of declining world agricultural prices does not necessarily threaten the viability of midwestern agriculture. Farmers in that area have lived in such a price environment for several decades, yet have managed to expand total output greatly and to increase their penetration of world agricultural markets.

They were able to do this because advances in technology and management increased their productivity enough to more than offset the decline in the prices they received for their output. Their continued success depends crucially on the maintenance of this technical and managerial progress. Progress in this respect is driven in good part by changing relative scarcities of land, water, and labor and by investments in the development of managerial skills and new technology (Hayami and Ruttan, 1985). Concerns about the environmental impacts of new technology have begun to influence the pattern of technical change as well. These various factors will condition the without-climate-change availability of new technology to midwestern farmers over the next fifty to seventy-five years. Three of these factors will be of special importance, as follows.

Rising Value of Labor

Crosson (1986) notes that the process of economic development has brought substantial increases in the value of labor in both agricultural and nonagricultural pursuits. If the productivity of the U.S. economy continues to increase, and we think it will, this will continue. The implication is a continuing incentive to substitute capital for labor in American agriculture, and the likelihood that new and more productive forms of capital will be available is high.

Continued Technological and Management Advances

Productivity-enhancing breakthroughs are expected to come mostly through genetic improvements in the yielding potential of crops and through more efficient use of emerging information technologies. Developments in biotechnology will contribute to genetic improvement, as will traditional crop breeding. Duvick (1986) sees no reason why recent yield increases (1 to 2 percent per year for corn, wheat, and grain sorghum, slightly less than 1 percent for soybeans) will not continue for some years to come. Other studies support this view (Office of Technology Assessment, 1986; English et al., 1984).

King (1986) argues that we can expect an increasing ability on the part of U.S. farmers and agribusiness to process large quantities of information. This will increase production efficiencies. The advent of cheap microcomputers and a growing software industry focused on agriculture will permit agriculturalists to maintain extensive information bases for decision making. At present, there are no firm estimates of the productivity impact of this expanding information technology, but it is expected to be significant.

Environmental Concerns

Crosson (forthcoming) argues that the growth of U.S. agricultural production over the last several decades incurred rising environmental costs, particularly in loss of habitat values, loss of open-space amenities, and increasing damage to water quality from sediment and associated agricultural chemicals. The demand for these environmental values is expected to grow much more rapidly over the next fifty to seventy-five years than the demand for food and fiber, and their supply likely will prove to be more inelastic. In this event, public pressure to protect these environmental values adequately would be likely to have an increasing influence on the technical and managerial options open to midwestern farmers in responding to rising demand for food and fiber.

Regional Comparative Advantage

Over the past forty years, comparative advantages of natural, human, and institutional resources have favored agriculture in the United States. Only a few other countries (for example, Canada, Australia, and, more recently, Argentina and Brazil) could come even close to matching the advantages of the United States in these respects.

Whether the United States will retain this comparative advantage over the long term is an open question. Some countries not among America's traditional competitors—China, India, perhaps even the Soviet Union—show signs of having the potential for considerably improved long-term agricultural performance. While these countries probably are not as well endowed with land and water as is the United States, their principal disadvantage has been a lack of the institutions and policies necessary for development of the increasingly science-based agriculture currently characteristic of the developed world. India now is well advanced in developing the needed research capacity (Plucknett and Smith, 1982), and recent reforms in China and the Soviet Union suggest that they may be ready to give more play to the initiative and creativity of their farmers and other members of their agricultural establishments.

There is another, essentially political, factor which also could weaken the competitive position of the United States in world agricultural markets. Food is so basic to national welfare and uncertainties about world political alignments are so great that it probably is fair to say that, net exporters apart, most countries would like to be agriculturally more self-sufficient than they are. Many probably already are more self-sufficient than they would be by a strict calculation of comparative advantage, Japan being the prime example. Apart from China, India, and the Soviet Union—which clearly have the potential to become major, if not dominant, actors in the world agricultural economy—many other countries may seek greater food self-sufficiency over the decades ahead. The cost of doing this should decline as their economies grow and their stock of human and other kinds of capital increases.

The implication of this scenario is decreasing international specialization in agricultural production. World trade in agricultural commodities would continue to grow in absolute amount, but it would decline relative to world agricultural production. Comparative advantage would be more diffused, and some countries now wielding little of it might emerge as principal actors. For midwestern farmers, export markets would grow significantly more slowly than in the 1950s, 1960s, and 1970s.

How the above without-climate-change scenario would be reflected in decisions of midwestern farmers can only be speculated. Yet the following propositions appear to be justified.

A midwestern farmer over time would come to expect:

- Gradually declining prices for farm output;
- Average annual increases in productivity of roughly 1 to 2 percent;
- Rising production, although dampened by relatively slow growth of export demand;
- Increasing pressure from outside the farm community to recognize and properly weight the environmental values affected by farm operations;
- Some help in responding to this pressure in the form of technologies aimed at lowering environmental costs, especially with respect to pesticides and sediment damage to water quality.

A FUTURE WITH CLIMATE CHANGE

We now consider how climate change may alter the expectations of midwestern farmers. We also consider some of the adjustment and adaptation possibilities that may be open to them for lessening the costs and seizing the benefits of climate change.

Expected Prices

Since midwestern farmers operate in world markets, we must consider how prices in these markets might be affected by climate change. Present understanding of the impacts of climate change gives no persuasive reasons to believe that the growth of world population or income will be significantly affected by changes resulting from a doubling of atmospheric CO_2. Schelling (1983, p. 481) speculates that such changes would not lower world living standards by more than a few percent. If he is right, then the income effect of $2 \times CO_2$ climate change would be trivial in a world where per capita income is higher by 260 percent (2 percent compounded annually from 1985 to 2050 [Crosson, forthcoming]). The population effect of climate change would be similarly small, to the extent that population growth is a function of income growth. There appears to be no reason to believe that $2 \times CO_2$ climate change would have other significant effects on world population growth.

If total world population and income growth are little affected by $2 \times CO_2$ climate change, then the effect on growth in world demand for food and fiber also should be small. It does not follow, however, that the effect would be similarly small on the world's ability to expand agricultural capacity in step with demand. Should the expansion fall short, economic and environmental costs would rise. Even if aggregate world capacity grows equally with demand, climate change likely will alter the configuration of world comparative advantage, a matter of considerable interest to midwestern farmers.

Possible effects of climate change on world agricultural capacity, therefore, must be considered. We begin with a review of results of studies that estimate the effects of climate change on crop yields and on locations of crop production. In this connection we also review studies of the yield effects of higher levels of atmospheric carbon dioxide. We next discuss briefly how climate change might affect unpriced environmental costs of agricultural production. We then consider how various technical adjustments and long-term adaptation strategies might alter the estimated effects of climate change on yields. We conclude by drawing the apparent implications of these studies for world agricultural capacity.

Climate Effects on Yields

Climate change is likely to affect crop yields adversely in some regions and favorably in others. Results of a $2 \times CO_2$ climate experiment with a model developed by the Goddard Institute for Space Studies (GISS) were used by Parry and Carter (1988) to study the effects of $2 \times CO_2$ warming on crop yields in several regions around the world.[4] They did this by comparing the average yields expected under current climate conditions with average yields as they might be with $2 \times CO_2$ warming, assuming in each case the use of the same agricultural technology.

In some parts of the mid-to-high latitudes the GISS $2 \times CO_2$ climate scenario implied adverse yield effects because higher temperatures increase evapotranspiration, thus reducing the water supply available to plants. In Saskatchewan, Canada, for example, wheat yields would be as low as those that occurred during the driest five-year period on record (that is, 18 percent below).

Bach (1979) notes that warmer and drier conditions are detrimental to corn production. He argues that United States corn production could decline by 11 pecent per degree increase in mean maximum summer temperature and by 1.5 percent for each 10 percent reduction in summer precipitation.

Waggoner (1983) used a regression-based crop-weather model to examine the effect on crop yields in the United States of a uniform 1°C warming in combination with a 10 percent decrease in growing-season precipitation. Yields of North Dakota spring wheat declined by 12 percent; yields of Iowa corn declined by 3 percent; and Iowa soybeans declined by 7 percent (from the 1978–80 average).

For regions where available water is not a major constraint—for example, in Northern Europe, the United Kingdom, and Japan—yields could increase. In northern Japan, for example, yields of present-day rice varieties would probably increase by 4 percent with the 35 percent increase in growing degree days estimated under the GISS $2 \times CO_2$ scenario (Yoshino et al., 1988).

The GISS $2 \times CO_2$ scenario shows substantial lengthening of the growing season at middle and high latitudes. Parry and Carter (1988) found that in most parts of the high latitudes the warmer conditions would be expected to give higher yields. In northern Japan yields were roughly equivalent to those of the best (that is, warmest) year recently experienced. In Finland yields were approximately double those experienced during the best decade of this century, and in Iceland they were approximately four times the average for the best ten years since 1930.

Effects of Climate Change on Crop Geography

Studies of the effects of climate change on crop yields suggest that, in the absence of adjustments and adaptations, such effects could alter the present geographic distribution of crops.

If we assume that in most regions of the world agriculture is closely tuned to its present-day climate, then present-day farming types are one indicator of a system (which includes crop types, expected yields, optimum levels of fertilization, and so forth) that operates effectively in that climate. Whether that system is optimal for the given climatic conditions would be a matter for further inquiry. Yet by mapping present-day analogues of (for example) a $2 \times CO_2$

climate, we may gain some insight into the range of adjustments that may be required to match new land uses to the new climate. Parry and Carter (1988) used this procedure of analysis by analogue. Some of the results are shown in figure 7-2. The figure illustrates present-day regional analogues for the GISS 2 × CO_2 climate estimated for six regions in the Northern Hemisphere. The climate of Finland is estimated to become similar to that of present-day northern Germany; of southern Saskatchewan, to northern Nebraska; of the Leningrad region, to the western Ukraine; of the central Ural mountains (Cherdyn region), to south-central Norway; of Hokkaido in northern Japan, to northern Honshu in southern Japan; and of Iceland, to northeast Scotland.

Following this approach we might, to adopt the Iceland-Scotland analogue, expect hay yields in Iceland to increase by about 50 percent to match levels which are more typical of northeast Scotland today. Model experiments tend to confirm this conclusion (Bergthorsson et al., 1988). However, losses to pests and diseases, which are at present limited by winter cold in Iceland, might increase up to the level of 10 to 15 percent now faced by Scottish farmers. As for land-use changes: while hay, potatoes, and turnips are the only crops grown extensively outdoors in Iceland and sheep are the dominant stock, in northeast Scotland barley (stall-fed to cattle) is the main cash crop. Are the indications, then, that Iceland will see a shift from sheep farming to cattle rearing and fattening? And will the Leningrad region supplant the Ukraine as the wheat granary of the Soviet Union? Nothing is likely to be so simple, but indications are that the extent of the land-use changes may be substantial.

Other studies also suggest significant impacts of climate change on the location of agricultural production. Blasing and Solomon (1982) found that an increase in average annual temperature of 3°C combined with a 10 percent decrease in summer precipitation would shift the American corn belt several hundred kilometers on an axis from southwest to northeast. This assumes that the agroclimatic factors (for example, growing-season length, amount of sunshine, moisture requirements) that now determine the geographic extent of the corn belt remain unchanged.

Rosenzweig (1985) performed a similar spatial analysis for wheat in the United States, using the GISS 2 × CO_2 climate scenario. She found that, in general, the predicted climate changes would encourage a northward expansion in wheat production and that the southern part of the present wheat belt might experience more frequent heat stress that could lower yields.

A 2 × CO_2 climate change could expand the area of rice production in Japan (Parry and Carter, 1988). This is illustrated in figure 7-3 for northern Japan, where the safely cultivable area for irrigated rice under the GISS 2 × CO_2 climate is more than double that under the present-day climate. These studies, of course, are only illustrative. It is unlikely, however, that present geographic patterns of agriculture would simply relocate themselves to match the new geography of climate. This is considered in further detail below.

Environmental Effects

Climate change likely will increase the unpriced environmental costs of production. For example, a warmer and wetter global climate likely will encourage insect pests, weeds, and plant diseases. Unless nonchemical pest control systems are developed, the result will be more pesticide use than in the without-climate-change scenario and an

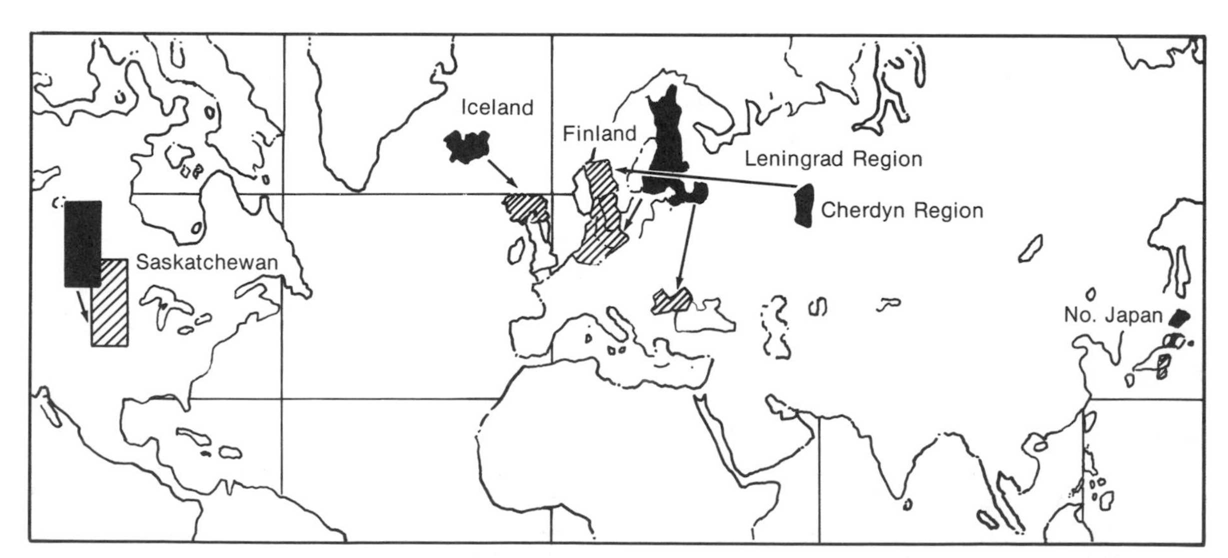

Figure 7-2. Present-day regional analogues of the Goddard Institute for Space Studies (GISS) 2 × CO_2 climate estimated for six regions: Saskatchewan, Iceland, Finland, Leningrad, Cherdyn (USSR), and Hokkaido and Tohoku (Japan). (*Source:* Redrawn, with permission, from Parry and Carter, 1988. © 1988 by Kluwer Academic Publishers.)

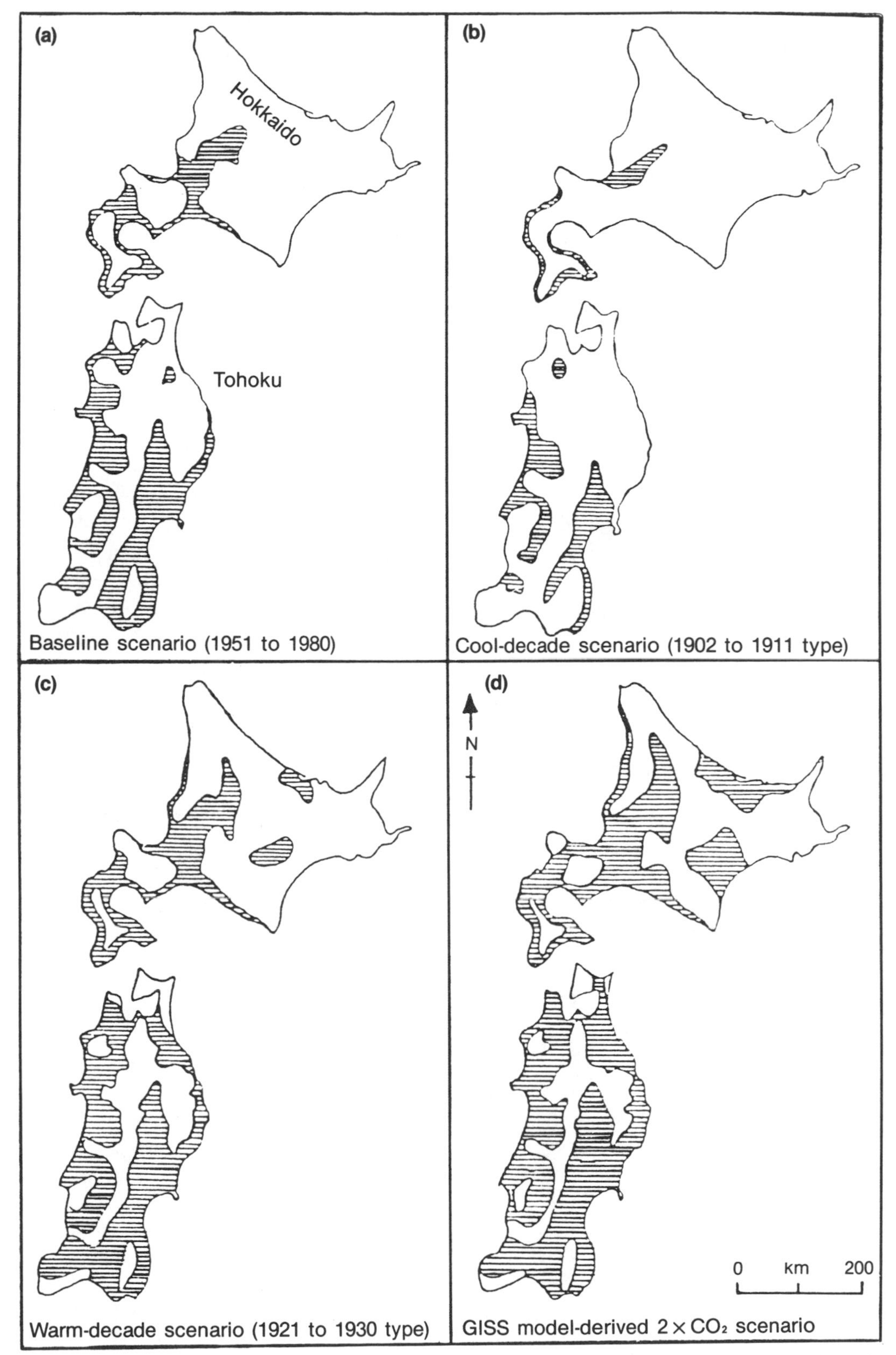

Figure 7-3. Safely cultivable area for irrigated rice in northern Japan under four climate scenarios. The safely cultivable area (shaded) is defined by the minimum level of accumulated temperatures during the growing season required for the crop to complete its normal life cycle. (*Source:* Reprinted, with permission, from Yoshino et al., 1988. © 1988 by Kluwer Academic Publishers.)

increased probability of pesticide damage to the environment. As another example, in an area that becomes hotter and drier, demand for water for irrigation likely would increase, putting more pressure on the water supply available for instream protection of ecological systems.

These are only illustrative of the types of environmental damage that may be abetted by climate change. It is not clear how these rising environmental costs will affect the productivity of global agriculture. If the costs are not borne by farmers—and many will not be—then they will have little effect on measured productivity. But, as we noted in our discussion of future agriculture without climate change, pressures from outside the agricultural sector are likely to force farmers to internalize more and more of the environmental costs generated by their operations. Additional costs imposed by climate change likely would strengthen this tendency. Unless farmers are provided with technologies and management practices to offset these higher environmental costs, global agricultural capacity may grow more slowly than it would in the absence of climate change.

The studies reviewed indicate that in some countries, and some regions within countries, the effects of 2 × CO_2 climate change on agriculture would be negative, but that in other countries and regions they would be positive. The implication is that in agriculture there would be both winners and losers from 2 × CO_2 climate change, with no clear indication of the net effect on global agricultural capacity.

Nevertheless, the studies tend to overstate the negative effects of climate change, for two reasons: (1) they fail to take into account the fertilization effects on crop yields of increased atmospheric CO_2; (2) they neglect the fact that farmers and institutions serving them will find adaptations and adjustments in technical and managerial practices that will offset some of the negative effects of climate change and enhance those that are positive. We consider each of these possibilities.

Effects of Carbon Dioxide Enrichment on Crop Yields

Results of studies in which plants are grown under controlled climatic conditions and in which atmospheres are enriched with CO_2 indicate increases in rates of photosynthesis and plant growth and yield and reductions in transpiration (see Kimball, 1983; Morison and Gifford, 1984; Goudriaan et al., 1984). These direct effects of CO_2 on crops have yet to be observed under actual field conditions, although they have been observed in chambers open to the sky. Factors other than CO_2, such as nutrient and water availability, may be more limiting to crop production than CO_2 levels. If this is so, the positive effect of CO_2 on plant growth would be reduced. However, controlled environmental studies indicate that CO_2 diminishes the negative impacts of soil-water shortage, soil salinity, and some nutrient shortages (Rosenberg et al., forthcoming). Clearly, the effects of CO_2 enrichment must be considered jointly with climate changes in estimating impacts on crop production.

Only a few studies have done this. Waggoner (1983) concluded that an increase in atmospheric CO_2 from the then current 340 parts per million by volume (ppmv) to 400 ppmv would increase yields of field crops in the American corn belt by some 5 percent, but that the associated warmer and drier climate (drawing on the global climate modeling in Manabe et al., 1981) would reduce yields by 3 to 13 percent, depending on the crop and the part of the region. The net effect of yields would thus be negative, but not as negative as the changes in temperature and precipitation alone would indicate. Decker et al. (1988), in a review of statistical climate-crop models, examined Missouri corn yield responses to a set of climate scenarios. The scenarios included changes in temperature and precipitation and a doubling of atmospheric CO_2 concentrations. They found that a scenario of 1°C warming and a 10 percent decrease in precipitation combined with a CO_2 concentration of 540 ppmv (double the preindustrial level) resulted in a net increase in corn yields over those under present climate and CO_2 conditions. With temperature increases greater than 1°C they found a net decrease in corn yields, despite the doubled CO_2 concentrations. However, the yield reduction was not as great as it would have been without the fertilization effects of CO_2.

Role of Adjustments and Adaptation

The research reviewed above has focused on how agricultural production in various regions around the world might be affected by climate change in the absence of changes in farm management and technology or in policies induced by the altered climatic regime. Since such induced responses are certain, the results of the research tend to overstate the negative effects of climate change on global agricultural capacity. Discussing adjustments by farmers to climate change, Easterling (forthcoming) argues that such simple technologial adjustments as changes in the timing of tillage, planting, and fertilization could turn a climate change from a net-negative to a net-positive production situation.

Over the longer term the vulnerability of agricultural systems to climate change would be reduced by increased investment in research aimed at developing new technologies and management practices better suited to the emerging climate conditions. Following are some examples of the adjustment and adaptation options that farmers around the world likely would count on.

Most scenarios of 2 × CO_2 climate change indicate higher temperatures, both seasonal and annual, at middle and higher latitudes. In many areas where low temperature during the growing season now is the major constraint on crop yield, this likely would mean increases in output. However, many of the early-maturing varieties currently grown in cool regions would fail to take full advantage of the longer, warmer growing seasons. In northern Japan, the substitution of late-maturing rice (now grown in central Japan) might increase yields by 26 percent. Similar increases have been estimated for spring wheat yields in Finland and the USSR. In each case only limited benefit would be derived from the present-day varieties, while substantial benefits are estimated for varieties with higher thermal requirements.

In the major Canadian grain-growing provinces (Manitoba, Saskatchewan, and Alberta), selection of different cultivars would almost eliminate crop losses from a 2 × CO_2 warming (Arthur et al., 1986). In fact, a switch from spring wheat, now the dominant crop, to winter-sown wheat in the southern prairies would avoid summer stress, thus reducing the threat of crop damage (Williams et al., 1988). An acceleration of this trend from spring to winter wheat could be an effective response to possible warmer growing seasons in the future.

Variations of fertilizer application may also be an appropriate technical response to climate changes over the long term. Experiments in Iceland indicate that fertilizer savings of up to 50 percent may be achieved in a warmer climate simulated for 2 × CO_2 conditions (Bergthorsson et al., 1988). Where long-term climate changes are unfavorable, it is possible that increased use of fertilizers may compensate for decreases in yield that otherwise would occur. For example, experiments indicate that yields of winter rye in the Leningrad area of the USSR may fall by more than a quarter under a warmer 2 × CO_2 climate, but could be increased by about 15 percent above present levels if fertilizer applications were increased by 50 percent (Pitovranov et al., 1988). Whether this would be economical would depend, of course, on relative fertilizer-crop prices.

In a study of farm-level response by Illinois corn producers to the GFDL[5] and GISS 2 × CO_2 midwestern climate scenarios, Easterling (forthcoming) suggests that several management strategies could keep corn production possible in Illinois from an agronomic standpoint. The timing of many operations could be altered to adjust production to climate changes. For example, more tillage and planting preparation activities in the drier fall season could offset potential difficulty in getting into fields with equipment during the wetter spring. Also, earlier planting, varietal changes (discussed above), and lower planting densities, though they may lower yields somewhat, might help offset heat and moisture stress during a warmer and, possibly, drier summer.

Many of the above management changes are not expensive, although they may not offset completely all of the negative impacts of climate change. More research is needed, especially research linking agronomy and economics, to determine the technical and economic feasibility of the various types of management responses.

In addition to adjustments to existing technologies at the farm level, policies aimed at developing new technologies and at promoting regional and natural adjustments to the changed climatic regime would tend to ease the negative impacts of the new regime on global agricultural capacity. It should be emphasized that few of these policies have yet been explored in any depth, and they are put forward here simply as an indication of the range of policy alternatives. Indeed, it may reasonably be argued that, at this early stage in our understanding of future climate change, it is more rewarding to concentrate on extending the range of policy options than on identifying and refining specific policies. Two types of options are considered here: development of new technologies and management practices and changes in land-use.

Policies to Develop New Technologies and Management Practices Investment in research to develop new cultivars, cultural practices, and machinery, all specifically adapted to changing climate conditions, is a promising policy option. The promise is suggested by experience with the spatial expansion of hard red winter wheat in the Great Plains of the United States. Rosenberg (1982) shows that between 1920 and 1980 the area of the Great Plains planted in hard red winter wheat expanded across spatial climate gradients that exceed the climate changes expected from 2 × CO_2 warming. Rosenberg argues that it was the development of new cultivars and new cultural practices that allowed this expansion to occur. A follow-up study now under way at Resources for the Future has found that this expansion has continued into southern Canada since 1980. This example suggests that agricultural technologies adapted to a variety of climatic conditions can be developed in a few decades, within the time frame expected for 2 × CO_2 warming.

Investments to expand the range of farm management practices appropriate for changing climatic conditions also would be a promising policy option. National farm policies that aim to increase resilience to short-term climatic variability (for example, drought, flooding, cold spells, and the like) incidentally will often improve resilience to possible changes of climate over the longer term. However, it may be necessary to adapt these policies specifically to encourage different types of farm management and different levels of production inputs to match climate-induced changes in agricultural potential. For example, further support may be necessary in traditional areas of agricultural extension work, such as water management (for example, improving efficiency of water use), land management (for example, instituting new soil management practices to improve control of soil erosion), and pest control (for example, adopting new crop varieties and cultivation practices to inhibit pests and diseases). Continued efforts in the reorganization of farm structuring and rural infrastructure (such as improvements to systems of transport, marketing, and credit) would be necessary, as they are today, to provide a suitable economic environment for the effective introduction of new technologies.

Change in Land Use Some of the studies cited above indicate that 2 × CO_2 climate change likely will bring shifts in the distribution of agricultural land use. As argued above, these shifts may be more complex than simply moving a particular mix of land uses to another location. Questions may arise as to what represents an optimal mix of land uses in a region after the climate change. Some changes of land use might be implemented not to optimize production (that is, to maximize total net returns) but to stabilize it. This strategy would seek to allocate land in such a way that the risk of heavy losses due to climatic variations (either short- or long-term) would be reduced, either by minimizing the variability or by striving for the best possible return under the worst possible conditions.

Implications for Global Agricultural Capacity

The material reviewed gives no clear evidence that 2 × CO_2 climate change would prevent the growth of global agricultural capacity at least in step with the relatively slow growth of demand stipulated in the without-climate-change scenario. Although some countries would be adversely affected by 2 × CO_2 climate change, others would benefit, even without accounting for opportunities for adjustment and adaptation. When these opportunities are taken into account, it appears likely that the adverse effects would be reduced and the beneficial ones enhanced. The likely positive-yield effects of atmospheric CO_2 enrichment strengthen this view. We conclude that the scenario of 2 × CO_2 climate change does not imply rising world prices of food and fiber. On the contrary, we believe that the declining price trend depicted in the without-climate-change scenario would hold also with 2 × CO_2 warming, although the rate of price decline might be lower.

A Note on Food Security Although 2 × CO_2 climate change does not appear to threaten the expansion of global agricultural capacity in step with demand, it does suggest significant shifts in regional comparative advantage in agriculture. Some countries losing advantage may see this as a threat to the security of their food supplies. If they respond by using protectionist policies to defend their agriculture from imports from countries favored by climate change, the effects on world agricultural capacity and prices would be problematical. Capacity and prices in the countries adopting protectionist policies to offset the loss of advantage would be higher than if the policies were not adopted. The policies would also limit the export growth which countries gaining advantage otherwise would enjoy. If the newly advantaged countries accept these limits, their capacity would be lowered. The effect on their prices would depend in large part on how much their potential for export expansion would be stifled by the protectionist policies adopted in disadvantaged countries.

What the effect of the protectionist policies would be on global agricultural capacity and prices is uncertain. However, the intercountry distribution of capacity would resemble more closely the without-climate-change scenario, and world prices would be distorted from what they would have been in the absence of protectionist policies in disadvantaged countries. The effect of this on the price expectations of midwestern farmers is uncertain.

If the countries disadvantaged by climate change are content to ensure food security by permitting increased imports from advantaged countries, then the distribution of world capacity would shift toward the advantaged countries, and prices in the disadvantaged countries would be lower than if they adopted protectionist policies. Whether prices in the advantaged countries would be higher or lower is, again, uncertain. Our guess, however, is that world prices would be lower in the absence of protectionist policies in the disadvantaged countries.

On balance, we think the effects of protectionist policies on global capacity and price would be unlikely to upset our conclusion that world agricultural prices would decline under 2 × CO_2 climate change, although perhaps not as much as they would decline without climate change. After all, because of technological advances, world agricultural prices have declined over the last forty or fifty years despite strongly protectionist policies in many countries (Crosson, 1986).

EXPECTATIONS OF MIDWESTERN FARMERS

The discussion above indicates that with 2 × CO_2 climate change midwestern farmers could expect declining prices for their output, as they could without climate change. However, 2 × CO_2 climate change may induce less favorable expectations about trends in farm productivity and income than in the without-climate-change scenario. Schneider and Rosenberg (chapter 2 in this volume), although noting the high uncertainty about how regional climates will change in response to 2 × CO_2 warming, nonetheless acknowledge the possibility that temperature in the Midwest likely will increase more than the global averages and that soil moisture in summer could be reduced from present levels. In the absence of adjustments, this suggests lower yields than in the without-climate-change scenario. The studies reviewed above by Bach (1979), Waggoner (1983), Blasing and Solomon (1982), and Rosenzweig

(1985) also suggest that, in the absence of adjustments, $2 \times CO_2$ climate change would have unfavorable effects on crop yields in the Midwest.

The unpriced environmental costs associated with midwestern agricultural production also might increase under $2 \times CO_2$ climate change. For example, a warmer and drier climate would decrease water supplies while the demand for irrigation water could increase. The water supply available for the protection of instream environmental values thus would be diminished.

The effect of higher environmental costs on the productivity and profits of midwestern farmers will depend on whether public policies force farmers to cover the costs—for example, by limiting the expansion of irrigation in order to protect in-stream environmental values. In this case, and in the absence of adjustments, the higher environmental costs would reinforce the adverse effects on farm productivity profits of lower yields.

In this event, farm income in the Midwest also would be reduced. The amount of the reduction is completely uncertain. Recent experience suggests it could be significant, however. The 1988 drought created conditions in the Midwest similar, in direction if not in magnitude, to those postulated with $2 \times CO_2$ warming. In the summer of 1988 the U.S. Department of Agriculture estimated that net farm income of all U.S. farmers (not just those most affected by the drought) would be down between $1 and $6 billion from 1987 because of the drought, a decline of 2 to 3 percent (U.S. Department of Agriculture, 1988).

Should $2 \times CO_2$ climate change reduce farm income in the Midwest, nonfarm income in the region likely would be depressed also. These off-farm income effects are being studied at Resources for the Future, but results are not yet available. The U.S. Environmental Protection Agency also is investigating this issue for the entire country, not just for the Midwest, but at this writing the study is not yet completed.

Studies in other countries, however, suggest that the negative off-farm effects of $2 \times CO_2$ climate change may be significant. For example, Williams et al. (1988) used an input-output model to estimate the effects of $2 \times CO_2$ warming (under the GISS climate scenario) on agricultural income in Saskatchewan and consequent feedback effects on provincial employment. They found that agricultural income in the province would decline from present levels by 3 percent and employment by 0.5 percent.

Other studies have sought insights into the regional economic effects of future climate change by studying recent experience. Magalhaes et al. (1988), for example, found that in northeast Brazil purchases of clothing, footwear, and processed food fell by about a third during 1983, following a two-year drought, as rural households shifted expenditures from these commodities to purchases of essential foodstuffs. And a study by Powell (1978) found that in some regions of Australia where agricultural output contributes 30 percent of total regional output, a drought-related fall in agricultural output of 10 percent can reduce regional output by 3 percent and regional income by 10 percent.

We take no position on how much off-farm income in the Midwest might be affected by reductions in farm output and income induced by $2 \times CO_2$ climate change. Common sense suggests that the whole regional economy would be less buoyant than otherwise. But this is as far as we are prepared to go.

All of this makes no allowance for the various adjustments midwestern farmers will have open to them for adapting to the productivity consequences of $2 \times CO_2$ warming. We already have discussed some of these possibilities and will not repeat them here. How effective the adjustments might be in offsetting the adverse productivity effects of $2 \times CO_2$ climate change is completely uncertain, although some offset is highly likely. Much will depend on the sensitivity of the U.S. agricultural research establishment in responding to emerging threats of climate change. Given the time frame for $2 \times CO_2$ warming, much undoubtedly could be done to develop technologies and management practices that would help midwestern farmers to cope. But attention and resources will have to be devoted to that end.

On balance, we believe that under $2 \times CO_2$ climate change midwestern farmers would come to expect the following:

- Prices, on average, to be declining, but perhaps not as much as in the without-climate-change scenario;
- Yields, on average, to be rising, but at a slower rate than without climate change;
- Environmental costs from farming operations to be rising at a faster rate than they would be without climate change, with environmental policies forcing farmers to bear some of the costs;
- Adjustments to existing technologies to offset some of the likely climate-induced yield reductions;
- Some help in dealing with the negative effects of climate change from new technologies better adapted to the new climatic regime.

We find it hard to avoid the conclusion that the expectations of midwestern farmers would be less buoyant under $2 \times CO_2$ climate change than without it. Agricultural production in the region likely would be reduced with climate change, reflecting shifts like those described by Blasing and Solomon (1982) for corn production and Rosenzweig (1985) for wheat production. Given the warmer and drier regime for the region depicted by Schneider and Rosenberg (chapter 2 in this volume), the great uncertainty tempering this outlook is the effectiveness of the various adjustment opportunities midwestern farmers might have available. Indeed, the future of midwestern agriculture would appear to hang decisively on the region's, and the nation's, success in providing adjustment opportunities appropriately and in a timely fashion, and success cannot be assumed. Yet failure likely would mean that the Midwest, and the United States, would lose position in world agricultural markets and the exodus of American farmers from agriculture would accelerate. In short, 2 × CO_2 climate change seems likely to pose severe challenges to U.S. agriculture even if its global effect is small. Those charged with responsibilities for U.S. agricultural policy cannot be indifferent to this possibility.

CONCLUSION

The main question asked in this chapter—How can future agriculture, especially midwestern agriculture, best adapt to climate change?—cannot be answered at this time. Throughout this paper we have emphasized the inadequacy of present knowledge. While it is important that the consideration of policy options should begin now, despite this inadequacy, it is quite clear that more information would help to identify the full range of potentially useful responses and would assist in determining which of those may be most valuable. Four specific areas need attention now.

1. *Understanding future climate change.* More information is required about the nature of future change—its spatial pattern (particularly of rainfall); its seasonal pattern (again, particularly of rainfall); and the rate of change in response to forcing factors such as CO_2 emissions, ozone depletion, and the like. We cannot assess with much accuracy the effects of climate change in the Midwest if we do not know how the midwestern climate will change. In the modelling studies described above, changes in mean climatic conditions are imposed instantaneously on the region and resource of interest. Changes in climatic variability and frequency of extreme events are generally ignored. Shifts in risks are examined as a set of comparative statistics (before and after change), not as a process of change. Yet it appears more likely that changes will occur continuously and that several significant climatic thresholds of agroclimatic significance are likely to be crossed at different points along the way. We may speculate about what kinds of adjustments and adaptations to transient climate changes will be made. Dynamic feedbacks onto production decisions and ultimately onto crop productivity will play a large role in determining net agricultural response to climate change. Simply put, farmers will change the rules by adjusting production practices to offset as much as possible the negative impacts of climate change and to take advantage of the positive impacts. Accordingly, research will produce more realistic insights by viewing climate change as a transient process.
2. *Assessing future contexts of agriculture.* Climate change will be one of many potential environmental, social, and economic problems facing world, and midwestern, agriculture in coming decades. Other problems will include (but not be limited to) increased competition for water, resource degradation (particularly of water), and the possibility of continued production surpluses in some regions combined with decreased food security in others. Meyer-Abich (1980) warns that climate change may be only "a particular darkness in the night," a barely noticeable problem among many problems of equal, if not more pressing, importance. It might also be a trigger that sets into motion impacts that are exacerbated by these other problems (Parry, 1986).

   More information is needed in order to portray climate change within the complex mosaic of future agriculture. We propose that this issue is best dealt with in a direct manner. What will future world and midwestern agriculture look like in the absence of major changes in climate? Answers to this question will provide a benchmark for approaching the further question, How might climate change alter this future?
3. *More knowledge of impacts.* There is a need for improved knowledge of the effects of climate changes on crop yields in different regions and under varying types of management. To date, less than a dozen regional studies have been completed, and these are insufficient for confident generalizing about effects on agricultural production at the regional or the global scale. Moreover, an improved ability is required to scale up our understanding of effects on crops to effects on farm production, village production, and national food supply. This is particularly important because policies must be designed to respond to impacts at the national level as well as at the regional level.
4. *Perception of climate change.* Improved understanding is needed of the perception of (actual or potential) impacts from climate change, particularly of the perception of risk and how this can be used to evaluate alternative policies which seek to reduce risk levels.

There are certainly a number of strategies that can be adopted to mitigate the effects of climate changes. The question is not "*Can* agriculture adapt?" but "*How best* can agriculture adapt?" Our task over the next decade is to explore, and seek to extend, the range of options available and to evaluate their relative efficacy with respect to different types and rates of climate change. We have stressed that much critical information is lacking concerning the regional details of changes in temperature and rainfall and possible changes in the distribution of rainfall through the year. However, it is evident that while we cannot, at present, forecast how the climate will change, we *can* estimate the potential consequences of each of a number of *possible* climate changes. In this way we can improve our understanding of effects and explore appropriate responses. Later, when sufficiently precise forecasts of climate change are available, we shall thus have acquired the ability both to assess their impact and to determine the best response.

NOTES

1. These are *equilibrium* temperature increases induced by the equivalent doubling of CO_2 concentrations. The *actual* temperature increases would lag, perhaps for decades, behind the time when CO_2 doubling occurs, primarily because of the thermal inertia of the oceans. See Schneider and Rosenberg (chapter 2 in this volume).

2. Throughout this chapter we will use the term "without-climate-change," although cumbersome, as the stationary or benchmark situation for contrast with the various scenarios of climate change that we consider.

3. The likelihood that increasing scarcity of land and water resources would negatively affect this optimistic scenario is considered at some length in Crosson (1986). The argument rests on the assumption of continued technological advance in agriculture. The grounds for the assumption are discussed briefly in the text above.

4. The GISS general circulation model (GCM) of the global atmosphere is one of several now in use to study climate change. For a comparison of the various models and the different estimates they yield, see Dickinson (1986).

5. GFDL refers to a general circulation climate model developed by the National Oceanic and Atmospheric Administration's Geophysical Fluid Dynamics Laboratory located at Princeton University.

REFERENCES

Arthur, L. M., V. J. Fields, and D. F. Kraft. 1986. *Towards a Socio-Economic Assessment of the Implications of Climate Change for Agriculture in Manitoba and Prairie Provinces: Phase I.* Report to Atmospheric Environment Services (Winnipeg, Department of Agricultural Economics and Farm Management, University of Manitoba).

Bach, W. 1979. "Impact of Increasing Atmospheric CO_2 Concentrations on the Global Climate: Potential Consequences and Corrective Measures," *Environment International* vol. 2, pp. 215–228.

Bergthorsson, P., H. Bjornsson, O. Dyrmundsson, B. Gudmundsson, A. Helgadottir, and J. V. Jonmundsson. 1988. "The Effects of Climatic Variations on Agriculture in Iceland," in M. L. Parry, T. R. Carter, and N. T. Konijn, eds., *The Impact of Climatic Variations on Agriculture* vol. 1, *Assessments in Cool, Temperate, and Cold Regions* (Dordrecht, The Netherlands, Reidel).

Blasing, T. J., and A. M. Solomon. 1982. *Response of the North American Corn Belt to Climatic Warming.* Oak Ridge National Laboratory, Environmental Sciences Division, Oak Ridge, Tenn., Publication No. 2134, pp. 1–16.

Crosson, Pierre. 1986. "Agricultural Development—Looking to the Future," in W. C. Clark and R. E. Munn, eds., *Sustainable Development of the Biosphere* (Cambridge, Cambridge University Press).

———. Forthcoming. "Climate Change and Mid-Latitudes Agriculture: Perspectives on Consequences and Policy Responses," *Climatic Change.*

Decker, W. L., V. K. Jones, and R. Achutuni. 1988. *The Impact of Climatic Change from Increased Atmospheric Carbon Dioxide on American Agriculture.* U.S. Department of Energy, Office of Energy Research, Carbon Dioxide Research Division, Report TR031. DOE/NBB-0077.

Dickinson, R. E. 1986. "Impact of Human Activities on Climate—A Framework," in W. E. Clark and R. Munn, eds., *Sustainable Development of the Biosphere* (Cambridge, Cambridge University Press).

Duvick, Donald N. 1986. "North American Grain Production: Biotechnology Research and the Private Sector," in C. F. Runge, ed., *The Future of the North American Granary: Politics, Economics, and Resource Constraints in North American Agriculture* (Ames, Iowa, Iowa State University Press).

Easterling, W. E. Forthcoming. *Farm-Level Adjustments to Climatic Change by Illinois Corn Producers.* Report to the U.S. Environmental Protection Agency, Office of Policy, Planning, and Evaluation.

English, B. C., J. A. Maetzold, B. R. Holding, and E. O. Heady. 1984. *Future Agricultural Technology and Resource Conservation* (Ames, Iowa, Iowa State University Press).

Goeller, H. E., and A. Zucker. 1984. "Infinite Resources: The Ultimate Strategy," *Science* vol. 223, pp. 456–462.

Goudriaan, J., H. H. van Laar, H. van Keulen, and W. Louwerse. 1984. *Photosynthesis, CO_2 and Plant Production,* NATO Advanced Workshop, Wheat Growth and Modelling, Long Ashton, U.K.

Hayami, Y., and V. W. Ruttan. 1985. *Agricultural Development* (Baltimore, Md., Johns Hopkins University Press).

Kimball, B. A. 1983. "Carbon Dioxide and Agricultural Yield: An Assemblage and Analysis of 430 Prior Observations," *Agronomy Journal* vol. 75, pp. 779–788.

King, R. P. 1986. "Technical and Institutional Innovation in North American Grain Production: The New Information Technology," in C. F. Runge, ed., *The Future of the North American Granary: Politics, Economics, and Resource Constraints in North American Agriculture* (Ames, Iowa, Iowa State University).

Magalhaes, A. R., H. C. Filho, F. L. Garagorry, J. G. Gasques, L. C. B. Molion, M. Neto, M. da S. A., C. A. Nobre, E. R. Porto, and O. E. Reboucas. 1988. "The Effects of Climatic Variations on Agriculture in Northeast Brazil," in M. L. Parry, T. R. Carter, and N. T. Konijn, eds., *The Impact of Climatic Variations on Agriculture*, vol. 2., *Assessments in Semi-Arid Regions* (Dordrecht, The Netherlands, Kluwer).

Manabe, S., R. T. Wetherald, and T. J. Stouffer. 1981. "Summer Dryness Due to an Increase of Atmospheric CO_2 Concentration," *Climatic Change* vol. 3, pp. 347–386.

Meyer-Abich, K. M. 1980. "Chalk on the White Wall? On the Transformation of Climatological Facts into Political Facts," in J. H. Ausubel and A. K. Biswas, eds., *Climatic Constraints and Human Activities,* International Institute of Applied Systems Analysis, Proceedings Series, vol. 10 (Elmsford, N.Y., Pergamon Press), p. 70.

Morison, J. I. L., and R. M. Gifford. 1984. "Plant Growth and Water Use with Limited Water Supply in High CO_2 Concentrations." I. Leaf Area, Water Use and Transpiration, *Aust. Journal of Plant Physiology* vol. 2, pp. 361–374.

Nordhaus, W., and G. Yohe. 1983. "Future Paths of Energy and Carbon Dioxide Emissions," in *Changing Climate* (Washington, D.C., National Academy Press).

Office of Technology Assessment. 1986. *Technology, Public Policy, and the Changing Structure of American Agriculture* (Washington, D.C., U.S. Government Printing Office).

Parry, M. L. 1986. "Some Implications of Climatic Change for Human Development," in W. E. Clark and R. E. Munn, eds., *Sustainable Development of the Biosphere* (Cambridge, Cambridge University Press).

———, and T. R. Carter. 1988. "The Assessment of Effects of Climatic Variations on Agriculture," in M. L. Parry, T. R. Carter, and N. T. Konijn, eds., *The Impact of Climatic Variations on Agriculture* vol. 1, *Assessments in Cool, Temperate, and Cold Regions* (Dordrecht, The Netherlands, Kluwer).

Pitovranov, S. E., Y. Iakimets, V. I. Kiselev, and O. D. Sirotenko. 1988. "The Effects of Climatic Variations on Agriculture in the Subarctic Zone of the USSR," in M. L. Parry, T. R. Carter, and N. T. Konijn, eds., *The Impact of Climatic Variations on Agriculture* vol. 1, *Assessments in Cool, Temperate, and Cold Regions* (Dordrecht, The Netherlands, Kluwer).

Plucknett, D., and N. Smith. 1982. "Agricultural Research and Third World Food Production," *Science* vol. 217, pp. 215–220.

Powell, R. A. 1978. "Stagnating Economic Growth: Its Effect on Agriculture and Regional Implications." Paper presented to the Fourth Advanced Studies Institute in Regional Science, Siegen, Federal Republic of Germany.

Rosenberg, N. J. 1982. "The Increasing CO_2 Concentration in the Atmosphere and Its Implication on Agricultural Productivity." II. Effects Through CO_2-induced Climatic Change, *Climatic Change* vol. 4, pp. 239–254.

———, B. A. Kimball, Ph. Martin, and C. F. Cooper. Forthcoming. "Climate Change, CO_2 Enrichment and Evapotranspiration," chap. II-C in P. E. Waggoner and R. Revelle, eds., *Climate and Water* (New York, Wiley).

Rosenzweig, C. 1985. "Potential CO_2-induced Effects on North American Wheat Producing Regions," *Climate Change* vol. 7, pp. 367–389.

Schelling, T. 1983. "Climate Change: Implications for Welfare and Policy," in *Changing Climate: Report of the Carbon Dioxide Assessment Committee* (Washington, D.C., National Academy of Sciences).

Sonka, S. T., and P. J. Lamb. 1987. "On Climate Change and Economic Analysis," *Climatic Change* vol. 11, no. 3, pp. 291–312.

U.S. Department of Agriculture. 1988. *Agricultural Outlook,* A0144, August.

Waggoner, P. E. 1983. "Agriculture and a Climate Changed by More Carbon Dioxide," in *Changing Climate: Report of the Carbon Dioxide Assessment Committee* (Washington, D.C., National Academy Press), pp. 383–418.

Williams, G. D. V., R. A. Fautley, K. H. Jones, R. B. Stewart, and E. E. Wheaton. 1988. "Estimating Effects of Climatic Change on Agriculture in Saskatchewan, Canada," in M. L. Parry, T. R. Carter, and N. T. Konijn, eds., *The Impact of Climatic Variations on Agriculture,* vol. 1, *Assessments in Cool, Temperate, and Cold Regions* (Dordrecht, The Netherlands, Kluwer).

World Bank. 1984. *World Development Report 1984* (New York, Oxford University Press).

Yoshino, M. M., T. Horie, H. Seino, H. Tsujii, T. Uchijima, and Z. Uchijima. 1988. "The Effects of Climatic Variations on Agriculture in Japan," in M. L. Parry, T. R. Carter, and N. T. Konijn, eds., *The Impact of Climatic Variations on Agriculture,* vol. 1, *Assessments in Cool, Temperate, and Cold Regions* (Dordrecht, The Netherlands, Kluwer).

8

Climate and Forests

Roger A. Sedjo
Allen M. Solomon

There is general agreement that the observed rise in global carbon dioxide (CO_2) levels in the atmosphere will, via the greenhouse effect, lead to a global warming. Humankind can respond to the prospective global change in several different ways. One possibility is through an increase in the size of the global carbon "sink." In this chapter we examine the effects of forests on climate change and the effects of climate change on forests. Forests can postpone the buildup of carbon in the atmosphere. Such a postponement may delay the advent of global warming and thus buy time for the substitution of nonfossil fuels as a more permanent solution to the global warming problem.

This chapter examines the mechanisms whereby changing CO_2 levels and global warming might affect forest growth and composition. It also examines the extent to which naturally generated climate-induced changes in forest area and forest cover mitigate or exacerbate the problem of rising atmospheric CO_2 levels. Increasing the forest stock by better management and decreased harvest from existing forestlands does not appear promising in light of our investigation. Neither does the natural generation of new forests in response to the climate change. Hence, the focus of our analysis is on the artificially established plantation forests.

The economic costs required to establish plantation forests in temperate and tropical regions are estimated. In addition, the economic uses to which the newly established forest stock might be put are investigated. These include industrial wood uses and also wood-fueled power generation. The possible effects of these higher stocking levels on world industrial wood markets and on investments in traditional industrial forestry activities are addressed and the broad economic implications examined. In neither case is the plantation likely to recover its costs. The question of the source of financing is addressed briefly and the possibility/necessity of foreign assistance is examined.

THE PROBLEM

The earth's climate may be undergoing a warming in response to the atmospheric accumulation of the greenhouse (radiatively important) gases (RIGs), at this time predominantly but not exclusively CO_2. The carbon content of the atmosphere is increasing annually by approximately 2.9 billion metric tons (gigatons [Gt]). This occurs in an environmental system where between 4.8 and 5.8 Gt of carbon in the form of CO_2 are released annually through fossil fuel combustion and cement production (Detwiler and Hall, 1988). Also contributing to this release are land-use changes that have eliminated large amounts of terrestrial biomass, particularly tropical forests, thereby freeing into the atmosphere the carbon that had been captive in the vegetation.

In principle, three basic strategies exist for responding to potential global warming: (1) prevent or limit the warming to nonthreatening levels; (2) allow the warming to occur and find low-cost ways to adapt; or (3) some combination of (1) and (2).

A number of ways have been suggested for preventing or moderating the emission of CO_2 into the atmosphere. In addition to the obvious approach of reducing carbon release into the atmosphere by dramatically reducing the use of fossil fuels and the conversion of forestlands, carbon sinks can be created or expanded. Such sinks could sequester the "excess" increments of CO_2 released that are currently responsible for the atmospheric buildup. Since forests experiencing net growth will sequester CO_2 from the atmosphere as part of the growing process, any buildup of forest biomass will reduce the buildup of atmospheric CO_2.

While the temperate forests may be roughly in carbon balance (Melillo et al., 1988) or in a modest carbon sink (Armentano and Ralston, 1980), the tropical forests are experiencing decline and are a source of carbon (Detwiler and Hall, 1988). The generation of net positive increments to the rate of accumulation of forest biomass would serve to sequester an increased portion of the free CO_2. Net additions in the carbon-sequestering ability of forests through a net buildup in forest timber inventories could be accomplished through (1) an expansion in the total forest biomass through massive investments in fast-growing plantation forests; (2) an increase in forest timber inventories, generated by increased management on existing timberlands, or a reduction in harvest, or both; or (3) an expansion of total forested area and biomass via a migration of natural forest into previously unforested land in response to higher temperatures, increased precipitation, and/or the fertilization effects of higher CO_2 levels. Any changes that result in increased forest biomass would contribute to a sequestering of at least a portion of the "excess" carbon released into the atmosphere. Similarly, any reduction in the rate of deforestation would reduce the amount of carbon accumulating in the atmosphere.

In this chapter we examine the mechanisms whereby climate changes forests and estimate the likely change in total forest area and biomass under the 2 × CO_2 scenario (see chapter 2 in this volume). This chapter also explores the potential of additional forest area and/or biomass growth to mitigate the buildup of atmospheric CO_2. Possible expansions of the forest area are explored. These could be caused by natural forest expansion generated by the warming process, more intensive management and/or reduced timber harvests, or the establishment of new plantation forests. Where plantation forests are involved, estimates are also made of the economic costs necessary to accomplish the desired sequestering of carbon. While it is obvious that many of the means to mitigate the buildup of CO_2 are not mutually exclusive and thus can be employed in concert, the analysis examines each of the options separately.

BACKGROUND

The Impact of Forests on CO_2 Levels

Over the centuries, the area of forestland has declined as populations have increased and lands have been converted from forests to other uses, particularly agricultural uses. Matthews (1983), for example, estimates the decline in global forestland at about 700 million hectares (ha) (1 hectare = 2.47 acres) or a reduction of about 15 percent from the preagricultural forest area. On stronger evidence, Olson et al. (1983) calculated a 50 percent decline in carbon stored in vegetation, most of it in forests. While the area of tropical forest is declining, the major declines in temperate forest area that occurred due to clearings in past decades and centuries have ceased. The temperate forest area is now stable. In fact, it may have increased modestly in recent decades. Data examined by Armentano and Ralston (1980) imply that net forest area has increased throughout the northern temperate zone, including North America, Western Europe, and Siberia.

The effect of the change in forestland area, however, need not translate directly into changes in the carbon sequestered by forests. Forest inventories vary with age and with tree species as well as site location and characteristics. Unmanaged forests may be thought to consist primarily of large mature trees. In fact, however, these natural systems range in age from infancy to senescence. Old forests, weakened by age, are ultimately destroyed by natural forces, only to regenerate once again. Fire, insects, and disease all contribute to the natural cycle of regeneration, growth, and decline. For example, until recently, several times more forest was destroyed by fire in the United States than at present and the natural pine forests of the U.S. South would not exist in the absence of a recurring fire cycle. Indeed, most forests have a measured fire return time, ranging from 10 to 1,000 years but normally 50 to 500 years. Hence, even a pristine landscape consists not simply of undisturbed, old-growth forests, but of a variety of forests at different stages of development and growth, thus embodying and sequestering very different amounts of carbon.

Initially, most forest clearing by humans was initiated by the desire to change use of the land, usually to agriculture. Hence, the biomass was largely destroyed, often by fire if feasible. The net stock of forest biomass was reduced thereby, liberating carbon into the atmosphere. However, over time, the production of industrial wood produced large volumes of poles, fences, rails, boards, timbers, and solidwood for construction of all types. Clearly, the wood used for these purposes and contained in structures, while no longer part of the forest, continues to sequester carbon. While the volume of wood in this form is obviously difficult to estimate, the stock of wood in this form has increased dramatically over time as has the volume of lumber and wood panels produced.

If we assume that one-quarter of the industrial wood harvested worldwide over the past thirty-five years remains in a solidwood form, then about 10 billion cubic meters (m^3) of wood exist in structures, furniture, and so forth. (Rotty, 1986, estimated 6 billion m^3 of solidwood remaining over a thirty-five-year period.) This volume is roughly equal to one-half the current total volume of growing stock in all the U.S. commercial forest inventory and would constitute a bank of roughly 2.6 Gt of carbon. However, this is only a very modest 0.17 Gt per year. Additional carbon is no doubt sequestered in wood and paper products and captured in landfills.

The relation of forest biomass to carbon sequestering is essentially a stock-flow problem. Net changes in the stock of forest biomass indicate the current flow of carbon being sequestered, while the total stock of forest indicates its carbon-sequestering capacity. Ultimately, the question of terrestrial carbon sequestering or release hinges on the total stock of carbon sequestered not only in the forest ecosystem biomass and soils, but also in the stock of industrial solidwood that remains. However, during transitions, the focus is on the net flux (flows) of CO_2 between forest ecosystems and the atmosphere (on the assumption that the carbon sequestered in industrial solidwood is small).

For forests and biomass to reduce the levels of CO_2 in the atmosphere, the inventory of forest biomass (and solidwood uses) must increase. The general approaches available are to increase the land area in forest biomass, increase the forest biomass per unit land area, or increase both. Mature fuelwood plantations will maintain equilibrium, absorbing and releasing offsetting amounts of CO_2. The creation of sustainable energy plantations with higher average biomass levels than those found in the previous land use, however, will result in a net sequestering of previously free carbon. Similarly, industrial forests could continue to produce the same volumes of harvested industrial wood (flows) while simultaneously having larger growing stock inventories. Ultimately, it is the size of the permanent change in the inventory stock that is important. The simple concept of plantation forests being used as a sink for carbon involves increasing the world's forest stock without a parallel increase in its harvests.

The Impact of Climate Change on Forests

There are a number of mechanisms whereby CO_2 and climate changes might affect forest growth and composition. These include the direct effects on physiologic, anatomic, ecosystematic, and geographic properties of tree growth induced by increases in ambient CO_2 concentration and air temperature and by shifts in precipitation patterns. The mechanisms also include the indirect effects of climate change on land use, that is, migration of high-intensity agriculture and concomitant release of current agricultural land for growth of forests. The eventual effects of such land-use shifts on the amount of carbon stored in vegetation are quite difficult to project and will not be considered further here.

Physiologic Mechanisms

The trees under discussion and, indeed, all green plants depend for life upon photosynthesis, the storage of solar energy when carbon from CO_2 in the atmosphere is united with hydrogen from water to form molecules of sugar (CH_2O) and elemental oxygen (O_2) (see chapter 2 in this volume). In addition to the CO_2 and water, photosynthesis requires warmth and nutrients, especially nitrates ($-NO_4$) and phosphates ($-PO_4$). The construction and life of tree cells occur during respiration in which oxygen combines with the previously synthesized sugar molecules to release stored energy. Carbon dioxide and water vapor are the byproducts of this process.

Too little warmth slows or stops photosynthesis and respiration. Required enzymes are inactive below a threshold temperature of 0°C to 10°C, depending on species (Steponkus, 1981). Photosynthesis and respiration increase with increasing warmth up to a temperature optimum, and then decline with further temperature increases. Temperature optima for respiration are normally much higher than those for photosynthesis. Thus, after a given temperature (compensation point) is passed, net photosynthesis (the difference between the energy fixed by photosynthesis and the amount lost by respiration) declines with continued increases in respiration. Photosynthesis essentially ceases altogether at about 38°C in boreal conifers, 43°C in temperate deciduous trees, and even higher in certain tropical trees (Kappen, 1981).

The availability of water in the soil also controls the processes of photosynthesis and respiration. When too little water is available, photosynthesis is slowed or halted. When soil is too wet, the flow of oxygen to the roots is impeded. Root respiration and the respiration-demanding uptake of nutrients is diminished.

Unlike annual plants that must survive the conditions extant during only one growth season, trees are adapted to surviving the wide environmental extremes encountered at one location for as long as 100 to 500 consecutive years. Thus, minor environmental fluctuations frequently have little lasting effect because trees are very conservative organisms. By definition, healthy trees at any given location are receiving adequate but not excessive warmth, water, and nutrients for survival and growth. Climate changes can be expected to shift this balance for individual trees, either with no effect when the broad requirements for survival and growth continue to be met or else with the effect of diminishing their photosynthetic potential, hence reducing the net uptake of atmospheric CO_2. Reductions of growth could force the death of trees during the 10 to 100 years that follow growth reduction (Solomon, 1986).

Although systematic changes from the current climate might reduce net photosynthesis, increases in atmospheric CO_2 concentration could increase it (see chapter 2 in this volume). Like warmth, water, and nutrients, CO_2 is required for photosynthesis. Greenhouse experiments indicate that photosynthetic rates of tree seedlings increase during short-term (minutes to days) CO_2 fumigations. The amount of water required to fix a given amount of carbon is also reduced as increased CO_2 concentrations close the leaf stomates through which gas exchange takes place (CO_2 into leaves, O_2 and H_2O vapor out of leaves). Unfortunately, there are few experiments on multiple-year fumigations and none that use mature trees. The few available results are ambiguous (B. R. Strain, Duke University, personal communication with A. M. Solomon, 1986) or suggest no permanent increase in photosynthetic rate or decrease in water use despite greater CO_2 concentrations (W. A. Oeschel, personal communication with A. M. Solomon,

1987). If such CO_2 "fertilization" does occur, it is unlikely to reach the values used by Solomon and West (1986) in modeling forest response to concurrent climate and CO_2 changes; for example, for a doubling of atmospheric CO_2 they assumed an 11 percent growth increase in conifers, 20 percent increase in deciduous trees, and an 18 percent decrease in water use in deciduous trees.

Anatomic Mechanisms

Trees of seasonal environments (the seasonally arid tropics; temperate and boreal zones) are also seasonal in their growth, sequestering most of the annual carbon fixed by photosynthesis at specific times of the year and thereby forming growth rings. Particularly in ring-porous wood (possessed by boreal conifers and about half of the temperate deciduous species), the growth rings are composed of light, thin-walled springwood (earlywood) cells and dense, thick-walled summerwood (latewood) cells. Although there is considerable variability in a given growth ring, latewood may contain three or four times as much material as earlywood (Panshin and DeZeeuw, 1964). The amount of material in earlywood cells—hence the carbon sequestered there—depends upon the amount of photosynthate available at the end of the previous growing season. In other words, it is a function of the total growth year (Fritts, 1976). In contrast, the latewood reflects growing conditions of the current spring, increasing with increasing spring warmth inducing greater spring photosynthetic rates, with earlier onset of summer weather inducing earlier shifts to latewood cell formation, and with mild summer drought to increase cell wall thickness (Conkey, 1986). Increasing spring warmth, earlier onset of summer weather, and mild summer drought are possible consequences of climate change (Manabe and Wetherald, 1986).

The amount of carbon storage that could result from enhancement of latewood density in temperate and boreal zone trees is difficult to project. Wood from most trees has a density of 300 to 800 kilograms per cubic meter (kg/m^3). Because latewood is two to three times as dense as earlywood (Panshin and DeZeeuw, 1964), little widening of the growth ring would be required to accommodate a significant increase in latewood carbon storage. Such a change might be first detected in an increase of the ratio of latewood to earlywood width and density.

Ecosystematic Mechanisms

From the beginning of tree growth on previously burned or farmed land, forest stands undergo a predictable succession of changes. Initially successful tree seedlings and saplings are tolerant of direct sunlight. They require few nutrients and grow rapidly, sequestering large amounts of carbon in short periods of time. As the trees mature, different seedlings and saplings appear. These grow more slowly, demand greater amounts of nutrients, and are less tolerant of sunlight. Ultimately, the most slowly growing species become dominant and retain dominance until mortality due to disease, or insects, or fires opening the sites to sunlight again. Intensive forest management is aimed at maintaining plots with these rapid-growth pioneer species and at harvesting these trees before they can be replaced by the slower growing shade-tolerant species. A climate change that kills large numbers of slowly growing late-successional trees will encourage growth of pioneer trees that store carbon at perhaps two or three times the rate of the shade-tolerant species (Solomon, 1986; Solomon and West, 1987).

Geographic Mechanisms

Tree growth is generally limited by lack of summer warmth at the high-latitude boundaries of species ranges. Northern boundaries of ecosystems and biomes are also thought to be warmth limited. As a result, and unlike most tree populations and communities, those growing near their northern range boundaries tend to experience increased growth during a warming of the magnitude projected by general circulation models of the atmosphere such as that of Manabe and Wetherald (1986). This effect should be expressed most clearly in the higher latitude forests (for example, northern boreal forests), where temperature increases are expected to be greatest, and where tree growth is now limited most by low growing-season temperatures.

The migration rates of tree species beyond their current boundaries in response to increasing temperatures and changing precipitation will be limited by supply of seeds. Solomon et al. (1984) calculated that the warming induced by a $2 \times CO_2$ scenario (see chapter 2 in this volume) during the next 100 years could occur at a rate ten times more rapid than that at which trees have migrated in North America during the past 10,000 years. Numerous barriers to migration in the form of agricultural fields, developed land, and road rights-of-way exist today. Hence, rates of migration would be much lower today than in the past. As a result, a rapid loss of tree species from landscapes can be projected in the absence of massive reforestation programs. The few tree species that might repopulate these areas would probably be those that possess small, light, wind-dispersed seeds that can be transported long distances (Davis, 1986). These are the pioneer species that grow rapidly. This feature of climate change could promote additional carbon storage in the short term.

Climate Change and Changes in Forest Types

Climate modelers generally agree that global warming will generate the greatest temperature increases in the high latitudes, while only modest temperature increases will be experienced in the tropics. Hence, the effect of warming on forests might be expected to be greatest on the high-latitude boreal forests of the northern regions (for example, those in Canada, Alaska, the Soviet Union, and the Nordic countries). Some observers expect the impact to be manifested through both the northerly expansion of forests into areas previously unforested and their higher growth rates (Kauppi and Posch, 1987; Hari et al., 1984). These high-latitude areas may be the most easily invaded by forests in the sense that competing human uses for the high-latitude lands are minimal.

The regional predictions of climate models are still very crude and must be treated with considerable skepticism. Nevertheless, some of the effects of global warming on North American forests can be projected. The response in eastern North America to a CO_2-induced warming, simulated by Solomon (1986), is a northerly shifting of the current forest zones with an overall expansion of forested area as the boreal forest migrates into unoccupied tundra. At the southern edges and on the western fringes, however, forestlands lose ground to nonforest vegetation. As a result, the volume of forest biomass increases by 30 to 50 percent in the North as the boreal forest expands and the southern edge of the initial boreal forest is eventually replaced by the cool temperate forest. Biomass is lost in the more southerly forests as they contract. According to this simulation, the total biomass for eastern North America ultimately declines.

Despite the regional projections, climate modelers and forest biologists have said little that is definitive as to whether global forests in total will expand, contract, or experience no net change under conditions of rising CO_2 levels and global warming. Later in this chapter we make a crude estimate of the possible effect of global warming on worldwide forest biomass and area.

FORESTS AS A CARBON SINK

In this section we examine the potential strength of global forests as a carbon sink. Three mechanisms for increasing forest stocks are examined: (1) natural increases in forest growth and biomass as the result of the climate change due to greenhouse warming; (2) an increase of forest stocking on existing forest either through increasing growth, decreasing harvest, or both; and (3) the establishment of rapidly growing plantation forests.

Natural Forest Expansion

The Gaia Hypothesis (Lovelock, 1979) proposes that the global biosphere, through natural processes, has the capacity to ameliorate changing pressures placed on the system. One mechanism whereby the biosphere might negate effects of CO_2-induced warming is by increased carbon sequestering in global forests. This could be accomplished if forests expand into regions where they do not now grow, and if photosynthesis rates and carbon storage are enhanced by warming, or perhaps in response to CO_2 enrichment of the atmosphere. The removal of this carbon from the atmosphere by these mechanisms could reduce warming, perhaps lowering global temperatures.

The potential for unmanaged forests to affect the carbon cycle in this way forms the subject of this section. First, we estimate the area new forests would have to cover in order to affect the rate at which atmospheric CO_2 concentrations change. Next, using the Holdridge Life Zone classification, we estimate how forest area and biomass might change in response to CO_2-induced climate changes independent of human interventions. We stress that forest ecologists have hardly begun to study whether global forested area and biomass would increase or decrease due to warming, much less to reach a consensus on this question. Based on a dynamic model of forest growth at a limited set of points on the landscape, Solomon (1986) and Solomon and West (1985, 1986, 1987) suggest that the most common forest response to initial warming will be tree mortality, with enhanced tree growth occurring in few places, and no change in growth or mortality in other places.

There is as yet only a very weak basis to support our reasoning. Changes unrelated to those considered by the specialists in each discipline are quite likely to determine the future course of forest growth. For example, technological change that produces agricultural land abandonment and subsequent forest growth may proceed at a rate great enough to mask any climate-induced changes in forest biomass. The only certainty is that "steady-state" climate and "equilibrium" forests are highly unlikely in the foreseeable future.

How Much Natural Expansion Is Required?

It is instructive to estimate the extent of natural forest expansion that would be required to mitigate the buildup of atmospheric CO_2 and to explore the likelihood that sufficient land area would be available should the forest-expanding effect of warming be greater than estimated. Assume for convenience that most of the expansion would take place in and around the current areas of the boreal forest—areas that are sparsely populated and have few competing human uses for land. Growth rates in boreal forests are relatively modest, averaging perhaps 2 m^3 per ha per year (Kauppi and Posch, 1987). If we accept an adjustment factor of 1.6 to account for the limbs and roots and an average of 0.26 tons of carbon sequestered per m^3 of forest growth, roughly 0.83 tons of carbon per ha will be sequestered annually. At this rate, a net addition of 3.5 billion ha of new forest would be required to sequester the requisite carbon. This represents more than a doubling of the world's area of closed forests and an increase in the world's total forest area to 7.5 billion ha from the present 4 billion ha. The requisite 3.5 billion ha of land is not available for naturally generated boreal forest expansion. Should the hypothetical expansion take place in a region with more rapid growth (for example, 10 m^3 per ha), only 700 million ha would be required, a large but more attainable area. However, such an area of productive land would probably be available worldwide only if large areas of other nonforested uses were to be suspended.

Change in Forested and Unforested Areas

An assessment of potential increased carbon storage in unmanaged vegetation must examine the global vegetational distributions and how these might change with CO_2-induced global warming and climate change. To date, Olson et al. (1983) have provided the most comprehensive and credible estimates of global distributions of large-scale vegetation groups (biomes) and of carbon in above-ground biomass. Forests cover about 42 percent of the earth's land surfaces, and they contain 483 of the 562 Gt (86 percent) of the globe's above-ground carbon (Olson et al., 1983). An estimated 73 percent of the carbon in the world's soil (927 of 1,272 Gt) is in forest soils (Post et al., 1982). Hence, forests dominate the dynamics of the terrestrial carbon cycle.

The question we wish to address in this section is how forest area might change in response to CO_2-induced climate changes independent of the direct human interventions described above. The only formal approximation of that answer currently available can be derived from the output of the Holdridge Life Zone classification. This is a system that uses temperature and precipitation data to predict the distribution of life zones. It can also be used to assess probable shifts in biological communities in response to different climate scenarios. These results are mapped in Emanuel et al. (1985a); data based on an improvement to the standard Holdridge calculations is provided in Emanuel et al. (1985b). The improved maps appear in the inside front and rear cover of Bolin et al. (1986). The Holdridge system is imperfectly suited to our purposes, but it is the only spatially explicit and comprehensive work now available.

Based on data from approximately 7,000 meteorological stations, life zone classes were assigned to all 1/2 degree by 1/2 degree latitude and longitude land areas (cells) of the globe. Then the climate for each cell was changed, according to the Manabe and Wetherald (1986) projections for a doubling of CO_2. When the reclassified areas are compared with the original classes, an estimate emerges of changes that could be expected in life zone areas under stable climate (Emanuel et al., 1985b, table 1). Results are shown in table 8-1, columns 1 and 2. (Also see Batie and Shugart, chapter 9 in this volume, for an additional application of the Emanuel et al., 1985a, analysis.) Although Emanuel et al. provided thirty-seven life zone classes, they mapped only eighteen of these (Emanuel et al., 1985a, figures 2, 4). We combined classes to match their eighteen-class maps of life zone distribution (table 8-1, column 1).

Emanuel et al. (1985b) point out that the greatest changes in distribution of life zones occur in the high-carbon forests at the highest and lowest latitudes: boreal forests decline by 37 percent and tropical forests increase by 28 percent. This is all the more notable considering that the majority of global forests occur near the equator or in the boreal zones. The boreal forest that remains after warming is largely new, having replaced about 42 percent of the cells designated as tundra in the base case. Overall forested life zone area declines in the simulations by about 6 percent, equivalent to 444 million ha of forested land. Thus, if climate determines life zone distribution, and if life zones coincide with forested and nonforested land, then the climate scenario used implies that slightly less forested land would be found on an earth warmed by 2 × CO_2. Forest losses are made up in this analysis by increases in almost all nonforest (and low-carbon storage) life zones except tundra and temperate desert bush.

We may contemplate the processes or phenomena that might increase the areal extent of forests on the globe (that is, what information not considered in the Holdridge-based analyses could nullify the conclusion that forested area would be reduced under warming conditions). As noted, four mechanisms—physiologic, anatomic, ecosystematic, and geographic—could be considered in examining effects of climate on forests. The physiologic and geographic processes are particularly relevant here. The Holdridge scheme contains no consideration of the process whereby physiological response to increased CO_2 in the atmosphere involves closing stomates and reducing water losses, which could allow trees to grow in steppe regions now limited to herbs and shrubs. The Holdridge projections indicate that current steppe areas (tropical thorn woodland, temperate thorn steppe, cool temperate steppe) cover about 23 billion ha—about 30 percent of the area now in forests. There is no way to estimate the possibility that such a mechanism would operate on a scale grand enough to affect such loss of steppe, but the possibility cannot be ruled out.

A geographically oriented process of potential importance that is absent from the Holdridge simulation is the role of moisture at lower latitude and dry boundaries of tree species and forest ecosystems. Emanuel et al. did not use the precipitation changes projected by Manabe and Stouffer (1980) for a 2 × CO_2 climate because of the inherent uncertainty of these projections. Although most climate models suggest decreased precipitation over large continental regions (see chapter 2 in this volume), enhancement of the hydrological cycle due to warming could also take place. If this occurs, and if the resulting increases in

Table 8-1. Forested Area and Stored Carbon: Current and Future by Holdridge Life Zones

Area	$1 \times CO_2$ area (10^6 km^2) [1]	$2 \times CO_2$ area (10^6 km^2) [2]	Olson biomass (kg m^{-2}) [3]	$1 \times CO_2$ carbon (Gt) [4]	$2 \times CO_2$ carbon (Gt) [5]	Carbon difference (Gt) [6]
FORESTS						
Tropical wet[a]	0.387	0.413	20	7.740	8.260	+0.520
Tropical moist[b]	8.647	9.888	14	121.062	138.432	+17.370
Tropical dry[c]	9.992	14.033	10	99.920	140.330	+40.410
Subtropical wet[d]	0.510	0.281	15	7.650	4.215	−3.435
Subtropical moist[e]	11.453	9.096	7	80.171	63.672	−16.499
Tropical totals	30.989	33.711	31	316.543	354.909	+38.366
Warm temperate[f]	0.591	0.515	10	5.910	5.150	−0.760
Warm/dry temperate[g]	15.222	14.175	5	76.110	70.875	−5.235
Cool temperate[h]	11.287	11.627	10	112.870	116.270	+3.400
Temperate totals	27.100	26.317		194.890	192.295	−2.595
Wet boreal[i]	4.609	1.499	11	50.699	16.489	−34.210
Moist boreal[j]	12.654	9.381	8	101.232	75.048	−26.184
Boreal totals	17.263	10.880		151.931	91.537	−60.394
Total all forests	75.352	70.908		663.364	638.741	−24.623
NONFORESTS						
Tropical thorn woodland[k]	7.066	9.189	3	21.198	27.567	+6.369
Temperate thorn steppe[l]	6.784	7.413	1.6	10.854	11.861	+1.007
Cool temperate steppe[m]	8.931	11.922	1.0	8.931	11.922	+2.991
Tropical desert bush[n]	13.654	16.290	0.4	5.462	6.516	+1.054
Temperate desert bush[o]	11.591	9.454	0.6	6.955	5.672	−1.283
Boreal desert[p]	1.309	2.591	1.0	1.309	2.591	+1.282
Tundra[q]	4.470	3.034	0.5	2.235	1.517	−0.718
Ice	2.218	0.567				
Total all nonforest	56.023	60.460		56.944	67.446	+10.502
TOTAL FOREST AND NONFOREST	131.375	131.368		720.308	706.187	−14.121

Note: The quoted classifications in notes a to q are from Olson et al. (1983).

Sources: For cols. [1] and [2], Emanuel et al., 1985b, table 1, and Emanuel et al., 1985a, figures 2, 4. For col. [3], Olson et al., 1983. Col. [4] is col. [1] × col. [3]; col. [5] is col. [2] × col. [3]; col. [6] is col. [5] − col. [4].

[a] Combines rain and wet tropical forest zones. "Evergreen equatorial forest."
[b] "Evergreen or deciduous forest."
[c] "Poor site or marginal tropical forest."
[d] Combines rain and wet subtropical forest zones. "Lowland and other tropical and subtropical broadleaved humid forest."
[e] Combines moist and dry subtropical forest zones. "Tropical dry forest and woodland."
[f] Combines rain and wet warm temperate forest zones. "Mostly temperate broadleaved forest."
[g] Combines moist and dry warm temperate forest zones. "Other semi-arid woodland or lowland forest."
[h] Combines rain, wet, and moist cool temperate forest zones. "Mixed woods, alternating evergreen and broadleaved forests."
[i] Combines rain and wet boreal forest zones. "Southern continental taiga."
[j] "Main taiga."
[k] Combines very dry tropical forest and tropical thorn woodland zones. "Tropical savannah and woodland."
[l] Combines subtropical thorn woodland and warm temperature thorn steppe zones. "Warm/hot grassland with more shrub."
[m] "Cool grassland."
[n] Combines tropical desert bush and desert with subtropical desert bush and desert zones. "Other desert and semi-desert."
[o] Combines warm temperate desert bush and desert with cool temperate desert bush and desert. "Cool semi-desert shrub."
[p] Combines boreal dry bush and boreal desert. "Low arctic tundra."
[q] Combines rain, wet, moist, and dry tundra zones. "Other high arctic tundra."

rainfall are not negated by increased evapotranspiration, then one might expect additional growth of trees in current areas of steppe and increased boreal forest growth in tundra, with much less concomitant boreal forest losses at southern forest boundaries.

Future land-use and technological changes could also increase the area of forests. For example, current trends to abandonment of marginal agricultural land (Delcourt and Harris, 1980), if continued, will certainly result in return of forests to previously forested land. The amount of land made available with improved fire suppression technology could also be important. In the absence of prairie fires, many tree species and communities can grow out onto the steppe. With increased development and population, national governments find fire suppression activities to be quite feasible and important to citizens. While forest fires burned some 2.5 million ha of U.S. forests in 1988, this is far below the average of 10 million ha per year in the 1920s. Thus, increasing effectiveness of wildland fire suppression seems likely in the future, further enhancing the spread of forests into unforested regions. Continuing destruction of tropical forests, however, opposes these trends.

Change in Carbon Density

Carbon sequestered as forest biomass may not decline when forest area declines because increases in area of carbon-rich forests, such as tropical wet forests, may offset or outweigh decreases in area of carbon-poor forests, such as the boreal forest zones. We present the estimates by Olson et al. (1983) of carbon density for each life zone in column 4 of table 8-1. Their values for areal cover of life zones or combinations of life zones are not included because of difficulties in matching Olson classes with Holdridge classes. However, Olson et al. estimates for areal cover are lower than the corresponding Emanuel et al. estimates. The Holdridge area estimates conform to "potential" vegetation, while Olson et al. provide areas of extant forests, minus areas in which man has replaced forests with crop, residential, and other uses. In addition, certain differences are impossible to reconcile in vegetation classes derived from differing criteria. However, the Olson and Emanuel classes are generally comparable for our purposes of determining the direction of change in carbon storage in the life zones.

Overall, forest carbon storage is predicted to decrease under a 2 × CO_2 climate from that of the base case climate (table 8-1, columns 5 and 6). The most important shifts in carbon densities due to warming occur again among tropical forests (57.784 Gt greater biomass in tropical dry and moist forests) and boreal forests (60.294 Gt less biomass in wet and moist boreal forests).

One problem with invoking the proposed CO_2 fertilization phenomenon is that the curvilinear and asymptotic relationship of photosynthetic rate with increasing atmospheric CO_2 concentrations should make any CO_2 fertilization decreasingly important with greater CO_2 concentrations (Regehr et al., 1975). As yet, the phenomenon of CO_2 fertilization has not been detected in trees, despite extensive searches for it in the field (LaMarche et al., 1984; Hari et al., 1984; Graybill, 1987; Parker, 1987) and in growth chambers (Doyle, 1987; Telewski and Strain, 1987). The most likely response, if any, would be an increase in the rate of tree growth in forest gaps (Solomon and West, 1986), thereby closing gaps more quickly and decreasing the "carbon voids" in forest canopies. To significantly affect atmospheric CO_2 concentrations, annual carbon storage by trees would need to increase by only about 10 percent over the annual 31.5 Gt of net production estimated by Olson et al. (1983, table 2). A change of this magnitude has not occurred.

The results summarized in table 8-1 indicate that there is likely to be little natural net expansion in forest area or volume as the result of a global warming. Certainly, a climate-induced expansion of the 700-million hectare natural forest area as required in our estimate above is unlikely.

Managed Forests and Industrial Harvests

Another potential option for sequestering the excess atmospheric carbon would be by increasing the biomass growth of the world's existing forests through application of increased management and improved silvicultural practices and/or decreased harvesting. To increase forest growth enough to sequester all of the excess atmospheric carbon (2.9 Gt per annum), biomass growth for the 3,000 million ha of the world's existing closed forest would need to increase by an average of 4 m^3 per ha per year, or about 2.5 m^3 per ha per year for stemwood. This option would require an approximate doubling of the average rate of growth in all of the world's closed forests—temperate and tropical. The prospects of generating such a growth increase are discussed below.

Foresters have demonstrated their ability to increase forest productivity by intensive management. However, for the existing forests to sequester the atmospheric excess of carbon over the next several decades would require a rapid increase in the total forest biomass possible only if current growth rate and accompanying biomass accumulation are doubled.

Much of the increase in forest productivity is due to intensive management early in the growing cycle—planting that ensures a fully stocked stand, the use of genetically improved trees, weed control for newly planted seedlings, and so forth. These practices are currently applied in the process of reforestation after a harvest. But to double the growth of existing forests and significantly increase biomass accumulation in the next few decades would require a dramatic increase in management. Most of the increase in forest growth that would be required must come from increases to be induced in already established stands.

Fertilization is probably the most effective way to increase growth during the growing cycle. Ford (1984) and Nambiar (1984) cite experiments that show significant responses to fertilization in older stands. However, even if the growth rate were to increase 80 percent in response to fertilization (Beverge, 1984), its widespread application to the world's existing timber stands to accelerate growth and carbon sequestering is of questionable practicality because much of the closed forest inventory of the world is now in a mature or "overmature" state (e.g., Holowacz 1985, for the Soviet Union; Richardson, forthcoming, for China;

Bonnor, 1982, for Canada). Even if fertilizer were added, little net growth would occur. Furthermore, a large fraction of the world's forests are in cool, high-latitude climates where the potential growth rate is limited by lack of warmth (Zavitkovski and Isebrands, 1983). Therefore, the burden of increasing overall world average forest growth would fall upon the young and middle-aged stands in warmer climates that are, in many cases, already growing relatively rapidly and may already be under management.

In summary, although forest management can help in increasing growth rate, forest management is most effective with initial forest establishment and in the early years of growth. Management has little to offer if applied to existing mature forest. Therefore, it appears that to double the growth rate of the world's existing closed forest while simultaneously maintaining and accumulating the increased biomass in a short period of time (one to two decades) is not technically feasible.

Dramatic increases in growth rates of many existing forests are unlikely, but forest biomass could be increased by lengthening industrial timber rotations through a temporary reduction of the harvest. As a limit, the maximum reduction from the industrial wood side would be accomplished through a complete cessation of commercial harvesting, thereby maintaining 1.5 billion cubic meters of timber that would otherwise be removed from the forests. The volume of biomass associated with 1.5 billion m^3 of stemwood is roughly 2.4 billion m^3 (using 1.6 factor for nonstemwood), and the carbon retained therein would be about 600 million tons (roughly 20 percent of the annual increment of carbon to the atmosphere). In point of fact, however, a substantial fraction of industrial wood already remains in solid form and decays only slowly over many years, and some nonstemwood parts of the harvested trees, such as the root system, also decay only slowly. Thus, the actual net retention of carbon in forest biomass following from a no-harvest policy would be considerably fewer than 0.6 Gt, perhaps in the neighborhood of 0.3 Gt, or roughly one-tenth of the annual accumulation of carbon in the excess atmosphere. Neither more intensive forest management nor reduced industrial harvests would increase the world's forest biomass enough to have more than a very modest impact in reducing buildup of carbon in the atmosphere.

Forest Plantations

Two estimates have been made recently of the forest area necessary to offset a portion of that buildup. Marland (1988, p. 53) estimates that as little as 500 million hectares of intensively managed rapidly growing plantation forests could sequester all of the 5 Gt or so of carbon, approximately the total amount released annually by the burning of fossil fuels. Marland's generally optimistic assumptions about growth rates, and particularly sequestering potential, lead to this relatively modest estimate of forestland required, especially given the immensity of the task of sequestering 5 Gt of carbon.

Attempting to address the same CO_2 sequestering problem, Woodwell (1987) estimates that 200 to 400 million ha of new, rapidly developing forest would be enough to withdraw between 1 and 2 Gt of carbon now accumulating in the atmosphere at current rates of emission.

To maintain the present level of atmospheric carbon, it is not necessary to capture all of the CO_2 generated by fossil fuel burning, as posited by Marland; 2.9 Gt would offset the increment now accumulating in the atmosphere at current rates of emission. However, maintaining the present level of atmospheric CO_2 requires capturing more than the 1 to 2 Gt of carbon posited by Woodwell.

The ability of forests to store large amounts of carbon and their importance in the global carbon cycle have led to the suggestion of massive afforestation to provide an additional carbon sink (Cooper, 1983; Brown et al., 1986; Marland, 1988). The ability of forests to sequester carbon is related very directly to biomass growth. Most growth figures used by foresters relate to stemwood and ignore branches, roots, leaves, and the like (Marland, 1988, p. 36). In our calculations, these are included in order to estimate carbon-sequestering capacity. Carbon is also sequestered in litter on the forest floor. The decomposition rate of litter is rather rapid, so after a few years the forest floor litter is likely to be in a steady state. Litter is ignored in our calculations.

Land Area Requirements

On reasonably good sites in the Pacific Northwest and the southern United States, a plantation forest can average growth of industrial wood (the stem) of about 15 m^3 per ha per year with a minimum of silvicultural inputs beyond planting (Farnum et al., 1983). Assume that 1 m^3 of stemwood is associated with 1.6 m^3 of tree biomass (including roots, branches, and so forth), and that 1 m^3 of biomass contains 0.26 tons of carbon (Brown et al., 1986). Then one ha of new forest will sequester about 6.24 tons of carbon annually. To sequester the 2.9-Gt annual increment of carbon in the atmosphere would require, therefore, about 465 million ha of new plantations. This is about 50 percent greater than the currently forested area in the United States and more than 15 percent of the current area of closed forest worldwide. Figure 8-1 helps put these numbers in perspective. The shaded area covers about 465 million ha. This is approximately 75 percent of the nonforested land area in the United States (figure 8-1). An area this size of suitably productive land would probably require a worldwide search.

The FAO (1982) projected the total area of plantation forests in the tropics by 1985 at only 17 million ha. Earlier,

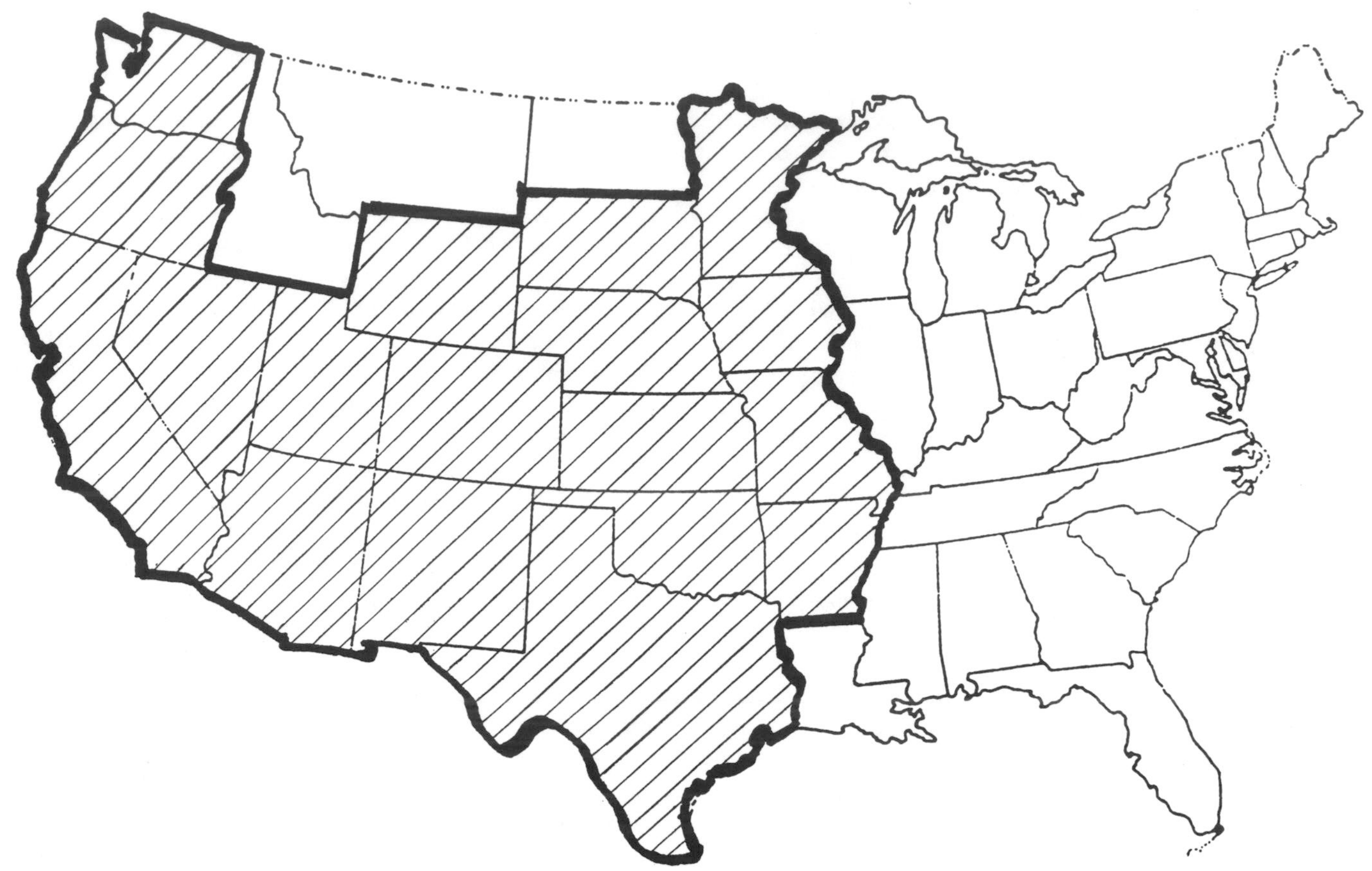

Figure 8-1. An area approximately equal to the 465 million hectares of newly planted forest that would remove the 2.9 Gt of carbon added annually to the atmosphere.

Lanly and Clement (1979) estimated that the total area of manmade forests worldwide, both industrial and nonindustrial, in the mid-1970s was 90 million ha. The most recent estimate, by Postel and Heise (1988), puts total industrial plantations worldwide in the mid-1980s at about 92 million ha. Most of these plantations are found in the industrial countries in the temperate zone of the Northern Hemisphere, such as the Soviet Union, Japan, Western Europe, and the United States. Protection and other noncommercial forest plantations would add significantly to this total.

New forest plantation large enough to sequester all of the free carbon currently being absorbed by the atmosphere would be on a scale that would dwarf all previous forest plantation efforts. Is there enough suitable land to accommodate planting of this magnitude anywhere in the world? Much of the forestland of the requisite productivity is located either in the tropics or in the more southerly parts of the Northern Hemisphere temperate zone. Much of this land is already forested.

Another alternative might be to locate the plantations on less productive sites, perhaps in more northerly regions. However, in this situation tree growth would be less rapid; hence, the land area involved, and associated plantation establishment costs, would be greater. If lower productivity sites were to be used, the total land area required could easily be several times greater than we estimate.

Table 8-2 summarizes the areas of new or more rapidly growing forest required to stabilize atmospheric carbon dioxide for naturally induced forest expansion, increased growth of existing closed forests, and new plantation forests.

Table 8-2. Areas of New or Rapidly Growing Forest Lands Required to Stabilize Atmospheric Carbon Dioxide at Current Levels (annual atmospheric increase of carbon: 2.9 Gt)

To sequester carbon requires either:		No. of hectares
A. Warming-induced expansion of forests northward (assumes no forest decline elsewhere)		3,500 million
B. Increased growth of existing global forests (closed); need to increase their growth by 2.5 m^3/ha/year (current U.S.-wide average growth is 3.15 m^3/ha/year)		3,000 million
C. New fast-growing plantations		465 million
Cost:		
Temperate zone	$372 billion	
Tropics	$186 billion	

Economic Costs

The economic costs of undertaking a plantation effort of the magnitude just described are truly enormous. Establishment costs for forest plantations vary with location between $230 and slightly over $1,000 per ha. An average establishment cost of $400 per ha (Sedjo, 1983, p. 161) involves an initial investment of about $186 billion to establish the minimum area of new plantations required, 465 million ha. This does not include the costs of land procurement. The price of land is also likely to vary with location. In the United States today the average cost would be at least $400 per ha for land of the requisite productivity and could be as high as $1,000 per ha (Graham, 1988). Hence, the minimum per ha start-up costs for CO_2 plantations in the United States could range from $800 to $1,400.

Using these minimal cost estimates, the total costs involved in establishing 465 million ha of carbon-sequestering plantations in the temperate zones might be $372 billion, almost 8 percent of the annual gross national product (GNP) of the United States. However, due to the tremendous land areas involved, valuable agricultural lands almost surely would be required and would demand higher prices. Thus, total costs would be significantly higher. To reduce land acquisition costs, plantations could be located on less productive forest sites. However, total plantation establishment costs would rise as more area would be required for planting and the sum of both acquisition and planting costs would increase. The costs of obtaining the land are not simply accounting costs that may be avoided by donations or expropriation. These costs represent real opportunities forgone. Land used as a carbon sink cannot be used to produce other socially useful products such as crops or livestock. However, such forest would generate some benefits other than carbon sequestering in the form of wildlife habitat, recreation, erosion control, and watershed stabilization.

The degraded lands of the tropics may also offer some opportunity for large-scale plantation. In many parts of the tropics such degraded lands have little if any alternative use. Thus, land costs are likely to be minimal. While the land area requirements are large, so too is the the land area potentially available. Grainger (1988) has estimated that 758 million ha of degraded tropical lands, including 203 ha of previously forested land in the humid tropics, have a potential for forest replenishment. However, the huge volume of land area required almost ensures that some low-productivity lands, perhaps in semiarid regions, will be required and thus drive up the total cost of land for establishing plantations. Plantation establishment costs in the tropics are similar to those in the United States. Recent data from Indonesia indicate that plantation establishment costs on degraded grasslands are about $400 per ha (JICA, 1986, p. 192).

While productivity on degraded lands is sometimes low, growth rates of 15 m^3 per ha per year and more are attainable on many tropical sites. If these cost and productivity figures are broadly representative of conditions in the tropics and if the land is essentially free, the total cost of establishing 465 million ha of plantation forests could be as low as $186 billion, about one-half the cost of a comparable carbon-sequestering plantation forest in the more temperate regions. Of course, these two options are not mutually exclusive. However, such plantation forests, no matter where they are located, would require significant follow-on costs of protection and maintenance.

Another consideration affecting the economic feasibility of the plantation approach is the question of what to do with the mature timber. Such a question would not need to be addressed if the purpose of the plantations were simply to buy time to develop a more adequate approach for dealing with carbon releases from fossil fuels. It becomes relevant only if the plantation approach were to be used for a period longer than a single rotation, which is typically thirty to fifty years.

If the plantation is to continue sequestering high levels of carbon over longer periods, the mature trees would need to be harvested to make way for a new, more rapidly growing second rotation. Cooper (1983), for example, suggests that shorter rotations may be necessary to keep the level of carbon sequestering high. However, he ignores the problem of disposing of the harvested timber. If the timber is simply treated as industrial wood, the large volumes harvested could severely disrupt world markets, depressing prices and generating a host of adverse impacts. In addition, much industrial wood becomes residuals or is used for products such as paper that quickly decompose. The carbon-sequestering function is lost thereby. If the timber is harvested and the timber then inventoried, the sequestering function continues but the costs rise appreciably. Additional costs would involve harvest, transport, and inventory costs. Assuming a harvest rotation of thirty years involving 450 m^3 per ha, and a harvest and transportation cost of $15 per m^3, the total harvest-related costs are $6,750 per ha, exclusive of significant costs for timber storage. This suggests harvest costs thirty years hence for the 465 million ha totaling $3.1 trillion, in addition to substantial storage and preservation costs.

Cost Recovery from Forest Investments

Prospects are poor for recovering a substantial portion of the costs of plantation establishment by releasing mature timber from these forests onto world industrial wood markets. Commercial harvest of even a fraction of the mature timber of the plantations envisioned could easily swamp these markets. For example, the sustained yield harvest from the 465 million ha of new plantations would produce almost 7 billion m^3 of industrial wood. This volume is over four times the total world industrial wood harvest in 1985 (1.5 billion m^3). Such a volume of industrial wood would certainly disrupt wood markets dramatically and sharply reduce the price of industrial wood. At markedly lower prices, harvest and transport costs probably would not be covered by the market price for much of the timber.

The critical point is that the amount of forest required for carbon sequestering would create far more timber than could be absorbed by the world's industrial wood markets. Timber markets no doubt would return to an equilibrium even with radically lower prices, but those lower prices would not compensate private investors for their investments in industrial forestry. The lack of financial incentives would be translated into a decline in forestry investment and ultimately in the growth and productivity of the world's traditional industrial forests. With this would come an associated decline in the carbon-sequestering ability of these forests. Hence, some of the increased carbon sequestered by the newly created plantation forests would be offset by reduced carbon sequestering in the traditional industrial forests.

In the case of the naturally generated forests, the same possibilities for disrupting world wood markets exist. However, should most of the growth occur in the northern boreal regions distant from major world markets, the market disruption features would probably be small simply because the forest would be economically inaccessible. Warming in the high latitudes could, however, improve accessibility.

ADDITIONAL PROBLEMS ASSOCIATED WITH CARBON-SEQUESTERING FORESTS

The earlier discussion indicated a number of ways that the forest stock might be significantly increased and how, in so doing, it would serve as a carbon sink to mitigate the buildup of atmospheric carbon and thereby offset, to some degree, the anticipated global warming. Other issues include (a) the extent to which costs might be recoverable at harvest if the timber is sold for industrial wood, used as fuelwood for power generation, or otherwise utilized and (b) the problem of preventing the carbon sequestered in the forest from reentering the atmosphere if the wood is utilized.

An increased global stock of biomass is (for our purposes) a sufficient condition to the sequestering of a higher level of carbon. To the extent that a higher biomass stock is maintained, more carbon will be permanently sequestered. If the carbon sequestered in the increased biomass "originates" from the burning of fossil fuels, the carbon is simply shifted from a fossil fuel to a biomass residence. In a physical sense, a greater permanent stock of forest need not preclude greater industrial and fuelwood harvests. For example, an increased level of industrial or fuelwood use need not exacerbate the atmospheric CO_2 problem if a greater stock of forest is established and maintained to "feed" the increased use. In other words, the burning or destruction of biomass will not, per se, increase free carbon if the total stock of biomass is rising. Therefore, to the extent that an enlarged stock of forest is created for energy generation and the permanent stock requirement results in a net worldwide increase of biomass, the effect of biomass energy production would simply be to recycle carbon from the biomass to the atmosphere and back to the biomass.

The substitution of plantation wood for fossil fuel combustion would have an advantage with regard to the carbon cycle: it would decrease the rate of carbon accumulation in the atmosphere resulting from energy production. This advantage could be maintained if the total stock of biomass did not decline as a result of utilizing the world's forest biomass stocks needed to feed the power generators.

The energy potential of the plantation-grown wood could be very large. We estimate that one year's wood energy potential is in excess of two years of energy generated by all of the coal-generated power plants in the United States. From an economic perspective, investments in energy forest plantations and power-generating equipment could provide some financial return. The output of the plantation becomes feedstock in the power generation process, replacing fossil fuels. But how financially competitive could wood-burning power generation be as a substitute for the current fossil fuel modes of energy production? Is it close enough to be justified on the basis of its positive climate-warming effects?

WHO PAYS AND HOW?

Market-generated financial incentives are unlikely to induce enough private investment in plantation forests to moderate significantly the buildup of atmospheric carbon. This is almost certainly true if the timber is to be used as industrial wood. Substituting wood for fossil fuels as feedstock to large boilers for power generation also is unlikely to be financially viable given current technology.

Probably the lowest cost locations for plantations are in the tropics. Should wood-fueled power stations be established, some of the costs could be offset by use of wood for power generation. Furthermore, foreign financial assistance could enable the industrial countries to finance directly a substantial portion of the plantation costs. This would provide an opportunity to contribute to Third World development while helping to mitigate the carbon buildup in the atmosphere and providing some relief to the global energy problem. However, even in concept, this approach has problems. Often the lands available for plantations are not situated close to the population centers needing power. Long-distance transportation of fuelwood or long-distance transmission of power would increase, immediately and dramatically, the costs of the useful power due to fuelwood's relatively poor energy density. Technology may eventually make low-cost, long-distance transmission of power feasible, but such technology is not currently operational.

Degraded tropical lands have few alternative uses. Hence, these locations are likely to be the low-cost locations for carbon-sequestering forests of the type suggested. In addition, the plantations will contribute to the rehabilitation of the degraded lands, lands that at some future time may be usable for other purposes.

CONCLUSIONS

In this chapter we examined the relationship between forests and the climate. In particular we investigated methods whereby increases in forest growth and the forest biomass might increase the terrestrial carbon sink and thus provide a means of delaying the buildup in atmospheric CO_2 and the onset and degree of global warming.

Plantation forests, especially if located on tropical lands with few alternative uses, may offer a relatively low-cost means for mitigating and postponing the buildup of atmospheric CO_2. A three- to five-decade postponement of global warming could provide society with enough time to develop energy source alternatives to carbon-releasing fossil fuels.

Three principal methods of increasing the global forest stock were examined in this chapter: (1) the establishment of massive areas of fast-growing plantation forests, (2) the use of forest management and a reduction in the harvest to allow a buildup of the stock of forest biomass, and (3) expansion of the forest system through natural generation of additional forests on previously unforested lands in response to environmental changes created by increased CO_2 levels and the attendant global warming. Based upon our preliminary assessment of these three possibilities, only the wide-scale establishment of forest plantations appears to have much potential.

Stabilization of atmospheric CO_2 at current levels would require a minimum of 465 million ha of new man-made plantations. A much larger area might be needed if some of the land was not productive, as is likely given the massive areas involved. Establishing tree plantations of this size would cost at least $372 billion—and probably significantly more in a temperate region like the United States where land has alternative productive uses. In principle, large areas of degraded lands in the tropics with few economic uses could be made available for afforestation. Thus, the tropics appear to offer a low-cost location, perhaps as low as $187 billion, for forest plantations sufficiently large to sequester the free carbon now accumulating in the atmosphere.

The problem of disposing of the plantation timber as it becomes mature was also considered. Two possibilities were examined: marketing the timber in the industrial wood market and using the wood as a substitute for fossil fuel in energy generation. In either case, much of the carbon sequestered in the wood would be liberated with its commercial use. However, the larger worldwide stock of forest associated with the newly created permanent plantations could ensure that a greater amount of carbon would be permanently sequestered. If significant portions of the new timber were made available to industrial wood markets, those markets surely would be disrupted, greatly depressing prices and inhibiting private-sector investments in forestry and forest plantations.

The option of using the timber in energy generation as a substitute for fossil fuels is examined briefly and appears promising. This approach has the advantages of reducing the use of fossil fuels and therefore the rate at which carbon is being freed from fossil fuels while simultaneously allowing for an increase in biomass stocks and the production of useful wood-generated power. The absence of this type of economic activity today suggests that the market does not currently provide the necessary financial incentives. Hence, subsidies almost certainly would be needed to achieve financial viability. The extent of such a subsidy warrants research.

The wood-growing possibilities for the tropics and Third World countries' increasing need for energy suggest the appealing possibility of linking these requirements together via multilateral economic assistance. A project of this type would address the world's carbon cycle problem, energy problem, and Third World development problem in a manner whereby the industrial countries of the temperate zones would also assist in alleviating all three. To be sure, there are a number of potentially serious problems including questions of plantation location vis-à-vis the major energy-using centers of the Third World. Nevertheless, these possibilities deserve careful consideration.

REFERENCES

Armentano, T. V., and C. W. Ralston. 1980. "The Role of Temperate Zone Forests in the World Carbon Cycle," *Canadian Journal of Forest Research* vol. 10, pp. 53–60.

Beverge, D. I. 1984. "Wood Yield and Quality in Relation to Tree Nutrition," pp. 293–326, in G. D. Bowen and E. K. S. Nambiar, eds., *Nutrition of Plantation Forests* (London, Academic Press).

Bolin, B., B. Döös, J. Jäger, and R. Warrick, eds. 1986. *The Greenhouse Effect, Climatic Change and Ecosystems* (New York, Wiley).

Bonnor, G. M. 1982. "Canada's Forest Inventory 1981," Forestry Statistics and Systems Branch, Canadian Forestry Service, Department of Environment, p. 79.

Brown, Sandra, Ariel E. Lugo, and Jonathan Chapman. 1986. "Biomass of Tropical Tree Plantations and Its Implications for the Global Carbon Budget," *Canadian Journal of Forest Research* vol. 16, pp. 390–394.

Conkey, L. E. 1986. "Red Spruce Tree-Ring Widths and Densities in Eastern North America as Indicators of Past Climate," *Quaternary Research* vol. 26, pp. 232–243.

Cooper, Charles F. 1983. "Carbon Storage in Managed Forests," *Canadian Journal of Forest Research* vol. 13, pp. 155–165.

Davis, M. B. 1986. "Climatic Instability, Time Lags, and Community Disequilibrium," pp. 269–284, in J. Diamond and T. J. Case, eds., *Community Biology* (New York, Harper and Row).

Davis, M. B., K. D. Woods, S. L. Webb, and R. P. Futyma. 1986. "Dispersal Versus Climate: Expansion of *Fagus* and *Tsuga* into the Upper Great Lakes Region," *Vegetatio* vol. 67, pp. 93–103.

Delcourt, H. R., and W. F. Harris. 1980. "Carbon Budget of the Southeastern United States: Analysis of Historic Change in Trend from Source to Sink," *Science* vol. 210, pp. 321–323.

Detwiler, R. P., and Charles A. S. Hall. 1988. "Tropical Forest and the Global Carbon Cycle," *Science* vol. 239, pp. 42–47.

Doyle, T. W. 1987. "Seedling Response to CO_2 Enrichment under Stressed and Nonstressed Conditions," pp. 501–510, in G. C. Jacoby and J. W. Hornbeck, eds., *Proceedings of the International Symposium on Ecological Aspects of Tree-Ring Analysis.* CONF-8608144 (Washington, D.C., U.S. Department of Energy).

Emanuel, W. R., H. H. Shugart, and M. P. Stevenson. 1985a. "Climate Change and the Broad Scale Distribution of Terrestrial Ecosystem Complexes," *Climatic Change* vol. 7, pp. 29–43.

———. 1985b. "Response to Comment: Climatic Change and the Broad-Scale Distribution of Terrestrial Ecosystem Complexes," *Climatic Change* vol. 7, pp. 457–460.

FAO (Food and Agriculture Organization of the United Nations). 1982. "Conservation and Development of Tropical Forest Resources." Forestry Paper 37, Rome.

Farnum, P., R. Timmis, and J. L. Kulp. 1983. "Biotechnology of Forest Yield," *Science* vol. 219, pp. 694–702.

Ford, E. D. 1984. "The Dynamics of Plantation Growth," pp. 17–52, in G. D. Bowen and E. K. S. Nambiar, eds., *Nutrition in Plantation Forests* (London, Academic Press).

Forest Service, U.S. Department of Agriculture. 1982. "An Analysis of the Timber Situation in the United States, 1952–2030." Forest Resource Report 23, December.

Fritts, H. C. 1976. *Tree Rings and Climate* (New York, Academic Press).

Graham, David. 1988. Personal communication, Forest Investment Associates, Atlanta, Georgia, May 24.

Grainger, Alan. 1988. "Estimating Areas of Degraded Tropical Lands Requiring Replenishment of Forest Cover," *International Tree Crops Journal,* vol. 5 (1/2).

Graybill, D. A. 1987. "A Network of High Elevation Conifers in the Western U.S. for Detection of Tree-Ring Growth Response to Increasing Atmospheric Carbon Dioxide," pp. 463–474, in G. C. Jacoby and J. W. Hornbeck, eds., *Proceedings of the International Symposium on Ecological Aspects of Tree-Ring Analysis.* CONF-8608144 (Washington, D.C.. U.S. Department of Energy).

Hari, P., H. Arovaara, T. Raunemaa, and A. Hautojarvi. 1984. "Forest Growth and the Effects of Energy Production: A Method for Detecting Trends in the Growth Potential of Trees," *Canadian Journal of Forest Research* vol. 14., pp. 437–440.

Holowacz, J. 1985. "Forests in the U.S.S.R.," *The Forestry Chronicle* vol. 61, no. 5, October, pp. 366–373.

JICA (Japan International Cooperative Agency). 1986. "Technical Guidance for Afforestation on the Grassland in Benekat South Sumatra," ATA-186, April.

Kappen, L. 1981. "Ecological Significance of Resistance to High Temperature," pp. 439–474, in O. L. Lange, P. S. Nobel, C. B. Osmund, and H. Ziegler, eds., *Physiological Plant Ecology I. Responses to the Physical Environment* (New York, Springer-Verlag).

Kauppi, P. 1987. "Forests and the Changing Chemical Composition of the Atmosphere," in M. Kallio, D. P. Dykstra, and C. S. Binkley, eds., *The Global Forest Sector* (New York, Wiley).

———, and M. Posch. 1987. "A Case Study of the Effects of CO_2-Induced Climatic Warming on Forest Growth and the Forest Sector: A. Productivity Reactions of Northern Boreal Forests," in M. L. Parry, T. R. Carter, and N. T. Konijn, eds., *The Impact of Climate Variations on Agriculture,* vol. 1, *Assessments in Cool Temperate and Cold Regions* (Dordrecht, The Netherlands, Kluwer).

LaMarche, V. C., D. A. Graybill, H. C. Fritts, and M. R. Rose. 1984. "Increasing Atmospheric Carbon Dioxide: Tree Ring Evidence for Growth Enhancement in Natural Vegetation," *Science* vol. 225, pp. 1019–1021.

Lanly, J. P., and J. Clement. 1979. "Present and Future National Forest and Plantation Areas in the Tropics," *Unasylva* vol. 31, no. 123.

Lovelock, J. E. 1979. *Gaia: A New Look at Life on Earth* (New York, Oxford University Press).

Manabe, S., and R. J. Stouffer. 1980. "Sensitivity of a Global Climate Model to an Increase of CO_2 Concentration in the Atmosphere," *Journal of Geophysical Research* vol. 85, pp. 5529–5554.

Manabe, S., and R. T. Wetherald. 1986. "Reduction in Summer Soil Wetness Induced by an Increase in Atmospheric Carbon Dioxide," *Science* vol. 232, pp. 626–627.

Marland, Gregg. 1988. *The Prospect of Solving the CO_2 Problem through Global Reforestation.* Report DOC/NBB-0082 (Washington, D.C., U.S. Department of Energy, Office of Energy Research).

Matthews, E. 1983. "Global Vegetation and Land Use: New High-Resolution Data Base for Climate Studies," *Journal of Climate and Applied Meteorology* vol. 22, pp. 474–487.

Melillo, J., J. R. Fruci, R. A. Houghton, B. Moore, and D. L. Skole. 1988. "Land-Use Change in the Soviet Union Between 1850 and 1980: Causes of the Net Release of CO_2 to the Atmosphere," *Tellus* vol. 40B, pp. 116–128.

Nambiar, E. K. S. 1984. "Plantation Forests: Their Scope and a Perspective on Plantation Nutrition," pp. 1–15, in G. D. Bowen and E. K. S. Nambiar, eds., *Nutrition of Plantation Forests* (London, Academic Press).

Olson, J. S., J. A. Watts, and L. J. Allison. 1983. *Carbon in Live Vegetation of Major World Ecosystems.* Report ORNL-5862 (Oak Ridge, Tenn., Oak Ridge National Laboratory).

Panshin, A. J., and C. DeZeeuw. 1964. *Textbook of Wood Technology* (New York, McGraw-Hill).

Parker, M. L. 1987. "Recent Abnormal Increase in Tree-Ring Widths: A Possible Effect of Elevated Atmospheric Carbon Dioxide," pp. 511–521, in G. C. Jacoby and J. W. Hornbeck, eds., *Proceedings of the International Symposium on Ecological Aspects of Tree-Ring Analysis.* CONF-8608144 (Washington, D.C., U.S. Department of Energy).

Post, W. M., W. R. Emanuel, P. J. Zinke, and A. G. Stangenberger. 1982. "Soil Carbon Pools and World Life Zones," *Nature* vol. 298, pp. 156–159.

Postel, Sandra, and Lori Heise. 1988. "Reforesting the Earth." Paper 83, April (Washington, D.C., Worldwatch).

Regehr, D. L., F. A. Bazzaz, and W. R. Boggess. 1975. "Photosynthesis, Transpiration and Leaf Conductance of *Populus deltoids* in Relation to Flooding and Drought," *Photosynthetica* vol. 9, pp. 52–61.

Richardson, Dennis. 1988. "Forestry in China—Revisited." Draft manuscript (February).

Rotty, R. M. 1986. "Estimates of CO_2 from Wood Fuel Based on Forest Harvest Data," *Climate Change* vol. 9, pp. 311–326.

Sedjo, Roger A. 1983. *The Comparative Economics of Plantation Forests* (Washington, D.C., Resources for the Future).

Smith, W. H. 1985. "Forest Quality and Air Quality," *Journal of Forestry* vol. 83, pp. 82–94.

Solomon, Allen M. 1986. "Transient Response of Forests to CO_2-induced Climate Change: Simulation Modeling Experiments in Eastern North America," *Oecologia* vol. 68 (Spring), pp. 567–579.

Solomon, A. M., and D. C. West. 1985. "Potential Responses of Forests to CO_2-induced Climate Change," pp. 145–170, in M. R. White, ed., *Characterization of Information Requirements for Studies of CO_2 Effects: Water Resources, Agriculture, Fisheries, Forests and Human Health.* Report DOE/ER-0236 (Washington, D.C., U.S. Department of Energy).

———. 1986. "Atmospheric Carbon Dioxide Change: Agent of Future Forest Growth or Decline?" pp. 23–38, in J. G. Titus, ed., *Effects of Changes in Stratospheric Ozone and Global Climate,* vol. 3, *Climate Change* (Washington, D.C., Environmental Protection Agency).

———. 1987. "Simulating Forest Responses to Expected Climate Change in Eastern North America: Applications to Decision Making in the Forest Industry," pp. 189–207, in W. E. Shands and J. S. Hoffman, eds., *The Greenhouse Effect, Climate Change, and United States Forests* (Washington, D.C., Conservation Foundation).

Solomon, A. M., M. L. Tharp, D. C. West, G. E. Taylor, J. M. Webb, and J. L. Trimble. 1984. *Response of Unmanaged Forests to Carbon Dioxide-Induced Climate Change: Available Information, Initial Tests, and Data Requirements.* Report TR-009 (Washington, D.C., U.S. Department of Energy).

Steponkus, P. L. 1981. "Responses to Extreme Temperatures: Cellular and Sub-cellular Bases," pp. 371–402, in O. L. Lange, P. S. Nobel, C. B. Osmund, and H. Zeigler, eds., *Physiological Plant Ecology I, Responses to the Physical Environment* (New York, Springer-Verlag).

Telewski, F. W., and B. R. Strain. 1987. "Densitometric and Ring Width Analysis of Three-Year-Old *Pinus taeda* L. and *Liquidambar styraciflua* L. Grown under Three Levels of CO_2 and Two Water Regimes," pp. 494–500, in G. C. Jacoby and J. W. Hornbeck, eds., *Proceedings of the International Symposium on Ecological Aspects of Tree-Ring Analysis.* CONF-8608144 (Washington, D.C., U.S. Department of Energy).

Woodwell, George M. 1987. "The Warming of the Industrialized Middle Latitudes." Paper presented at the Workshop on Developing Policies for Responding to Future Climate Change, Villach, Austria.

Zavitkovski, J., and J. G. Isebrands. 1983. "Biomass Production and Energy Accumulation in the World's Forests." The Seventh International FPRS Industrial Wood Energy Forum '83 1:12–22. Forest Products Research Society, Madison, Wis.

9

The Biological Consequences of Climate Changes: An Ecological and Economic Assessment

Sandra S. Batie
Herman H. Shugart

The alteration of the earth's atmospheric carbon dioxide (CO_2) concentration by the combined action of clearing land, tilling virgin soils, and burning fossil fuels since the industrial revolution in the 1850s and 1860s has been termed an "uncontrolled experiment" (Baes et al., 1977). The direct effect of increased atmospheric carbon dioxide levels on plant growth could be considerable if the performance of plants in a greenhouse is an indication of their performance in nature (Bazzaz and Carlson, 1984; Strain, 1985; Schneider and Rosenberg, chapter 2 in this volume). It is also possible that the increase in carbon dioxide has altered the functioning of today's plants with regard to their use of carbon dioxide and water (Woodward, 1987). These direct biological effects complicate the evaluation of another potentially large effect of carbon dioxide on the biosphere—global warming from carbon dioxide working as a greenhouse gas, or the greenhouse effect.

Large uncertainties surround the nature and magnitude of the greenhouse effect, but at a minimum it appears that it may no longer be valid to assume that the weather patterns of past centuries will continue into the future. Furthermore, there is justification for public concern because the climatic changes induced by greenhouse warming may be irreversible. Predictions as to the impact of climate change on biological diversity are confounded by delays in warming caused by the slow temperature response of the soils and oceans to air temperature change (that is, by delays in the response time of ecosystems to climate warming). There also appears to be a long lag in human response to climate impacts. Together these long delays may well imply a "dearth of time for effective action" (Kerr, 1988, p. 23).

To take advantage of whatever time is available for effective action and to make informed policy decisions, it is necessary to evaluate the consequences of the alterations in ecological systems caused by climatic change and also the effects on these systems caused directly by increased atmospheric carbon dioxide. We must attempt to answer two types of questions: First, what sorts of changes will take place? Second, what consequences will such changes hold for society? An evaluation of this sort is confounded not only by the uncertainties already mentioned but also by the difficulty of estimating the societal value of natural ecosystems and the diversity of species that they harbor.

Economists have developed several methods for evaluating decisions that involve nonpriced goods and services. These methods attempt to arrive at the costs and benefits that would result if private markets for the nonpriced goods and services actually existed and were well functioning (Bromley, 1986). These values tend to be relatively easy to obtain under certain conditions: that is, when the usefulness of the good and service is well known, possible alternatives represent only a small change from the status quo, there are not likely to be questions of intergenerational equity (discussed below and by Crosson, chapter 5 in this volume), the consequences of a choice are fully understood, and no major externalities such as pollution are likely to attend the provision of the good or service. These conditions are not met in the case of global warming impacts on biological diversity, and yet decisions—perhaps affecting many future generations—will be made. This chapter addresses the uncertainties surrounding the physical and ecological impacts of climate change and explores the use of economic methods of analysis for evaluating such changes in the face of uncertainty. It concludes with a discussion of public policies to reduce gas emissions and protect biological diversity.

WILL CLIMATE CHANGE ENOUGH TO MATTER?

The relationship of greenhouse warming to the increased concentration of atmospheric carbon dioxide is discussed in chapter 2 of this volume. For present purposes a key point in that discussion is that predictions of CO_2-induced climate change made by general circulation models (GCMs) do not agree about how regional climates may change. Yet evaluation of the consequences of greenhouse warming for ecological systems requires a scenario of likely changes in regional climates.

The mean global temperature change predicted by the models for a doubling of atmospheric carbon dioxide ($2 \times CO_2$), about 3.5 degrees Celsius (°C), may lead to temperatures higher than any that are thought to have occurred for hundreds of thousands of years. (See Schneider and Rosenberg, chapter 2 in this volume.) Such a change is ecologically significant by several rules of thumb. For example, Kerr (1988) estimates that for each 1°C of warming, climatic zones can shift 100 to 150 kilometers poleward. "By the mid-21st century," he says, "the climate that nurtures Yellowstone National Park could be well into Canada. The tundra of the Arctic National Wildlife Refuge could be pushed into the sea" (p. 23). The range of average monthly temperatures that occurs during the growing season of the boreal forest that occupies vast regions of Alaska, Canada, Fennoscandia, and the Soviet Union is itself of the order of 4°C. This suggests that the boreal forest might be shifted from its present boundaries by a 4°C warming. Of course, such evaluations are themselves subject to considerable uncertainty and to assumptions about the nature (and time scale) of biotic responses to change. Despite the many uncertainties, the magnitudes of climatic change being discussed would in some cases be large enough to induce considerable change in ecological systems.

Thus, in answer to the question "Will climate change be enough to matter?" we conclude that it may, but that we are uncertain about how to use GCMs to evaluate the ecological implications of these changes for the following reasons:

1. The high level of uncertainty in GCM parameters (for example, in their treatment of ocean currents, of heat storage by the sea, and of their coupling to vegetation);
2. The lack of agreement of output among GCMs that are constructed under different assumptions as to the magnitude (or the direction) of changes in moisture conditions associated with greenhouse warming;
3. The lack of comparability among GCMs with regard to the regional patterns of climate change associated with greenhouse warming;
4. The lack of fundamental ecological data at appropriate spatial scales and, in particular, the lack of modelers' knowledge of how to "scale-up" from short-term and small-scale processes to longer-term and larger-scale consequences;
5. The fact that the effects of climate change on ecosystems will be superimposed on those of other equally profound environmental agents of change such as forest clearing, air pollution, erosion, and species extinctions. The separation of climatic effects from these other effects through controlled experimentation is difficult if not impossible.

One might conclude that these deficiencies would completely proscribe any attempt to predict the ecological consequences of climatic change. Certainly they suggest that any such predictions should be taken as reasoned speculations at best. However, the recent advances in paleoecological reconstructions of responses of ecosystems to climatic changes of the past, the development of longer-term ecological studies aimed at characterizing change, and the application of tested ecological models (in both paleoecological and long-term studies) have all contributed to our improved capability to judge climatic effects on ecosystems. Using all of these tools, defective as they may be, we undertake an evaluation of the impacts of climate change on ecological systems.

EVALUATING IMPACTS OF CLIMATE CHANGE ON ECOLOGICAL SYSTEMS

We know from paleontological evaluations that, over long time periods (say, millions of years), species can respond to a wide range of climatic and other environmental changes. The panorama of species evolution that one sees when walking through a museum is testimony to the resiliency of biological systems in response to changes that are easily as profound as the climatic changes that might follow a carbon dioxide-induced climatic warming. Since 15,000 years ago, when the continental glaciers began to recede from what is now the contiguous forty-eight states, major plant and animal species have moved their ranges thousands of miles in response to the recession of the glaciers and the consequent climatic change. It would appear that many species are still expanding their ranges in response to the displacement caused by the last glaciation. While it is comforting to realize that there is considerable resiliency in the response of certain species to climate change, it is also important to remember that these responses have occurred on millennial time scales and over unmanaged landscapes. The present situation involves climate change on the scale of a century or less and on landscapes vastly changed by human beings.

In evaluating the effects of climate change on ecological systems, then, it is necessary to employ temporal and spatial scales that apply to the present situation and to recognize the phenomena that operate to control ecosystem function at these scales. For example, the migration of the range of tree species over thousand-year time scales (figure 9-1) in response to a glacial recession may be largely controlled by the spread of seeds (dispersal). An understanding

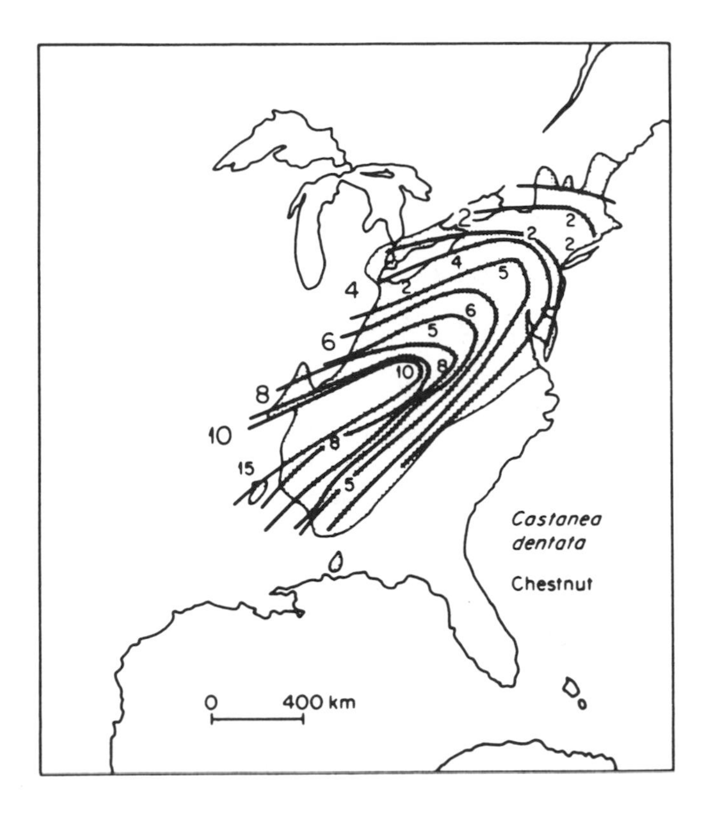

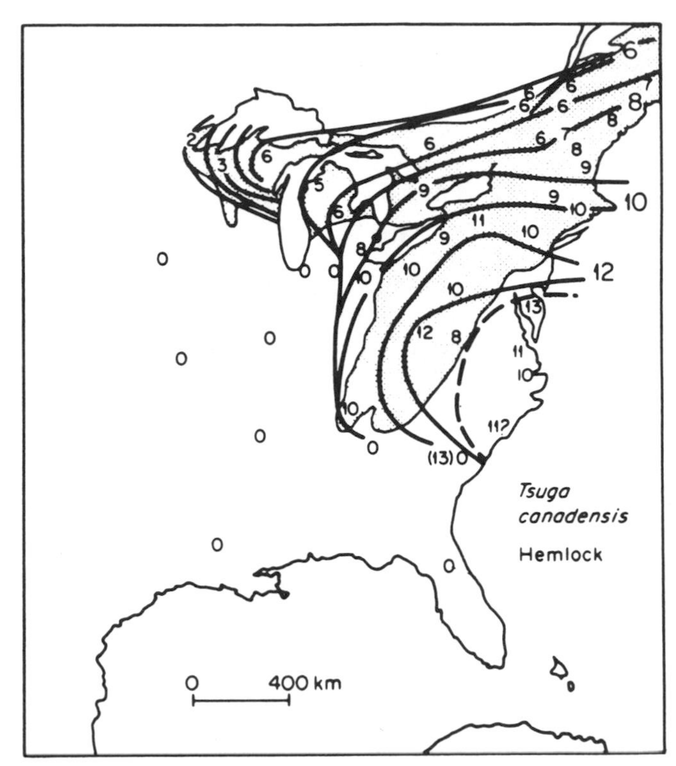

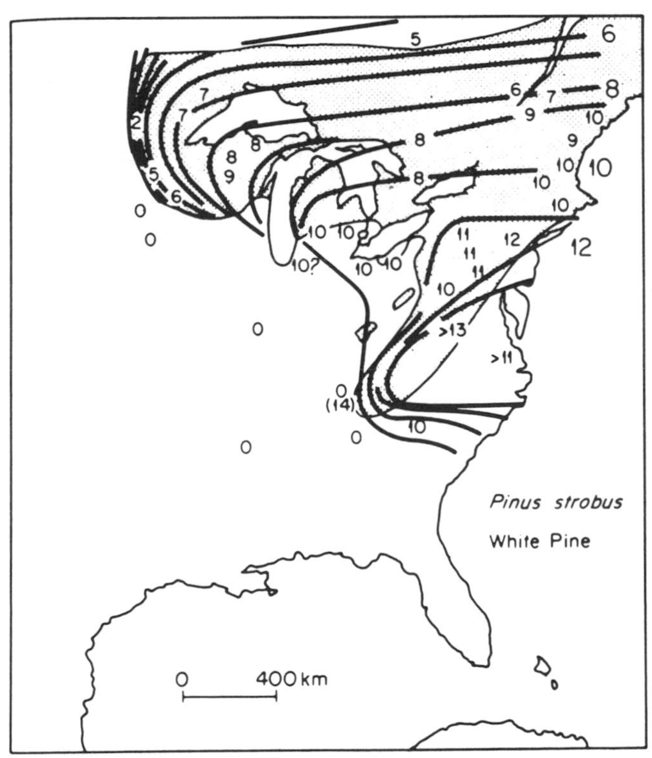

Figure 9-1. The response of three tree taxa over the past 15,000 years to the changes associated with the recession of the continental glaciers of the ice ages. The stippled lines represent the present-day ranges of the taxa; the isoclines indicate the position of the species' northern ranges (as evidenced by the presence of fossil pollen found in dated lake sediments and other paleontological evidence) in increments of 1,000 years. Different trees move across the landscape at different rates and in different directions. It appears that, to a degree, the ranges of tree species are still changing in response to the waning of the last glaciation. (*Source:* Reprinted, with permission, from M. B. Davis, "Quaternary History and the Stability of Forest Communities," chap. 10 in D. C. West, H. H. Shugart, and D. B. Botkin, eds., *Forest Succession: Concepts and Application.* © 1982 by Springer-Verlag, New York.)

of this process would be a prerequisite to predicting changes in these species' ranges. In contrast, the effects of a climate change on a small, forested park or forest reserve over the course of a few years would be strongly dependent on the age structure of the tree populations in the park or on the amount of potential fuel material that might feed a forest fire during a drought (such as occurred in the summer of 1988 in Yellowstone Park).

This relation between the temporal and spatial scales under consideration and the dominant phenomena that control the response of the ecosystems at those scales is an essential construct in modern ecology. Figure 9-2 illustrates the temporal and spatial scale of the relationships between ecosystem patterns and exogenous "disturbances" such as climate change. The implication of the figure is that it is inappropriate to speak of the effects of stress on an ecosystem without first considering the temporal and spatial context of the stress. The next subsection identifies some of the possible ecological consequences of climate change at differing spatial and temporal scales.

Consequences for Terrestrial Ecosystems

The consequences of a global warming of the magnitude and pattern predicted by current GCMs run under doubled CO_2 conditions will be discussed at three different temporal scales: short-term (decadal), medium-term (centenary), and long-term (millennial). Typically, the spatial scales associated with these ecosystem responses will increase as the time scales do.

Short-Term (Decadal) Effects

Evaluation of the short-term effects of a particular climatic change depends on the temporal nature and the magnitude of the change. Even in the short term, however, many important conservation systems such as parks and reserves and management plans respond to fluctuations in weather that are difficult to characterize as anything more than chance variations of the current climatic regime. For example, duck populations depend on breeding habitat in

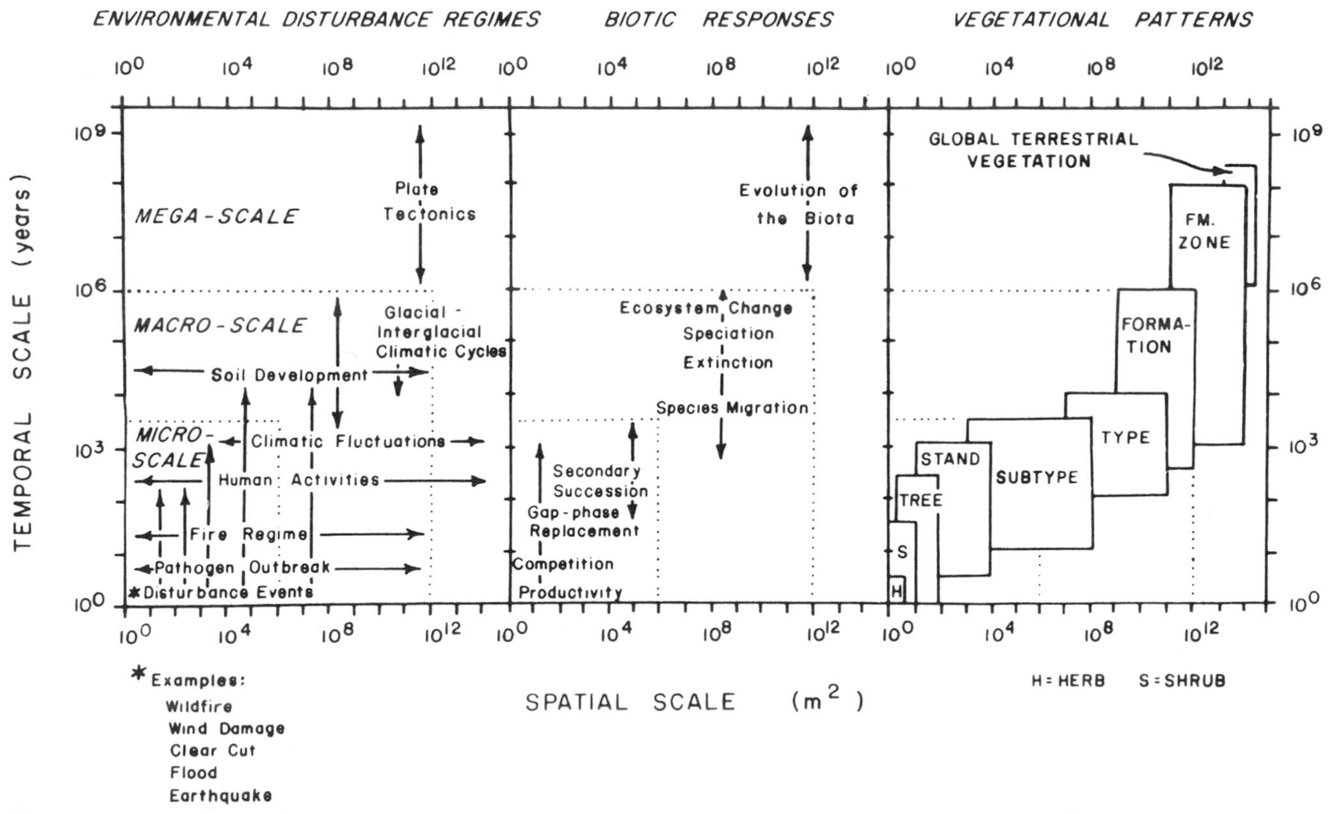

Figure 9-2. Environmental disturbance regimes, biotic responses, and vegetational patterns viewed in the context of space-time domains in which the scale for each process or pattern reflects the sampling intervals required to observe it. The time scale for the vegetational patterns is the time interval required to record their dynamics. The vegetational units are graphed as a nested series of vegetational patterns. (*Source:* Reprinted, with permission, from H. R. Delcourt, P. A. Delcourt, and T. Webb III, "Dynamic Plant Ecology," *Quaternary Science Review* vol. 1, pp. 153–175. © 1983 by Pergamon Press, plc.)

small wetland areas that can be completely dried by one or two years of drought. At this writing a second year of drought in central Canada is drying the small potholes that provide the nesting habitat for many of the North American ducks, and a second calamitous year for fledgling production seems likely. The probable result will be a continental-scale reduction in duck populations. Many of our conservation systems—for example, parks, refuges, and preserves—have been designed to protect populations under current climatic conditions but may fail under different conditions. Such failures are not due to misjudgment on the part of conservation scientists. Rather, it appears that the public loses interest in preserving species when they move from the endangered to the marginal category.

Other short-term responses to climatic change can be expected in ecological systems subject to large disturbances related to weather. The intensity and spread of wildfire is one obvious case. Outbreaks of insect pests and of fungal pathogens and other agents of disease capable of killing dominant species over large areas represent another likely response in many ecosystems.

Forest ecosystems with a substantial proportion of older trees (often termed *overmature forests* by commercial foresters) tend to be sensitive to short-term fluctuations in climate and are prone in some cases to episodic dieback of trees. Many plant and animal species are dependent on habitat conditions associated with these forests. Thus, species could be reduced or even eliminated if these habitats were modified by climatically induced increases in mortality.

In evaluating these and other short-term responses to climatic change, it is difficult to separate the conditions associated with a climatic change from those due to unusual (but statistically feasible and precedented) conditions characteristic of the current climate. Of course, ecosystems may be sensitive to short-term deviations from normal conditions less directly associated with climate such as pest outbreaks or fires.

Medium-Term (Centenary) Ecological Effects

At the time scale of a century, a climate change could produce fairly substantial effects on many terrestrial ecosystems. A 100-year period exceeds the longevity of animal species and falls within the range of life spans of most of the longer-lived plant species. The population dynamics of trees and grasses in the 100-year time range depend upon successful regeneration. Relic populations living within a range of environmental conditions under which they can survive, but cannot reproduce, suffer irreversible declines at this time span. Present-day examples of such populations are timberline forests that, once cleared, do not regenerate as well as forests in the tundra-to-taiga transition. Species with attributes that reduce their likelihood of successful reproduction (for example, by low production of propagules, complex regeneration requirements, or protracted maturation periods for juveniles) would likely diminish.

Shifts in the composition of species and alterations of several fundamental ecosystem processes and characteristics, such as nutrient cycles, decomposition rates, and total biomass, should be expected in century-scale responses to climatic change. It is difficult to assess the degree to which such changes would produce less stable ecosystems, in the sense that internal feedback processes fostering homeostasis would be reduced. In general, ecosystems recently stressed by a relatively severe exogenous change should be simplified by an increased redundancy of organisms performing particular ecosystem processes. Such ecosystems should be dominated by "generalist" species with the ability to survive and reproduce over a wide range of environmental conditions. However, ecologists do not have a great body of experience with such systems.

Long-Term (Millennial) Effects

We know from paleontological evidence that during and after the last glaciation, ecosystems for which there are no modern-day analogs existed over large areas. There has been widespread extinction of the megafauna of North America—the mastodons, mammoths, horses, giant sloth, giant beaver, and others—since the last glaciation. Entire ecosystems, as well as conspicuous taxa, have disappeared in fairly recent times, indicating the mutability of ecological systems in the face of changing conditions.

Over relatively long periods of time, one would expect the global vegetation to equilibrate to a change in climate. The duration of the transition could be quite long and will likely depend on the rate at which organisms can migrate to suitable environments (if they survive long enough to do so). The relatively slow rate at which soils equilibrate to climatic changes is another important factor that would delay the long-term response of ecosystems to a changed climate.

General circulation models predict that the buildup in atmospheric CO_2 will result in substantial warming in the higher northern latitudes. If this is the case, ecosystems in the boreal zones are most likely to be affected. Figure 9-3 shows the predicted change in potential vegetation in response to the temperature changes forecast by one GCM under doubled CO_2 conditions. This tabulation was developed by Emanuel et al. (1985), who used temperature and precipitation data from 7,000 weather stations to extrapolate an expected world vegetation map based on the Holdridge (1947) vegetation classification system. Emanuel and his colleagues then modified the temperature data to reflect the changes in temperatures predicted by Manabe and Stouffer of the Geophysical Fluid Dynamics Laboratory

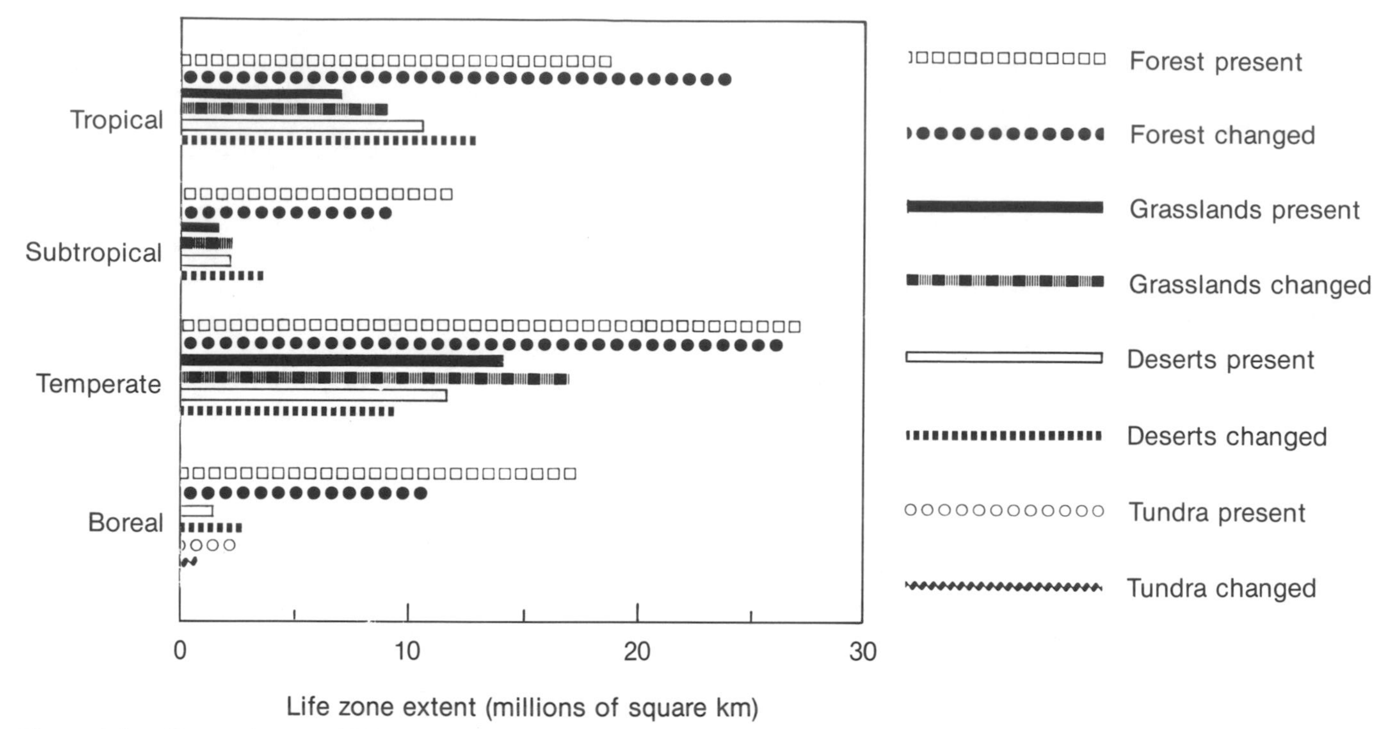

Figure 9-3. Changes in potential vegetation of the world under the temperature change predicted by Manabe and Stouffer (1980) for the climate under a CO_2 doubling. (*Source:* Redrawn, with permission, from W. R. Emanuel, H. H. Shugart, and M. P. Stevenson, "Climatic Change and the Broad-Scale Distribution of Terrestrial Ecosystem Complexes," *Climatic Change* vol. 7, pp. 29–43. © 1985 by Kluwer Academic Publishers.)

(1980) in a GCM-simulated climate response to a doubling of carbon dioxide. Changes in precipitation were not considered in this analysis, although vegetation responses are likely to be quite sensitive to changes in global moisture regimes. Since precipitation at each of the 7,000 meteorological stations was assumed unchanged, the increased evapotranspiration that is predicted by the higher temperatures in the 2 × CO_2 scenario implies a somewhat drier world than that of today.

GCMs do not agree well with respect to the regional changes in moisture conditions that would follow a carbon dioxide doubling (see chapter 2 in this volume). In a recent review Dickinson (1986, p. 262) notes, "There is qualitative agreement among all current three-dimensional models that the largest equilibrium temperature increases would occur in regions of high latitude winter. There is some but not compelling evidence for mid-latitude mid-continent summer drying. High latitude precipitation increases in winter are likely." With these predictions and the inattention to precipitation changes in the Emanuel et al. (1985) analysis borne in mind, it is not surprising that the vegetation changes predicted by the Holdridge model are quite substantial. They show vegetation changes on 34 percent of the terrestrial surface of the earth. Ecosystems in northern latitudes are shifted north considerably. The total areal extent of several northern ecosystems is changed significantly.

In theory it is possible to mitigate the transitional effects of vegetative response to climate change by actively moving plant material (seeds and seedlings) to appropriate locations. Such transplants would require well-developed techniques of plant propagation and for this reason would, realistically, be an option only for commercially important plant species. Even for these species problems would abound. For example, pronounced warming in the higher latitudes might open land for forestation, but southerly genotypes moved north to respond to a warmer climate might not be adapted to the northern light regime.

ECOLOGICAL CONSEQUENCES OF CLIMATE CHANGE: EVALUATING THE SOCIAL COSTS

The preceding discussion should convince us that the climate changes associated with 2 × CO_2 global warming could have profound implications for ecological systems all around the world, both in the short term and long term. How can we evaluate the social costs of the changes that may occur?

For many ecosystems the consequences of climatic change will be affected by human responses to perceived alterations in existing conditions. In a managed environment such as agriculture, for example, climatic change may induce changes in the mix of crops, location of planting, and even in farm programs. Biotechnological research may be directed toward potential agricultural adjustments to climate change.

Human responses to climate change, however, need not be solely reactive; it may be desirable to implement decisions to reduce the probabilities of negative impacts of climatic change. For example, there could be a deliberate policy of planting forests on idle croplands to reduce the rate or degree of global warming (see Sedjo and Solomon, chapter 8 in this volume). However, any decision—including the decision not to decide—involves valuation. How much should society spend to reforest? Should any crop production be forgone? Are the probable net benefits from reforestation large enough to warrant such a public policy?

With respect to the effects of changing climate on biological diversity, there is much we do not know that limits the contribution of conventional economic methods to evaluating alternative climate "protection" strategies. Randall (1986a) notes, for example, that in most cases we do not know whether there are such strong dependencies between species within a community that the loss of some will influence the survival of others. Will the loss of any or of many species lead to the systematic collapse of the community? Are some species more dispensable than others?

It is clear that the loss of biological material is, at a minimum, a loss of opportunity for future benefits to society. "For example, oral contraceptives for many years were produced from Mexican yams; muscle relaxants used in surgery worldwide come from an Amazonian vine; the cure [sic] for Hodgkin's disease comes from the rosy periwinkle, a native of Madagascar; and the gene pool of corn has recently been enriched by the finding, on a small area in the mountains of Jalisco, Mexico, of a perennial wild relative" (Raven, 1988, p. 13). It is less clear, however, whether we can sustain a high standard of living with a more simplified ecosystem. China, for example, is managing to feed and clothe 1 billion people on a relatively small arable area at the expense of natural biodiversity. What future potential is being sacrificed by this simplification of the ecosystem? We do not know.

Nevertheless, decisions will be made that affect biological diversity, and they will be made with a paucity of information—not only as to the probable outcomes of the decisions but also as to the values attached to these outcomes. As renowned geographer Gilbert F. White observed:

> To deal effectively with the whole range of environmental problems that are evident or emerging would call, ideally, for perfect knowledge of the natural systems to be affected. The knowledge would be so nearly complete that for any proposed intervention in the environment it would be possible with a high degree of confidence to estimate the expected effects on people and ecosystems, and to specify the full costs and benefits of any measures that might be devised to enhance resource use . . . this ideal never will be reached. At best, new technologies will present new solutions and new complications, and further investigations will reveal processes and conditions previously unsuspected. . . . Yet, the effort to estimate impacts will be pursued. The sobering prospect is that most of the major public decisions about resource use and environmental management will be made in the face of large uncertainty deriving from ignorance of physical and biological systems and from evolving techniques and social values. (White, 1980, p. 185)

Having to evaluate the ecological consequences of climate change and to decide what to do about them is indeed a sobering prospect. The time and space scales and the complexity of the task pose major problems for conventional benefit–cost analysis, the technique on which economists have come to rely for evaluation of resource management policies. The problems are of two main types: (1) attaching social values to the ecological consequences of climate change; and (2) dealing with the fact that the consequences are spread over decades, even centuries, thus involving the welfare not only of the present generation but of generations to come.

The Problem of Valuing Consequences

To an economist, social value is a reflection of society's preferences with respect to a good or service. According to this view, the values of an ecosystem are those reflecting human satisfactions derived from the system. Society's preferences for biological diversity are thought to be based on perceptions of three types of value: (1) use value, (2) option value, and (3) existence value.

Use value refers to the human satisfaction obtained from using the good or service, or in this case the ecosystem, in any manner and at any time—in the present or in the future. Uses can include those of a nonconsumptive nature such as bird-watching or receiving aesthetic satisfaction as well as consumptive activities such as making wood products.

Option value is associated with risks influencing future choices—risks that will have an effect on future demand for or supply of goods and services from the ecosystems. Option value is measured as the willingness to guarantee the availability of a resource in the future. Economists also distinguish a category called *quasi-option value*, which is the value of preserving an option to use the resource in the future, given the expectation of growth in knowledge about the resource (Arrow and Fisher, 1974). The term quasi-option value relates to decisions that are irreversible; that is, there is a value to preserving a resource in the present if one expects new information to emerge about the value of the resource in the future.

Existence value is the satisfaction derived from the knowledge that the ecosystem continues to exist. Its destruction may mean considerable loss of satisfaction to individuals—even to those who have never used its resources and never intend to do so. All goods and services can have existence value, not just unique natural systems that might be threatened with irreversible damage. However, the natural system's existence value is likely to be much higher than that from more common goods and services. For example, after reviewing many arguments with respect to the preservation of species, Sober (1986) notes that much of the concern with respect to environmental diversity relates to existence value and is aesthetic in nature. The damage done when a species is destroyed is likened to the destruction of a great work of art.

Value estimates are relatively easy to obtain when one is investigating a marketed service or a commodity such as bread. In contrast, the use, option, and existence values provided by biological diversity typically are not measured in markets, and this makes them exceptionally difficult to estimate. Consequently, knowledge of these values is quite imperfect. Moreover, choices among them frequently represent nonincremental changes from the status quo, involve questions of intergenerational equity, and, because of the interconnectedness of ecosystems, generate numerous unforeseen consequences that have ramifications well beyond the system directly manipulated.

When prices do not accurately reflect value (as is the case with most biological diversity values), economists base their measurements of value on revealed preferences as reflected in past choices. Unfortunately, past choices may be a poor guide to determining the effects of climate change on the social values of biological diversity, in part because the range of prospective climate change, while highly uncertain, is outside human experience. Furthermore, even past choices are dependent not only on preferences but also on the opportunities associated with access to resources or technology. Under some circumstances, small changes in either preferences or opportunities can result in large changes in choices made (Boulding, 1969). It may well be that the choices made with respect to the protection of biological diversity will be different in the future from those in the past as knowledge is gained, as technology changes, as biological diversity diminishes, and as preferences with regard to environmental quality are altered.

The problem of estimating, even roughly, the future value of biological diversity—and the possible consequences of climate change in reducing it—is compounded also by our ignorance about the current stock of species. Randall (1986a) notes that the information base for valuation of species is very weak since almost 70 percent of the earth's species remain uncataloged. Furthermore, even for the species that are known, there is little appreciation of their roles in life-support systems or of their use, option, and existence values. There are gross uncertainties with respect to the probability, nature, and consequences of losses with respect to climate change. As Randall notes:

> The economic analyst lives rather high on the information food chain. He/she functions by ingesting information about baseline conditions and consequences of change developed by practitioners of many other disciplines, and metabolizing it according to economic principles. When information about consequences is highly speculative (as it often is in preservation cases) the economist can do little to fill the information gap. (Randall, 1986a, p. 95)

The Intergenerational Problem

Even if valuation of the ecological effects of climate change were no obstacle, conventional benefit–cost analysis would still have major limitations as a criterion for setting policies that would attempt to avert or adapt to the effects of greenhouse warming. Benefit–cost analysis is a technique that embodies a particular way of viewing the world, and it reflects particular ethical premises. Philosophically, the technique rests on utilitarianism—the view that the objective of public policy should be to maximize the national income, regardless of how it is distributed among the population. This indifference of utilitarianism—and of benefit–cost analysis—to the distributional consequences of public policies has long brought the doctrine under criticism. The criticism has particular point when distributional consequences are spread across generations. The use of conventional benefit–cost analysis to evaluate these consequences is an exercise in what Page (1977) has termed a "dictatorship" of the present generation over future generations. The reason is that conventional benefit–cost analysis involves discounting future consequences to express them in "present value" equivalents. But discounting can make present molehills out of even the biggest future mountains. At the usual discount rates reflecting the opportunity cost of money, almost any benefit or cost occurring as much as twenty or thirty years in the future is reduced to irrelevancy. The ethics of our generation's discounting of benefits and costs, which by our actions we impose on future generations, is questioned by many. See, for example, Schulze and Kneese (1981); Page (1977, 1978); and Randall (1986b).

In summary, conventional benefit–cost analysis is best and most accurately used when market-generated prices are available for valuing incremental changes from the status quo over relatively short periods of time. The consequences of climate change for ecological systems promise to be unpriced, long-term, and nonincremental. Indeed, some of them—for example, species extinction—may be irreversible. It follows that benefit–cost analysis as conventionally applied is ill-suited to evaluation of the ecological consequences of climate change and to choices among policies for dealing with the consequences.

It does not follow, however, that we should abandon the economic way of thinking as we address the issues involved. There is more to economics than benefit–cost analysis. It has much to offer as a discipline for organizing thinking about the consequences of alternative courses of action and for making use of information about the consequences. Most generally, economics clarifies the truth that all decisions have costs. For example, strong global restraints on present greenhouse gas emissions, while reducing the rate of global warming, would come at a cost of forgone current global income. Economists also understand the enormity of the transaction costs of orchestrating a coordinated international response to reduce emissions or to reforest vast areas. They also argue that there are self-correcting mechanisms in the global social system: for example, incentives for global agreements to limit emissions of greenhouse gases will be strengthened as the costs of adapting to the effects of climate change rise.

Economics also can help to clarify the issues at stake in the intergenerational sharing of the ecological costs of climate change. Deciding on an equitable sharing of the costs is a political issue, but economists can contribute to the decision by clarifying the nature, if not the magnitude, of the trade-offs involved.

Safe Minimum Standard Strategies and Policy

One possible way of evaluating the trade-offs among alternative courses of action is by using what in the economics literature is called the Safe Minimum Standard (Bishop, 1978). The Safe Minimum Standard is a risk-averse, conservative criterion which states that society should ensure the survival of species, habitats, and ecosystems unless the costs of so doing are "unacceptably large." What is unacceptably large is a social decision to be made through the political process. However, it can be thought of as more than the maximum premium society is willing to pay to ensure against worst-case outcomes. The standard, thus, does not focus on the use, option, or existence benefits of ensuring survival nor on the probabilities of worst-case outcomes. Instead, it requires the identification of technically feasible protection strategies and the measurement of their attendant costs accurately enough to permit judgment as to their social acceptability.

The Safe Minimum Standard presumes but does not document benefits; it shifts any burden of proof to opponents of protection strategies; and it turns creative talents to identifying least-cost protection alternatives. Further, in cases where the opportunity costs of protection are trivially low, as with the snail darter—this small fish was endangered by an economically unattractive dam (Randall, 1986a)—the preservation choice is easily defended.

With regard to climate change and its consequences, the Safe Minimum Standard argues for strategies to reduce global warming even though the costs of doing so may appear high. The reason is the vast uncertainty about the long-term consequences of climate change and the likelihood that some of these eventualities—not just those that concern ecosystems—could be catastrophic for human society. Direct strategies to reduce gas emissions could include reducing the use of fossil fuels and reforestation (see chapters 3 and 8, respectively, in this volume). The costs of these strategies may fall with disproportionate severity on the low-income developing nations. Consequently, "compensation strategies" by which more wealthy nations would assist poorer nations in reducing greenhouse gas emissions and enlarging forested areas are worthy candidates for future policy design.

It should be noted here that whatever strategies are developed for reducing greenhouse gas emissions, economic incentives should be an integral part of them. Providing incentives to adopt more energy-efficient end-use technologies has received much attention (e.g., Goldemberg et al., 1987). But the opportunities for harnessing economic incentives to the task of slowing and eventually halting global warming are legion and may turn up in the most unexpected places. (For example, Hong Kong has been experimenting with a road-congestion tax where, through the use of computerized monitors at key highway intersections, car drivers pay a graduated tax that rises with the number of times their car enters a congested area. The scheme is not designed to reduce CO_2 emissions, and at most its contribution to solving the global problem would be trivial. But it illustrates the many opportunities for using economic incentives to achieve that goal, even when the immediate objective is something quite different.)

The impacts of climate change on ecological systems will be superimposed upon changes brought about by other forces. Substantial changes are currently occurring in habitat and biological populations. With respect to biological diversity, the short-term climate changes and variability coupled with current human activities appear to be the dominant concern. Many of the earth's ecosystems are already experiencing progressive simplification as a result of expanding human populations. It is sobering to realize that the world's population has doubled since 1950. When the twentieth century began there were 1.6 billion people; when it ends there will be more than 6 billion. Rapid growth in human population is an immense ecological force, and it is without precedent in world history. It is also true that the rapid population increase in this century has been accompanied by a sevenfold rise in real per capita income (Brown et al., 1987, as quoted by Krahl and Cook, 1988). However, there is great disparity in the distribution of gains in the quality of life. The rise in food per capita has been accomplished by the conversion of wild habitat to managed and urbanized habitat and, quite apart from climate change, the question remains as to whether global development is sustainable unless environmental deterioration is ameliorated (Raven, 1988).

The changes in ecosystem diversity caused by resource use, pollution, and deforestation thus could well mute the importance of any changes in biological diversity due solely to climate change. Therefore, any decisions with respect to climate change strategies to protect ecosystems and biological diversity must necessarily consider current activities as they relate to these objectives.

The Safe Minimum Standard provides guidance in the development of such strategies. With respect to biological impacts of climate change, it suggests the importance of protecting diverse types of habitat. The strategies might also include mechanisms by which rich nations would compensate the poor for protecting habitat that otherwise would be lost in the development process. Widespread habitat protection can be thought of as insurance that maintains future options by reducing decisions that may prove to be irreversible. Included in such insurance strategies are, for example, expanded seed banks and gene pool protection in zoos and preserves. For the United States, the implications of the Safe Minimum Standard would mean the abandonment of crisis management on a species-by-species basis in favor of the pursuit of overall habitat protection. For example, in the northwestern United States mature forests—as a habitat—are partially protected because of the crucial role they are thought to play in providing habitat for the spotted owl. But if the spotted owl should be found flexible enough to survive without mature old-growth forests, the Safe Minimum Standard would begin with the assumption that mature forests should be protected for possible but uncertain values in addition to those of the spotted owl—unless the costs of lost timber harvests are deemed unacceptably large. Wide-scale habitat protection reduces the risks associated with single preserves of unique habitat. Should the climate suitable for North American ducks, for example, move northward but the boundaries of bird refuges remain stationary, the future of the ducks would be jeopardized. Expanded habitat protection can improve the chances for survival not only of the North American duck but of many other species as well. Similar arguments can be made for protection of the habitat of pandas and other species.

The Safe Minimum Standard strategy means the protection of some species not widely thought, according to current standards of knowledge and valuation, to be worth protecting. For example, the federal endangered species list contains only a few insects: "It is easier to argue for funding to save a charismatic species than to save the abstractions we call ecosystems. It is no accident that there are only 13 insects among the 967 taxa listed in 1988 as threatened or endangered by the U.S. Fish and Wildlife Service" (Michael Scott, Fish and Wildlife Service, U.S. Department of the Interior, as quoted by Booth, 1988). This phenomenon—"that only a small minority of people possess much concern or empathy for the plight of endangered invertebrates"—has been referred to as "vertebrate chauvinism" (Kellert, 1986, p. 59). Norton (1986) also details what he feels are the dangers of such "undervaluing of species," arguing that human welfare may be more dependent on robust ecosystems than we know. He then concludes that the dangers of undervaluing species strengthens the case for use of the Safe Minimum Standard as a criterion for issues involving endangered species.

SUMMARY AND CONCLUSIONS

If a doubling of atmospheric CO_2 brings about the climate changes currently predicted by global climate models, there would be significant consequences for natural ecological systems. The essential nature of ecological systems is to display a range of the possible responses to a given stress; the responses are largely a function of the magnitude, spatial extent, and duration of the stress. Because ecosystems such as forests are multilevel or hierarchical systems with different processes controlling the system response at different time scales, it is difficult to predict the response of the system to a novel stress such as climate change.

There is uncertainty not only about the consequences of climate change for ecological systems and biological diversity but also about the social values that should be attached to the consequences. We have argued that because of these uncertainties and because of the intergenerational scale of the consequences of climate change, conventional benefit–cost analysis will not provide sufficient criteria for policies to avert climate change or to adapt ecosystems to it. The economic way of thinking is nevertheless valid, and, in particular, the Safe Minimum Standard provides a useful approach to decisions. It avoids a case-by-case incremental analysis and proceeds with the assumption that all species and increased biological diversity are prima facie valuable. The policy focus then becomes developing cost-acceptable, cost-effective ways of protecting species and diversity.

REFERENCES

Arrow, Kenneth J., and Anthony C. Fisher. 1974. "Environmental Preservation, Uncertainty, and Irreversibility," *Quarterly Journal of Economics* vol. 55, pp. 313–319.

Baes, C. F., H. E. Goeller, J. S. Olson, and R. M. Rotty. 1977. "Carbon Dioxide and Climate, the Uncontrolled Experiment," *American Scientist* vol. 65, pp. 310–320.

Bazzaz, F. A., and R. W. Carlson. 1984. "The Response of Plants to Elevated CO_2. I. Competition Among An Assemblage of Annuals at Different Levels of Soil Moisture," *Oecologia* vol. 62, pp. 196–198.

Bishop, Richard. 1978. "Endangered Species and Uncertainty: The Economics of a Safe Minimum Standard," *American Journal of Agricultural Economics* vol. 60, pp. 10–18.

Booth, William. 1988. "Reintroducing a Political Animal," *Science* vol. 242, pp. 156–158.

Boulding, Kenneth. 1969. "Economics as a Moral Science," *American Economic Review* vol. 59, no. 1, pp. 1–9.

Bromley, Daniel W., ed. 1986. *Natural Resource Economics: Policy Problems and Contemporary Analysis* (Boston, Kluwer Nijhoff).

Brown, Lester R., et al. 1987. *State of the World* (Washington, D.C.: Worldwatch Institute).

Davis, M. B. 1982. "Quaternary History and the Stability of Forest Communities," chap. 10 in D. C. West, H. H. Shugart,

D. B. Botkin, eds., *Succession: Concepts and Application* (New York, Springer-Verlag).

Delcourt, H. R., P. A. Delcourt, and T. Webb III. 1983. "Dynamic Plant Ecology: The Spectrum of Vegetation Change in Space and Time," *Quaternary Science Review* vol. 1, pp. 153–175.

Dickinson, R. E. 1986. "How Will Climate Change?" Chap. 5 in B. Bolin, B. R. Döös, J. Jaeger, and R. A. Warrick, eds. 1986. *The Greenhouse Effect, Climatic Change, and Ecosystems*. SCOPE 29 (Chichester, U.K., Wiley).

Emanuel, W. R., H. H. Shugart, and M. P. Stevenson. 1985. "Climatic Change and the Broad-Scale Distribution of Terrestrial Ecosystem Complexes," *Climatic Change* vol. 7, pp. 29–43.

Goldemberg, J., T. Johansson, A. Reddy, and R. Williams. 1987. *Energy for a Sustainable World*. (Washington, D.C., World Resources Institute).

Holdridge, L. R. 1947. "Determination of World Plant Formations from Simple Climatic Data," *Science* vol. 105, pp. 367–368.

Kellert, Stephen R. 1986. "Social and Perceptual Factors in the Preservation of Animal Species," pp. 50–76 in Bryan G. Norton, ed., *The Preservation of Species: The Value of Biological Diversity* (Princeton, N.J., Princeton University Press).

Kerr, Richard A. 1988. "Report Urges Greenhouse Action Now," *Science* vol. 241 (July 1), pp. 23–24.

Krahl, Lane, and Margaret E. Cook. 1988. "Population and Environment Experts Explore Linkages," pp. 1–8 in *The Conservation Foundation Letter* no. 3 (Washington, D.C., Conservation Foundation).

Manabe, S., and R. J. Stouffer. 1980. "Sensitivity of a Global Climate Model to an Increase of CO_2 Concentration in the Atmosphere." *Journal of Geophysical Research* vol. 85, pp. 5529–5554.

Norton, Bryan G. 1986. "On the Inherent Danger of Undervaluing Species," pp. 110–137 in Bryan G. Norton, ed., *The Preservation of Species: The Value of Biological Diversity* (Princeton, N.J., Princeton University Press).

Page, Talbot. 1977. "Equitable Use of the Resource Base," reprint 144 (Washington, D.C., Resources for the Future).

———. 1978. "A Generic View of Toxic Chemicals and Similar Risks." *Ecology Law Quarterly* vol. 7, no. 2, pp. 207–244.

Randall, Alan. 1986a. "Human Preferences, Economics and the Preservation of Species," pp. 79–109 in Bryan G. Norton, ed., *The Preservation of Species: The Value of Biological Diversity* (Princeton, N.J., Princeton University Press).

———. 1986b. "Valuation in a Policy Context," pp. 163–200 in Daniel W. Bromley, ed., *Natural Resource Economics: Policy Problems and Contemporary Analysis* (Boston, Kluwer Nijhoff).

Raven, Peter H. 1988. "State of the World: 2000—What We Should Do to Attest It." Paper presented at the Wingspread Conference, Racine, Wisconsin, April 15.

Schulze, William D., and Allen V. Kneese. 1981. "Risk in Benefit–Cost Analysis," reprint 188 (Washington, D.C., Resources for the Future).

Sober, Elliott. 1986. "Philosophical Problems for Environmentalism," pp. 173–194 in Bryan G. Norton, ed., *The Preservation of Species: The Value of Biological Diversity* (Princeton, N.J., Princeton University Press).

Strain, B. R. 1985. "Physiological and Ecological Controls on Carbon Sequestering in Terrestrial Ecosystems," *Biogeochemistry* vol. 1, pp. 219–232.

White, Gilbert. 1980. "Environment," *Science* vol. 209, pp. 183–190.

Woodward, F. I. 1987. "Stomatal Numbers Are Sensitive to Increases in CO_2 From Pre-industrial Levels," *Nature* vol. 327, pp. 617–620.

10

Water Resources and Climate Change

Kenneth D. Frederick
Peter H. Gleick

A greenhouse warming is certain to have major impacts on both water availability and water quality. Temperature, precipitation patterns, evapotranspiration rates, the timing and magnitude of runoff, and the frequency and intensity of storms will be affected by increasing concentrations of carbon dioxide and other trace gases. A rise in sea levels associated with a global warming could threaten the freshwater supplies of coastal communities. And changes in temperature and rainfall levels will affect the demand for water, especially for irrigation.

There is now a consensus among climate modelers that a global warming will increase global precipitation. An increase of 1.5 to 5.5 degrees Celsius (°C) in average global temperatures (the range expected to result from an equivalent doubling of atmospheric carbon dioxide) is expected to accelerate the hydrologic cycle and thereby increase global precipitation by an average of 7 to 15 percent (Schlesinger, 1988).

RISKS TO WATER SUPPLIES ASSOCIATED WITH CLIMATE CHANGE

Water problems, whether they involve floods or drought, are regional in nature. Consequently, changes in regional precipitation patterns are far more important than changes in the global average. Unfortunately, regional changes are difficult to estimate because global climate models have only coarse spatial resolution and relatively simplistic hydrologic parameterizations (World Meteorological Organization, 1988; chapter 2 in this volume). General circulation models do indicate some regional changes, such as increases in precipitation toward the poles, but the range of likely changes in average annual precipitation for any given region could be on the order of plus or minus 20 percent.

Changes in runoff are the direct result of changes in evaporation (which is strongly influenced by temperature) and in precipitation. Relatively small changes in temperature and precipitation can have large effects on runoff in arid and semiarid regions. A simulation study of the Great Basin in the United States, for example, showed that with a 2°C increase in temperature, a decrease in precipitation of only 10 percent led to a 17 to 28 percent decrease in runoff, while an increase in precipitation of 10 percent caused runoff to increase by 20 to 35 percent (Flaschka, Stockton, and Boggess, 1987). In another simulation Nemec and Schaake (1982) showed for the Pease River Basin in Texas that a temperature increase of 3°C and a 10 percent decrease in precipitation led to a decrease in runoff of 50 percent, while an increase in precipitation of 10 percent led to a 35 percent increase in runoff. Similarly, Gleick (1986, 1987) modeled California's Sacramento Basin and found that a temperature increase of 2°C and a 10 percent decrease in precipitation caused an 18 percent decrease in annual average runoff. These results are summarized in table 10-1.

Although detailed regional hydrologic predictions are not yet available from general circulation models, some additional generalizations can be made. There is evidence that runoff may increase in high latitudes due to increased high-latitude winter precipitation but may decrease in summer in mid and low latitudes as temperatures and evapotranspiration increase (Shiklamanov, 1987; World Meteorological Organization, 1988). Perhaps more importantly, regions where runoff comes largely from snowmelt are likely to see a distinct shift in the relative amounts of snow and rain and in the timing of snowmelt due to higher temperatures. The resulting changes in runoff patterns could greatly alter the likelihood of flooding and the availability of water during peak-demand periods such as the irrigation season. This type of seasonal effect was identified by Gleick (1988a) for basins in the western United States and has now been identified in China (Shi Yafeng, Director, Nanjung Institute of Geography, Nanjung, China, personal

Table 10-1. Effect on Runoff of Climate Changes in Arid and Semiarid Basins

Basin/study	Temperature change (degrees C)	Precipitation change (percent)	Annual runoff change (percent)
Great Basin	+2	−10	−17 to −28
		+10	+20 to +35
Pease River	+3	−10	−50
		+10	+35
Sacramento	+2	−10	−18
		+10	+12

Note: For details on methods and model calculations refer to individual authors: Flaschka, Stockton, and Boggess (1987) for the Great Basin; Nemec and Schaake (1982) for the Pease River; and Gleick (1986, 1987) for the Sacramento.

communication with Gleick, 1987), Canada (Cohen, 1986), and Europe (Bultot et al., 1988).

Gleick (1988a) showed, for example, that in the Sacramento Basin in California, a 2°C increase in temperature (with no change in precipitation) increased winter runoff nearly 10 percent and decreased summer runoff by 22 percent. An increase of 4°C increased winter runoff by 34 percent and decreased summer runoff by 62 percent. Such changes will have extremely important ramifications for both winter flooding and summer irrigation. Further studies are needed to explore the impacts of these runoff changes on water resource management.

Threats to water resources will also come from rising sea levels due to thermal expansion of the oceans and increased melting of glaciers and land ice. A sea-level rise of 0.2 to 0.8 meters (m) over the next fifty years is likely (Hoffman, 1984; Vellinga, 1987). As sea level rises, salinity levels in bays, inland waterways, and coastal aquifers will rise, threatening the water supplies of such coastal cities as San Francisco and Miami. In some regions rapid local subsidence from groundwater pumping or other causes will worsen the problem. In San Francisco Bay, an apparent sea-level rise of 0.17 m over the last century is the result of an actual sea-level rise of approximately 0.12 m and ground subsidence of about 0.05 m, and the problem of subsidence is considerably worse in other areas such as New Orleans. Yet, as chapter 4 (in this volume) by Gjerrit Hekstra suggests, the water supply problems associated with rising sea levels pale in comparison to the societal and environmental costs that would result from flooding in densely populated, low-lying coastal areas such as Bangladesh and Indonesia.

In summary, water resource planners and managers must expect global climate changes to cause significant changes in the overall hydrologic cycle: shifts in annual precipitation and runoff, altered seasonality of hydrologic processes, changes in the extremes of storms, and, perhaps, new threats to water quality. In the absence of clear regional predictions, it is only prudent to examine ways to increase our capacity to respond to such changes by both increasing the resilience and flexibility of our existing physical and management structures and decreasing our dependence on levels of water use developed under supply conditions that may prove to have been unusually favorable and to be impractical to sustain.

COPING WITH HYDROLOGIC CHANGE

The existing infrastructure for storing and transporting water was planned and designed and continues to be operated under the assumption that future levels and patterns of precipitation, runoff, and evaporation will be similar to those experienced in the past. The prospect of future climate change, however, raises questions as to whether this is still a useful assumption to guide the design and operation of water supply systems.

Infrastructure

The long construction periods and lifetimes of many water projects suggest that early planning for the hydrologic impact of climate change can prevent project failure when conditions change and thereby can save expensive redesign or reconstruction. But, as noted above, there is great uncertainty as to how the hydrology of specific regions will be affected by a greenhouse warming. In many instances, even the direction of change is unknown; the normal uncertainties about climatic extremes are compounded under conditions of changing climate. Yet knowing if streamflow or other hydrologic factors will exceed or fail to reach certain values is critical to the economic and social evaluation of a project and to estimates of the consequences of the failure of that project. Both building for changes that never materialize and failing to build facilities to deal with changes that do occur are costly prospects. As the uncertainties increase, the stakes associated with this dilemma grow.

Most water resource systems are designed to be "robust" and "resilient." Robustness refers to the ability to respond to the range of uncertainties associated with the variability of future events. Resiliency refers to the ability both to operate under a range of conditions and to return quickly to designed performance levels in case of failure (Matalas and Fiering, 1977).

The vulnerability of water supplies to future climate change will in many ways be similar to existing vulnerabilities to climatic variability and extremes. Climatic extremes are most often manifested as unusual flows or altered timing of flows, such as seasonal floods, persistent droughts, major storms, and the early or delayed onset of a range of meteorological events. Although reservoirs and aqueducts are designed to be robust and resilient, they still fail occasionally due to extreme events. Such failures are undesired, but they are recognized to be a result of normal uncertainties about hydrologic events. The planning goal should be to build projects to protect against extreme events up to the point where an additional dollar spent on protection no longer reduces the expected cost of failure by at least a dollar.

Critics have blamed the U.S. Army Corps of Engineers for overdesigning their water projects. In its defense, three Corps officials contend, "The design criteria merely reflected the conventional engineering approach of focusing on an extreme event or condition, e.g., as the critical drought period of record or the probable maximum flood or hurricane" (Hanchey, Schilling, and Stakhiv, no date, p. 6). This approach makes allowance for climatic as well as other uncertainties. As a result of these practices, the officials note, water resources agencies have effectively been designing large hydraulic structures and water control systems almost as if anticipating climate change.

This optimistic view of the ability of the existing infrastructure to handle the impacts of climate change has support among researchers. Matalas and Fiering (1977, p. 109) concluded: "It is comforting to recognize that most large systems contain so much buffering and redundancy that resilient design can be operationally achieved without recourse to sophisticated or elaborate projections about the climate." Nevertheless, the magnitude of future climate change could well produce alterations in the hydrologic cycle that overwhelm the safety margins of existing systems.

In addition, a strategy of building such redundancy into future water supply and control projects will become increasingly costly. Rising water project costs would appear to be inevitable for three reasons. First, the best reservoir and dam sites available for increasing a stream's safe yield (the quantity of water that can be supplied with some high degree of probability) are developed first. Thus, subsequent additions to a system's capacity require utilizing less desirable sites with higher costs per unit of storage. This trend is evident in table 10-2, which suggests that reservoir capacity per unit volume of dam declined thirty-five-fold in a comparison of large reservoir projects constructed in the United States before 1930 with those constructed in the 1960s.

Table 10-2. Decadal Reservoir Storage Capacity per Unit Volume of Dam for the 100 Largest Dams in the United States

Period	Index[a]
1920s or earlier	100.0
1930s	20.2
1940s	5.0
1950s	4.3
1960s	2.8

Source: Modified from Mermel (1958) as reported in U.S. Geological Survey (1984, p. 33).

[a]Index sets the average for the period of the 1920s and earlier at 100.0.

Second, Langbein (1959, p. 4) demonstrated that "water control by storage follows a law of diminishing returns. Each successive increment of control (safe yield) requires a larger amount of reservoir storage space than the preceding increment." Moreover, at some point the net yield actually declines as evaporation losses from the additional storage more than offset any gains from the additional reservoir capacity. Within the contiguous United States, Hardison (1972) found that safe yield peaks when storage is in the range of 160 to 460 percent of a region's average annual renewable supply. But the gains in yield from additional storage may be very low even when capacity is well below 160 percent of the average annual flow.

Third, the opportunity costs of storing and diverting water (for example, the recreational and wildlife benefits forgone by the loss of instream flows) rise as the amount of water left in the nation's streams declines. The traditional practice of ignoring the impacts of water projects on instream water uses ended with the National Environmental Policy Act of 1969, which requires all federal agencies to give full consideration to the environmental implications of their actions. The higher value placed on instream water use has been evident over the last two decades in the strength and success of the resistance to new storage and diversion projects.

While the costs of adding to supplies and increasing control over water through structural measures are rising, the benefits of additional infrastructure may also increase if future climate change produces climatic extremes far greater than those evident in the historical hydrologic record. Expectations of climate change are likely to affect the design of new facilities; they could also increase the demand for new water projects. Thus, reliance on structural approaches in the face of climate change presents the grim prospect of paying much higher costs to meet growing water demands and to limit risks of system failure. Alternative approaches, especially improving the management of the existing infrastructure, will become increasingly attractive and essential for providing secure water supplies in the future.

Managing the Infrastructure for More Water and Greater Security

Even in the absence of any major climate change caused by the greenhouse effect, the high financial and environmental costs of developing new supplies are forcing the United States to look beyond the traditional structural solutions for meeting growing water demands. Uncertainties as to how an anthropogenically induced climate change will alter the supply and demand for water strengthen the case for seeking ways to improve the management of existing water supply facilities.

Reservoirs have traditionally been managed independently, each under its own set of objectives and operating rules. Yet operating water supply projects as integrated systems may provide a low-cost way to increase safe yield in some areas without large additional investments in infrastructure. Joint management of the facilities of independent water suppliers may make it possible to improve significantly the supply capabilities of each individual supplier within a region with relatively little new investment. Such benefits have been achieved in the Washington, D.C., metropolitan area where joint operation of the systems of the three principal water supply agencies increased drought-condition water yields by more than 30 percent while saving between $200 million and $1 billion compared to the proposed structural solutions. Two studies by Sheer (1986) suggest that major benefits from improved management are also possible in regions with very different water supply characteristics. According to Sheer, conjunctive use of surface and groundwater supplies in Houston, Texas, could increase system yields by 20 percent; joint management of supplies in the North Platte River Basin could reduce water shortages by about 30 percent.

Despite the potential benefits, water supply projects are rarely operated as integrated systems. Institutional factors such as separate ownership of facilities, multistate jurisdictions, and water laws are often obstacles to the consideration of integrated solutions to water problems. Yet the success of integrated management in Washington, D.C., and a growing recognition that structural solutions alone will no longer suffice should help overcome such obstacles. The nation's principal federal agencies in charge of constructing water projects are at least talking about the need to give greater emphasis to improved water management. The key conclusion of *Assessment '87* by the U.S. Department of the Interior's Bureau of Reclamation (1987, p. i) is that "the Bureau's mission must change from one based on federally supported construction to one based on effective and environmentally sensitive resource management."

A recent paper by senior officials of the U.S. Army Institute for Water Resources stresses the need to place greater emphasis on management measures to complement their infrastructure in meeting the problems caused by extreme events and the uncertainties stemming from the prospect of climatic change. These officials concluded:

> Society cannot afford to build "fail-safe" projects anymore. It isn't economically efficient, either for water supply, flood control, or hurricane storm protection on coastal shorelines. Notwithstanding the Corps' past history of a reliability-based analysis for extreme events, their design concept has increasingly become "safe-fail". It is a strategy of designing flood protection for less than the extreme event of record. A structural measure (levee, channel) is complemented by emergency management plans that mitigate the damages of designed . . . failure of structural solutions. In other words local flood protection could conceivably be designed for a 20 to 50 year level of protection, as long as there is a complementary flood warning and evacuation plan. These complementary mechanisms represent an important contributing factor to the robustness and resiliency of water resources management systems, and are specially relevant to adaptation under climate uncertainty. . . . While hydrologists may not be able to explicitly factor in the anticipated variability of climate change, especially if they rely on empirical rather than causal models, water resources planners and engineers can implement a wide range of management measures that increase the robustness and resiliency of solutions that can operate under an anticipated range of climate uncertainty. . . . (Hanchey, Schilling, and Stakhiv, no date, p. 7)

Institutional factors are hardly the only obstacles to developing more efficient water-management practices. Climate-induced changes in the time of flows complicate the development of proper operating rules for reservoirs. In winter in California, for example, some reservoirs are operated to maintain sizable flood control storage. Yet summer irrigation demands require that these same reservoirs be as full as possible to provide a reliable supply for the dry season. This produces a conflict between the risks of flooding if the reservoirs are filled too early and the risks of shortage if the major runoff season ends before the reservoirs can be filled (Gleick, 1988b). Extreme weather conditions have led to adjustments in the operating rules for some of California's reservoirs in the past. For example, the severe 1976–77 drought prompted changes designed to make the system more resilient to droughts. These rules, however, led to a higher risk of flooding during high-runoff years in the early 1980s, and the operating rules were changed again. An increase in the variability of precipitation or a change in its seasonality would exacerbate the problem of deciding when to begin filling a reservoir and how large a flood-storage volume to maintain.

Demand Management

Water planners have traditionally taken a supply-side approach to meeting the growing demands for municipal, industrial, and irrigation uses. Water use was projected to grow virtually in step with population and income levels, and the resulting projections were treated as virtual requirements to be met regardless of cost. Prices were believed to have no significant effect on water use. Thus, until recently, concerns about the adequacy of future water supplies invariably led to plans for new dams, reservoirs, pumps, and canals. But, as noted above, exclusive reliance on new infrastructure to meet water demands that are unconstrained by price has become infeasible and environmentally unacceptable. The previous section described how improved management might increase the reliable yield of a given supply system. This section considers the use of demand management to balance water use with available supplies.

The need for demand-side management is gaining acceptance, but there are differing views as to how it should be implemented and the role it should play. The basic approaches to curbing water use involve, first, regulations (that is, command-and-control measures) and, second, prices and markets. The water planner who expects supplies to meet unconstrained demand levels under all but the most severe drought conditions would only reluctantly resort to demand-side management. Regulatory measures are likely to be viewed as a necessary, but last-ditch, tool for bringing water use into line with supplies during periods of extreme drought. Temporary drought-emergency measures would probably start with appeals for voluntary conservation and, if necessary, include restrictions on nonessential uses such as lawn watering and car washing. If it is necessary to curb the long-term growth of demand, this might be done by imposing water conservation standards for appliances and showerheads under the regulatory approach.

Economists, on the other hand, are likely to view demand-side management as a means of both allocating increasingly scarce water resources among competing uses (even during normal hydrologic periods) and eliminating construction of water projects for which the social costs exceed the expected benefits of the additional water. Marginal-cost pricing of publicly supplied water and market transfers of water rights are the means proposed for achieving these objectives. Prices, whether set by monopolistic suppliers to reflect the marginal costs of supply or determined in competitive markets, provide the incentives to encourage conservation, develop new supplies, and direct scarce water resources to their most valuable uses.

Despite the professed advantages of the economists' view, skepticism, and even outright rejection, of their prescriptions for using prices and markets to allocate water resources are widespread. Underlying these doubts are beliefs that water use is not responsive to price and that the characteristics of the resource make it difficult, if not impossible, to establish competitive markets in water. The rarity of well-functioning water markets reinforces these doubts. Yet, in view of the potential efficiency and flexibility of market solutions to economic problems, it is useful to consider whether the absence of water markets reflects fundamental factors rooted in the nature of the resource or institutional constraints to their operation.

Efficient markets must satisfy at least two conditions: there must be well-defined, transferable property rights in the commodity being traded, and the buyer and the seller must bear all the costs of the transfer. These conditions are not easily fulfilled for at least four reasons. First, it is difficult to establish clear and transferable property rights to groundwater and surface water supplies that vary over time and flow from one property to another. Second, even when streamflow rights are established, transfers of water to different uses and locations are likely to affect people other than the buyer and seller because of resulting changes in the quantity, quality, location, or timing of return flows. Third, some of the products of water resources, such as the amenities associated with a free-flowing stream, are not marketed. If left to private incentives and markets alone, there will be underinvestment in these outputs. And finally, since it is generally impractical to have more than one supplier, the price of water to domestic and commercial users is not set by market processes but by utility managers or regulatory agencies.

In spite of the importance of natural market failures, water marketing does occur and, with appropriate policies and institutional arrangements, could play a more important role in adjusting to changes in the supply and demand for water. Indeed, the flexibility of markets to adjust to unanticipated changes in supply and demand would be particularly helpful in coping with the hydrologic impacts of climate change. Currently, the relevant policies and institutions often discourage rather than encourage water marketing. The riparian doctrine, which grants the owner of land adjacent to a water body the right to use water as long as the use does not unduly inconvenience other riparian owners, provided the earliest basis for state laws controlling surface waters. By attaching the water right to the land, the riparian principle of water rights prohibits the establishment of water markets. The seventeen western states have abandoned, or at least modified, this doctrine in favor of the doctrine of prior appropriation, which establishes the basic principle of "first in time, first in right."

This step was important for creating definable property rights in water. But other features of western water law confuse the character of these appropriative rights and lessen their transferability (Frederick, 1984). For example, these rights are commonly encumbered by provisions restricting use to beneficial purposes or to certain locations. Sale of a water right has in a few instances even been interpreted under the beneficial-use provision as grounds for forfeiting the right. Thus, owners may be reluctant even to consider selling water rights for fear that beneficial-use provisions of state water codes would be used to invalidate their rights.

All western states permit water transfers under certain conditions so long as the rights of third parties are not impaired. Usually, however, third-party impacts are important, and satisfying the third-party provisions of state water laws involves costly and lengthy judicial or administrative procedures that are incompatible with the operation of efficient markets. Creating more efficient ways of handling third-party interests is perhaps the most difficult and important consideration for the emergence of more effective markets in water.

Traditional water-pricing practices and opportunities for introducing demand-side management in the urban and agricultural sectors are discussed below.

Urban Water Use

Planning and pricing within the urban water industry have emphasized increasing supplies at the lowest possible cost in order to meet projected levels of water use. Traditionally projected to increase in line with population and economic growth, water use has been treated as a necessity that is unresponsive to price. Water rates have been set just high enough to cover average costs plus some capital for future expansion and a fair return to investors.

This average-cost pricing policy has important implications in an industry where the costs of developing new supplies are rising sharply. Efficiency requires the expansion of supplies up to the point where the marginal cost of supplying another unit is just equal to the additional value of that last unit. Consumers will continue to use more of a good until the value they receive from another unit declines to their cost (that is, the price) of purchasing another unit. Thus, the price reflects the marginal value of water. In the water industry, where the costs of developing new supplies are relatively high and the price is based on average costs, the marginal costs of new supplies exceed the marginal benefits. Consequently, average-cost pricing practices have resulted in the construction of water projects for which the costs have far exceeded the benefits.

Marginal-cost pricing is generally rejected by the water industry on two grounds—a belief that use is not responsive to price and an expectation that the implied price levels would be unacceptably high. Adoption of marginal-cost pricing could indeed result in some sizable price increases just because the costs of developing new supplies are generally so much higher than the average costs, which are heavily weighted by the low costs of the earliest developed supplies. But contrary to the assumption traditionally employed by the industry, there is considerable evidence, much of it reviewed by Gibbons (1986), that such price increases would have significant impacts on water use.

The potential role of marginal-cost pricing in balancing water supplies with projected demands has been examined for southern California. Applying marginal-cost pricing to the additional supplies that the Metropolitan Water District hopes to get from an expansion of the State Water Project would increase consumer prices by about 75 percent. (The Metropolitan Water District is a wholesale supplier of water to six southern California counties; the State Water Project is a long-term effort to construct a major aquaduct and storage system to move large quantities of water from the well-watered northern parts of the state to the arid central and southern parts.) Conservative estimates of the price elasticity of demand (the percentage change in the quantity demanded resulting from a percentage change in price) suggest that a price increase of this magnitude would reduce water use by enough to offset the entire projected growth in demand for the region from 1985 to 2000 (Wahl and Davis, 1986, p. 117).

Agricultural Water Use

Four of every ten liters of fresh water withdrawn from the nation's streams and groundwater aquifers in 1985 were used for irrigation. In the arid and semiarid seventeen western states, irrigation accounts for nearly eight of every ten liters withdrawn and nine of every ten liters of consumptive use (the portion of the water withdrawn that is not available for reuse due to contamination, evaporation, or other losses) (Solley, Merk, and Pierce, 1987). Their prominence in current water use makes irrigators particularly susceptible to the impacts of climate change on future supplies.

Easy access to very low-cost water was an important factor in the expansion of irrigated agriculture during the first two-thirds of this century, and the increasing costs and scarcity of water have contributed to the recent decline in total irrigated acreage in the West. Indeed, past changes in regional irrigation developments can be explained in large part by changes in the availability and cost of water (Frederick, 1988). There are several reasons for expecting irrigation water use to be particularly sensitive to the cost of water. First, water costs are likely to comprise a higher proportion of total costs for irrigators than for most water users. Second, in comparison to most domestic, commercial, and industrial uses, irrigation water often goes for relatively low-value uses that are likely to be abandoned or modified if water prices rise significantly. And third, irrigators commonly have a variety of alternatives for reducing water use in response to rising costs. For instance, water losses might be reduced by lining canals or adopting more efficient irrigation systems. Crop yields might be maintained or increased even though less water is applied by scheduling irrigations so that water is applied only when it can be used most effectively by the plant. And switching to crops or varieties that use less water may become profitable as water becomes more expensive.

Although irrigators have many options for conserving water short of abandoning irrigation, they often have little incentive to do so. Rights to use surface water have been granted free by the western states to anyone making "beneficial" use of the resource. The principle of "first in time, first in right" gives the earliest appropriators the right to withdraw their full share even during periods of shortage. As the largest and commonly the earliest water users, irrigators own high-priority rights to much of the West's water for as long as they continue to use it. Since they pay nothing for the water itself, their water costs reflect distribution and perhaps storage costs. But even these costs are subsidized for federally supplied irrigation water. For many irrigators water costs do not depend on the amount actually used. Consequently, irrigators with senior water rights are likely to be isolated from the increasing scarcity and value of the resource unless they are able to sell some or all of their water, for even if the water is free, an opportunity to sell all or part of the right creates an opportunity cost reflecting the potential sale price. Opportunity costs introduce incentives to conserve. Increased opportunities to transfer water within and among regions and better incentives to conserve would facilitate adjustments to climate change, especially in regions where water will become scarcer.

Technological Change

The development and adoption of technology are strongly influenced by the prices of the principal factors of production (for example, land, labor, and water) as producers seek to substitute relatively abundant and inexpensive inputs for scarce and costly ones (Hayami and Ruttan, 1985). At least until the last decade or so, water was treated as the most abundant and inexpensive resource; saving water was not an important consideration in either the development or the adoption of production techniques and management practices. This attitude still prevails where costs do not depend significantly on the quantities of water used and the holders of the water rights lack opportunities to sell conserved water.

As the largest user, the agricultural sector is the prime candidate for saving water in the nation's arid and semiarid areas. Moreover, as noted above, irrigators have a number of alternatives for curtailing use when there is sufficient incentive to do so, and research is expanding their options.

As recently as two decades ago, researchers as well as irrigators saw little reason to conserve water. Consequently, emphasis on water conservation within the agricultural research community is relatively new. The increased concern over water use within the agricultural research community is illustrated by changes in the objectives of plant breeding in arid areas. In the 1960s, researchers sought mainly to develop seed varieties that would maximize yields under conditions of plentiful water. More recently, considerable effort has been devoted to increasing yields under conditions of less or uncertain water supplies. Investments in research are helping to develop inputs, such as more water-efficient crop varieties, and management techniques that will better enable farmers to adapt to more expensive and scarcer water. Other developments include chemicals to reduce transpiration without yield loss and irrigation and agronomic management practices that increase yields per unit of water under a range of circumstances. Many such developments have already been adopted where water is expensive.

Plant varieties that help individual irrigators adapt to scarcer and more expensive water will also help entire water resource regions adjust to the hydrologic impacts of climate change. More generally, technologies that increase yields per unit of applied water or reduce crop losses under low-water conditions tend to produce social benefits commensurate with the private returns (assuming that the water is not used to produce surplus crops). For some water-conserving technologies, however, there may be large differences between their net private and social benefits.

Because one farmer's runoff or losses may be another farmer's supplies, more efficient use of water on individual farms may not lead to equivalent increases in a system's or a region's efficiency. From the perspective of an entire water resource region, the important inefficiencies result in losses that are not recovered for later use through return flows or groundwater recharge. By contrast, any water that is not effectively utilized by crops is lost as far as the individual irrigator is concerned. As water becomes more expensive to the irrigator, the incentive to reduce on-farm water losses rises, even if that water is fully recovered by others through return flows. In this situation, the private benefits of applying water-conserving technologies exceed the social benefits.

On the other hand, environmental impacts tend to narrow, and may even reverse, the gap between the net private benefits and the net social benefits of technological alternatives. Concentrations of salts, sediment, and agricultural chemicals are higher in return flows than they are in water applied to a field. Thus, technologies that improve on-farm efficiency are likely also to improve downstream water quality. But the environmental benefits and costs of the alternatives are borne by downstream users rather than by the irrigator.

Even though computations of farm-level water savings generally overstate the net additions to a region's water supply, the value of the savings to the irrigator determines whether an individual will invest in water conservation. Based on recent sale prices, water sold for municipal uses in the West might command more than $80 per 1,000 cubic meters (about $100 per acre-foot). Permanent rights to an annual supply commonly sell for well over $800 per 1,000 cubic meters ($1,000 per acre-foot) (Western Network, 1988).

Irrigators can generally conserve water at costs well below these levels. Significant water savings at comparatively low cost are often possible on farms using gravity irrigation, the system employed on more than 60 percent of the West's irrigated lands. With an unimproved gravity system only about half of the water delivered to the farm—and perhaps much less if the terrain is uneven or the soil is sandy—actually reaches the plant. Under conditions characteristic of many irrigated areas, investments in laser-leveling of fields to reduce runoff or tailwater recycling to capture runoff for subsequent reuse might improve on-farm efficiency by 20 to 30 percent at costs of $40 per 1,000 cubic meters or less ($50 per acre-foot) (Frederick, with Hanson, 1982). Irrigation efficiencies of about 80 percent can be achieved even on sandy, hilly soils with sprinkler irrigation systems. While the combined investment and energy costs are considerably higher for sprinkler systems, they often produce significant labor savings as well as water savings. Drip or trickle irrigation systems virtually eliminate water loss, but their investment costs are likely to be at least three times those of most sprinkler systems.

Other than the minimum requirements for survival, water use can be reduced through water-conserving investments and technologies. The potential impacts on overall water supplies within the nation's arid and semiarid regions of water conservation by nonagricultural users are likely to pale in comparison to those of water conservation by the agricultural sector. Yet conservation by municipal and industrial users, which are discussed below, may be essential for meeting long-term demands within the nation's urban and industrialized areas. Even the mining sector, which accounts for less than 1 percent of the nation's total withdrawals, can be important for balancing local supply and demand.

Municipal water serves a variety of uses, including drinking, cooking, bathing, lawn watering, swimming, and firefighting. Some of these uses (especially those outdoors) are responsive to water prices, but very little of the response is likely to stem from technological change. At the prevailing levels of use, consumers' adjustments to price increases are likely to represent making do with less water rather than adopting new water-conserving technologies. Water-efficient dishwashers, showers, and toilets can reduce domestic water use, but water prices alone are unlikely to offer consumers sufficient incentive to demand them or producers incentive to develop and produce them. A strong conservation ethic or, more likely, use of building and manufacturing regulations may be required to ensure the development and widespread adoption of such technologies.

Where climate change results in increased water scarcity, technological changes that reduce desalting and recycling costs can help municipalities adapt. Desalinization of seawater is currently too expensive for all but the most water-scarce regions and the most high-value uses, and in the absence of a major decline in energy costs or an unexpected improvement in desalinization technologies, it is likely to remain so for the foreseeable future. On the other hand, some regions have large quantities of brackish water with salt concentrations that are well below those found in the oceans but are still too high for municipal use. As desalting and recycling technologies improve, they are likely to become increasingly important means of supplementing municipal water supplies. Where health or psychological factors make this water unsuitable for drinking, it might replace other supplies used for irrigating golf courses, public roadways, and crops. Farmers, however, are likely to find such water too expensive if it is priced to recover the costs of desalting and recycling. More promising technological developments for using low-quality water in agriculture may lie in developing crop varieties with a high tolerance for salt and improving water management practices for irrigating with such water.

Self-supplied industrial water accounts for about 45 percent of the nation's total withdrawals of fresh water but less than 10 percent of the consumptive use. Cooling water for thermoelectric plants accounts for about 85 percent of all industrial withdrawals, but only about 2 percent of that water is used consumptively. And only about 4 percent of the water withdrawn for other industrial water uses (such as manufacturing, refining, and food processing) is consumed (Solley, Merk, and Pierce, 1987; Solley, Chase, and Mann, 1983). Consequently, even though evaporative cooling technologies and the recycling of industrial process water greatly reduce withdrawals for industrial use and are relatively common, these changes have little impact on overall consumptive water use. The principal impacts of water-using technologies within the industrial sector are on the quality rather than the quantity of water. Moreover, water costs have little influence over the choice of industrial technologies because they are rarely more than a small part of a firm's total costs. Adoption of water-conserving technologies by industry is apt to be motivated more by environmental regulations or institutional obstacles to securing adequate water supplies than by strictly economic considerations.

VULNERABILITY TO CLIMATE CHANGE

The preceding discussion suggests a wide range of options for dealing with climate-induced changes in water supplies. With perfect foresight, wise planning, and flexible institutions, many regions should be able to adapt to likely changes in supplies with little economic or social pain. Improved management of existing supplies, marginal-cost pricing and water marketing to encourage conservation and reallocation of supplies in response to changing conditions, and development of water-saving technologies can provide considerable flexibility for adaptation. Unfortunately, foresight is invariably flawed, and water planners and institutions have not yet fully absorbed the significance of, nor adjusted appropriately to, the changes in the supply and demand for water that have occurred over the last several decades. These shortcomings suggest that it may be useful to consider the capacity of various water resource regions to adapt to large changes in water supplies in the absence of new infrastructure or institutional changes or technological developments.

Table 10-3. Indicators of Vulnerability to Climatic Conditions

River basin or region	Measure of storage[a]	Measure of demand[b]
New England	.15	.01
Mid-Atlantic	.10	.02
South Atlantic-Gulf	.16	.03
Great Lakes	.08	.02
Ohio	.12	.02
Tennessee	.23	.01
Upper Mississippi	.14	.03
Lower Mississippi	.31	.09
Souris-Red-Rainy	.93	.07
Missouri	1.12	.29
Arkansas-White-Red	.45	.17
Texas-Gulf	.61	.23
Rio Grande	1.89	.64
Upper Colorado	2.61	.33
Lower Colorado	4.22	.96
Great Basin	.35	.49
Pacific Northwest	.19	.04
California	.42	.29
Alaska	.00	<.01
Hawaii	.01	.05
Caribbean	.05	.06

Source: U.S. Water Resources Council (1978) and U.S. Geological Survey (1986).

[a]Ratio of maximum basin storage volume to total basin annual mean renewable supply (1985). Regions with values below 0.6 have small relative reservoir storage volumes. Large reservoir storage volumes provide protection from floods and act as a buffer against shortages.

[b]Ratio of basin consumptive depletions (including consumptive use, water transfers, evaporation, and groundwater overdraft) to total basin annual mean renewable supply (1985). Water is considered a decisive factor for economic development in regions with values above 0.20 (Szesztay, 1970).

Two indicators of the vulnerability to hydrologic change of the major American river basins are presented in table 10-3. The first indicator, the ratio of maximum storage volume to mean annual renewable supply within the basin, is one measure of a region's ability to withstand prolonged drought or flood. Large reservoir capacity relative to average annual streamflow provides protection against short-term droughts and flooding. Conversely, where this ratio is small (perhaps below 0.6), climate-induced reductions in water or changes in the frequency or intensity of droughts and floods may be strongly felt.

This measure alone can be misleading, however, since it neglects the demand side of the picture. Thus, the second measure in table 10-3, the ratio of demand to available supply, also provides important information. Where this ratio is high, existing supply is sensitive to growing population, growing industrial and commercial demand, and climatic fluctuations. Alaska, for instance, has essentially no storage. But since demand for water is negligible relative to supplies, Alaska is not currently susceptible to drought. On the other hand, even though the Lower Colorado Basin has by far the highest ratio of storage to average supply, it is vulnerable to the impacts of drought because consumptive use in the basin is 96 percent of average renewable supply. Szesztay (1970) suggests that water is a critical factor for development in regions where the ratio of demand to supply is 20 percent or higher. Basins where consumptive use exceeds 25 percent of average renewable supply, such as the Upper and Lower Colorado, California, Rio Grande, Great Basin, and Missouri, are particularly susceptible to further reductions in supply caused by climate change.

Neither of these measures captures more than a small part of the factors relevant for understanding a region's vulnerability to climate-induced changes in water supplies. Many water problems are local, and aggregating river basins into large water resource regions hides individual problem areas. A more fundamental shortcoming of these measures is that they do not take account of instream water uses (such as hydropower, fish and wildlife habitat, and recreation) that require large quantities of water and provide services of great importance to some areas, especially the Pacific Northwest.

Vulnerability to short- and long-term changes in water supplies also depends on how water is currently being consumed and the institutions in place to reallocate water when supplies are insufficient to meet all traditional uses. The more unused capacity or inefficiency there is in an existing system and the greater its ability to direct scarce supplies to higher value and more vulnerable users, the less susceptible to climatic conditions is the region as a whole. For instance, in times of shortage it would be much less costly and disruptive to curtail water use for irrigating lawns, pastures, and low-value annual crops than it would be to allow tree crops to perish, factories to close, or municipal water service to be rationed. With irrigation accounting for 90 percent of the consumptive use of water and with hay and pasture accounting for nearly one-third of irrigated acreage, it appears that there is at least the potential to have major reductions in water use at modest social costs in much of the West.

SUMMARY AND CONCLUSIONS

The quantity and quality of freshwater resources is of critical importance for both natural ecosystems and human development. Pressures on water resources are already rising due to population growth and industrial and commercial development. Any new threats to existing water resources must, therefore, be viewed with concern, especially where water is already limited in availability.

Water supply systems are naturally vulnerable to climatic variability and climate change. As a result, they are designed with a certain degree of resilience and robustness that permits them to operate over a wide range of conditions. When failures do occur, they are caused by unusually severe weather events or, occasionally, by the failure of system components to perform as designed.

Under some circumstances, climate change caused by the greenhouse effect will increase the risk of failure or the cost of maintaining or building facilities. In other cases, climate change will make changes in existing facilities imperative or will require new facilities. Lacking credible predictions of regional hydrologic changes, information about vulnerabilities and sensitivities can permit changes in design or operation in order to reduce the risk of failure.

The appropriate responses to permanent climate change are often different from those to short-term climatic variability. Unless long-term climate change can be distinguished from temporary variability, incorrect design and operation decisions could be made. A short drought, for instance, is often mitigated by surplus water from outside the basin or by temporary changes in water use. A long-term shortage, however, requires permanent changes in demand or changes in the physical infrastructure itself.

Of the various potential hydrologic consequences of climate change, one of the most difficult for most regions to handle would be increased variability, which translates into more frequent extremes—higher peak flood flows, more persistent and severe droughts, greater uncertainty about the timing of the rainy season. Few such changes would be beneficial. Changes in variability often prompt the most costly societal responses, such as flood control, large water transfers, and storage.

The existing infrastructure is generally both robust and resilient within the context of past hydrologic events, but its capacity for dealing with events that might result from a greenhouse warming is questionable. Uncertainties as to how climate change might alter regional supplies, as well as the high costs of building additional redundancy into water systems, increase the importance of nonstructural alternatives for coping with future scarcity and variability. These alternatives include improving the management of the existing infrastructure, encouraging conservation, marketing water to facilitate the movement of scarce supplies in response to changing supply and demand conditions, and developing and promoting technologies that increase the options for and decrease the costs of responding to changes in water supplies. Such measures not only may be needed but also may be more cost-effective approaches for coping with future hydrologic changes than would be developing major new facilities.

Institutional factors are often obstacles to realizing the potential of these alternatives. Multiple ownership and competing jurisdictions may stymie opportunities for integrated management of a region's water systems. Pricing policies of the urban water industry, subsidies for and restrictions on the use of federally supplied irrigation water, and state laws and administrative practices all discourage conservation. Federal and state laws and policies that obscure the nature of and reduce the transferability of water rights make water marketing unnecessarily difficult and expensive. Thus, incentives to develop and adopt water-conserving technologies on a timely basis are lacking.

There are no simple measures of a region's vulnerability to hydrologic events. The ratios of both storage and consumptive use to average renewable supply can be used to indicate whether or not a region is likely to be vulnerable to drought and flooding. However, these ratios alone do not incorporate such important considerations as instream water uses, the management of the water system, and the institutions that allocate and provide use incentives for scarce supplies. The flexibility to improve efficiency and allocate supplies to highly valued and vulnerable uses in response to adverse changes in water supply and demand will greatly reduce a region's vulnerability to the impacts of climate change on its hydrology.

REFERENCES

Bultot, F., A. Coppens, G. L. Dupriez, D. Gellens, and F. Meulenberghs. 1988. "Repercussions of a CO_2-Doubling on the Water Cycle and on the Water Balance: A Case Study for Belgium," *Journal of Hydrology*, in press.

Cohen, S. J. 1986. "Impacts of CO_2-Induced Climatic Change on Water Resources in the Great Lakes Basin," *Climatic Change* vol. 8, pp. 135–154.

Flaschka, I. M., C. W. Stockton, and W. R. Boggess. 1987. "Climatic Variation and Surface Water Resources in the Great Basin Region," *Water Resources Bulletin* vol. 23, pp. 47–57.

Frederick, Kenneth D. 1984. "Current Water Issues," *Journal of Soil and Water Conservation* vol. 39, no. 2.

———. 1988. "The Future of Irrigated Agriculture," *Forum for Applied Research and Public Policy* vol. 3, no. 2.

———, with James C. Hanson. 1982. *Water for Western Agriculture* (Washington, D.C., Resources for the Future).

Gibbons, Diana. 1986. *The Economic Value of Water* (Washington, D.C., Resources for the Future).

Gleick, P. H. 1986. "Methods for Evaluating the Regional Hydrologic Impacts of Global Climatic Changes," *Journal of Hydrology* vol. 88, pp. 97–116.

———. 1987. "Regional Hydrologic Consequences of Increases in Atmospheric CO_2 and Other Trace Gases," *Climatic Change* vol. 10, pp. 137–161.

———. 1988a. "Regional Hydrologic Impacts of Global Climatic Changes," in E. E. Whitehead, C. F. Hutchinson, B. N. Timmermann, and R. G. Varady, eds., *Arid Lands: Today and Tomorrow*. Proceedings of an International Research and Development Conference, Tucson, Arizona, October 20–25, 1985 (Boulder, Colo., Westview/Belhaven Press).

———. 1988b. "Climatic Changes and California: Past, Present, and Future Vulnerabilities" in M. Glantz, ed., *Societal Responses to Regional Climate Change: Forecasting by Analogy* (Boulder, Colo., Westview Press).

Hanchey, J. R., K. E. Schilling, and E. Z. Stakhiv. No date. "Water Resources Planning Under Climate Uncertainty" (unpublished paper, U.S. Army Institute for Water Resources, Fort Belvoir, Va., 22060).

Hardison, C. H. 1972. "Potential United States Water-Supply Development," in *Irrigation and Drainage Division Journal* vol. 98, pp. 479–492. Cited in U.S. Geological Survey (1984, p. 31).

Hayami, Yujiro, and Vernon W. Ruttan. 1985. *Agricultural Development: An International Perspective* (revised and expanded, Baltimore, Johns Hopkins University Press).

Hoffman, J. S. 1984. "Estimates of Future Sea Level Rise," in M. C. Barth and J. G. Titus, eds., *Greenhouse Effect and Sea-Level Rise* (New York, Van Nostrand Reinhold).

Langbein, W. B. 1959. "Water Yield and Reservoir Storage in the United States," U.S. Geological Survey Circular 409. Cited in U.S. Geological Survey (1984, p. 30).

Matalas, N. C., and M. B. Fiering. 1977. "Water-Resource Systems Planning," in National Research Council, *Climate, Climatic Change, and Water Supply* (Washington, D.C., National Academy of Sciences).

Nemec, J., and J. Schaake. 1982. "Sensitivity of Water Resource Systems to Climatic Variation," *Hydrological Sciences 17*, pp. 327–343.

Schlesinger, M. E. 1988. "Model Projections of the Climatic Changes Induced by Increased Atmospheric CO_2," in *Proceedings of the Symposium on Climate and Geosciences: A Challenge for Science and Society in the 21st Century* (Holland, Reidel Publishing, in press).

Sheer, Daniel P. 1986. "Managing Water Supplies to Increase Water Availability," in U.S. Geological Survey, *National Water Summary 1985—Hydrologic Events and Surface-Water Resources*, Water Supply Paper 2300 (Washington, D.C., Government Printing Office), pp. 101–112.

Shiklamanov, I. 1987. "Changes in Runoff in Soviet Rivers Due to Climate Change." Paper presented at the International Union of Geodesy and Geophysics Symposium, Vancouver, British Columbia, August 1987.

Solley, Wayne B., Charles F. Merk, and Robert R. Pierce. 1988. *Estimated Use of Water in the United States in 1985*, U.S. Geological Survey Circular 1004 (Washington, D.C., Government Printing Office).

Solley, Wayne B., Edith B. Chase, and William B. Mann IV. 1983. *Estimated Use of Water in the United States in 1980*, U.S. Geological Survey Circular 1001 (Washington, D.C., Government Printing Office).

Szesztay, K. 1970. *The Hydrosphere and the Human Environment: Results of Research on Representative and Experimental Basins*, Proceedings, Wellington Symposium, UNESCO Studies and Reports in Hydrology, no. 12.

U.S. Department of the Interior, Bureau of Reclamation. 1987. *Assessment '87 . . . A New Direction for the Bureau of Reclamation* (Washington, D.C.).

U.S. Geological Survey. 1984. *National Water Summary 1983—Hydrologic Events and Issues*, Water Supply Paper 2250 (Washington, D.C., Government Printing Office).

———. 1986. *National Water Summary 1985—Hydrologic Events and Issues*, Water Supply Paper 2300 (Washington, D.C., Government Printing Office).

U.S. Water Resources Council. 1978. *The Nation's Water Resources: 1975–2000* vol. 2, Second National Water Assessment (Washington, D.C., Government Printing Office).

Vellinga, P. 1987. "Sea Level Rise, Consequences and Policies," background paper no. 9 for the Workshop on *Developing Policies for Responding to Future Climatic Change*, Villach, Austria, September 28–October 2, 1987.

Wahl, Richard W., and Robert K. Davis. 1986. "Satisfying Southern California's Thirst for Water: Efficient Alternatives," in Kenneth D. Frederick, ed., *Scarce Water and Institutional Change* (Washington, D.C., Resources for the Future), pp. 102–133.

Western Network. 1988. *Water Market Update* (Santa Fe, N.M.) published monthly since 1987.

World Meteorological Organization. 1988. *Water Resources and Climatic Change: Sensitivity of Water-Resource Systems to Climate Change and Variability*, World Climate Applications Program-4 (Geneva, Switzerland).

Part III
Perspectives

11

Potential Strategies for Adapting to Greenhouse Warming: Perspectives from the Developing World

N. S. Jodha

Global warming due to increasing concentrations of greenhouse gases poses threats to human society by changing the living and working environment to which society has adapted over many generations. The rates of warming and other climatic changes predicted by some may be too rapid to permit the luxuries of purposeful experimentation and adaptation by trial and error that past generations have enjoyed. Experimentation can be made more efficient, and its output more rapid, through the use of modern scientific tools such as simulation modeling. However, uncertainty as to how climate change will play out, not only at the global but also at the regional and local levels, makes it difficult even to conceive of appropriate adaptation options (Jaeger, 1988). Agriculture, industry, and other activities adapt to concrete situations and real changes. Since neither the regional climates of the future nor their impacts are yet known, designing adaptation strategies is, obviously, not going to be easy.

Fully recognizing this limitation, I discuss in this chapter the question of how developing countries, particularly those in the semiarid tropical zone, may adapt to greenhouse warming. Large parts of the semiarid tropics are already exposed to climatic hazards which changes due to greenhouse warming may accentuate.

In many ways developing countries differ greatly among themselves. Nonetheless, there are certain ways in which they can be contrasted as a group with the developed countries. The relatively large size of the agricultural sector is one common feature that makes the developing countries particularly sensitive to climatic changes. I will attempt here to illuminate their specific agricultural vulnerabilities.

Because of agriculture's great importance in the developing countries and its significant direct dependence on natural factors such as precipitation levels, solar radiation, and the like—as against man-made factors like technology and market—the vulnerability of the developing countries to climatic changes is probably greater than that of the developed countries. The uncertainty of future climatic change and a general indifference to issues of greenhouse warming on the part of the governments of developing countries require an approach that may not be appropriate in developed countries. Specifically, I speculate below on possible adaptation strategies that would build on developing countries' past experiences of adjusting to risk and crises both at the farm level and at the policy and program levels. Adjustments to risks and crises caused by factors other than climate change are also illustrated in order to indicate how those experiences provide analogs useful in assessing options for adapting to any future impacts of climatic change.

Chapter 2 in this volume, by Schneider and Rosenberg, gives some details about the global and regional climate changes that may occur as a result of greenhouse warming. For the semiarid tropical regions (35°S to 35°N) regional temperature increases on the order of 0.5 degrees Celsius (°C) to 4.0°C are predicted by various general circulation models (GCMs). Increases in precipitation are predicted for most, but not all, of the semiarid tropics. This increase is expected to take the form of convective rainfall, which could imply a higher intensity, but not necessarily an increased frequency, of precipitation. Thus the seasonally dry tropics would have potentially high rainfall, high runoff, and high evaporation, without necessarily having a lengthened or improved growing season. Because these areas are now characterized by a large variability in seasonal and annual rainfall, any trend toward decreased precipitation could have a significant negative impact.

Despite the many studies of greenhouse warming and its relation to future climate change, a number of uncertainties and crucial gaps in our knowledge persist, even in the output of GCMs. These relate to the rate and timing of climatic changes and to details about specific climatic variables such as temperature, precipitation, and sunshine, and the regional dimensions of climatic changes (see chapter 2 in this volume). These uncertainties tend to reduce the utility of GCM results for the purposes of policymaking and program development aimed at either limiting the growth of greenhouse gases or developing strategies to adapt to potential climate changes. However, despite these uncertainties, there is no reason to assume that the tropical regions in which most of the developing countries are found will be exempt from change.

THE DEVELOPING WORLD AND ITS HIGH-RISK SECTOR

Agriculture, defined broadly to include all land-based and biomass-producing activities such as annual and perennial cropping, animal husbandry, and forestry, will likely be the sector most immediately affected by changing climate in the developing world. It is the sector whose resource base, production environment, and infrastructural support system (irrigation, for example) are most directly and closely linked with climate. What will happen to agriculture depends, of course, on the nature of the climate change and on the sensitivity of different components of agriculture to these changes. The changes could be positive or negative, but some degree of adaptation will be necessary in either case.

In the event of positive impacts, adaptations will be necessary to harness the new opportunities. In the case of negative impacts, adaptation will be required to manage new risks and losses. Tables 11-1, 11-2, and 11-3 show how agriculture, particularly in the developing countries, is related to climate and how it is, or could become, sensitive to climatic change. Table 11-1 identifies the components of the production environment and resource base most likely to be influenced by changes in climatic variables. Kates (1985) has called these influences "first-order impacts." The linkages between components of the production environment and resource base that may be affected by climate change and the components of the farming system are identified in table 11-2. These linkages identify "second-order impacts." Table 11-3 refers to "third-order impacts" and identifies the consequences of changes in farming systems for macrolevel activity. The magnitudes of these impacts in a given nation will depend, on the one hand, on the size of the agricultural sector, the extent of its dependence on natural and man-made factors, and its place in the total economy and, on the other, on the nature and magnitude of climatic change.

Agriculture in developing countries differs from that in the developed world in several ways, ranging from its capital base to its crop productivity (Hrabovszky, 1985), as shown in table 11-4. First, agriculture accounts for a much larger share of national income, and especially of employment, in developing countries (World Bank, 1987). Hence, any climatic change that adversely affects the natural resource base and production environment of agriculture would have more severe economic and social consequences in developing countries than in developed countries. In other words, the relative size of the unit exposed to climatic change is much larger in developing countries than in developed countries. Secondly, agricultural production in developed countries depends directly on such variables as precipitation and runoff, which strongly affect agricultural performance. Thirdly, the developing countries do not possess technical and social support systems to cushion shocks

Table 11-1. Likely First-Order Impacts of Changes in Climate and Other Variables on the Agricultural Resource Base and Production Environment

Major components of physical resource base and production environment	Variables							
	Temperature	Solar radiation	Precipitation	Humidity	Evapotranspiration	Soil moisture	Runoff	Sea level
1. Moisture regime	*	*	*		*		*	*
2. Length of growing season			*	*		*	*	*
3. Microclimatic stress	*	*	*	*	*	*	*	
4. Frequency of weather aberrations	*		*					
5. Seasonality	*		*			*		
6. Disease-pest complex	*	*	*	*		*		
7. Biomass productivity potential	*	*	*	*	*	*		*
8. Photosynthetic response	*	*		*	*	*		
9. Plant-input interactions[a]		*			*	*		
10. Soil chemistry	*	*	*		*	*	*	*
11. Soil erosion hazard			*		*		*	*

[a]Refers to plant-nutrient interactions and water–vegetation–soil interactions.

Table 11-2. Likely Second-Order Impacts on Components and Features of Farming Systems if the Resource Base and Production Environment Are Affected by Climate Change

	Components of resource base and production environment[a]				
Components/features of farming systems	Moisture regime, growing season, microclimatic stress (1-3)	Seasonality, instability of weather (4-5)	Disease-pest complex (6)	Biomass potential photosynthesis, plant-input interaction[b] (7-9)	Soil chemistry, erosion hazard (10-11)
1. Adapted cultivars	*	*	*	*	*
2. Agricultural enterprise combinations/linkages[c]	*	*	*	*	*
3. Moisture management systems/structures	*	*			*
4. Agricultural activity calendar	*	*	*		
5. Agronomy, input use practices	*	*	*		*
6. Risk adjustment mechanisms	*	*		*	*
7. Production flows	*	*	*		
8. Yields and returns	*	*	*	*	

[a]Components are taken from table 11-1; figures in parentheses refer to row numbers in table 11-1.

[b]Practices like mixed farming, intercropping, and sequential and relay cropping designed to diversify agriculture, to guard against risks, and to take advantage of complementarities and linkages between different activities.

[c]See note a to table 11-1.

generated by climatic or other natural crises. They lack the strong infrastructure, including irrigation systems and other technologies, to guard against natural hazards. Further, they lack widespread institutional arrangements, including responsive markets and insurance systems, to spread agricultural risks to other sectors. The comparative picture of agriculture in the semiarid tropics of Australia (Hobbs et al., 1987) and India (Jodha et al., 1987)—presented in the International Institute for Applied Systems Analysis/United Nations Environmental Programme volume on the impact of climatic variations on agriculture (Parry et al., 1987)—brings out these differences very sharply.

A complicating factor in the agriculture of many developing countries is the coexistence of a vast subsistence sector and a small, high-potential, surplus-producing modern sector. The subsistence sector usually encompasses the bulk of the marginal areas and marginal, or poor, people who are highly vulnerable to recurrent scarcities and risks due to climate variability, such as accompany droughts and floods. Their situation could conceivably worsen following climatic changes brought about by greenhouse warming. On the positive side, through a variety of practices aimed at diversification and flexibility, the farmer in this sector has developed adjustment mechanisms to cope with climatic risks (Jodha and Mascarenhas, 1985; Jodha et al., 1987). These adjustment strategies are sources of resilience in the system and might offer some clues as to how adaptations can be made in the face of climatic change.

The high-potential modern sector of agriculture in developing countries is often a source of surplus production of food and industrial raw material for domestic use and export. This sector is often well supported by infrastructure

Table 11-3. Likely Third-Order Impacts on Macrolevel Agricultural Systems and Activities if Components of Farming Systems Are Affected by Climate Change

	Potentially affected components of farming systems[a]				
Agricultural support systems and activities at macrolevel	Adapted cultivars, enterprise combinations (1-2)	Moisture management, water security (3)	Activity calendar, input use practices (4-5)	Risk adjustment mechanisms (6)	Production flows, yields and returns (7-8)
1. Irrigation systems strategies		*		*	
2. Relief strategies	*	*		*	
3. Agricultural infrastructure	*				
4. Input supply systems	*	*	*		*
5. Research & development strategies	*	*	*		*
6. Marketing, trade, food systems	*				*
7. Intersectoral linkages	*				*
8. Agricultural planning strategies	*	*		*	*
9. Employment, income distribution	*	*		*	*

[a]Components are taken from table 11-2 in this chapter; figures in parentheses refer to row numbers in table 11-2.

Table 11-4. Some Distinguishing Features of Agricultural Sectors in Developing Countries and Their Implications for Greenhouse Warming Impacts and Adaptations to Them

Feature	Implications
Agriculture as the dominant sector of developing countries	Large size of exposure unit, greater second- and third-order (farm-level and national-economy-level) impacts
Greater direct dependence on natural factors	Higher extent of first-order impacts caused by vulnerability to changes in physical–biological production environment
Inadequate institutional support system (banking, insurance, and marketing)	Limited scope for spreading risk through backward and forward linkages of agriculture; low risk-absorption capacity of agriculture
Weak physical infrastructure	Limited protection through irrigation systems and other physical facilities
Significant extent of climatically unstable subsistence sector with little institutional and technological protection against risks and high dependence on farmers' adjustment measures	Possible accentuation of climatic risks in already vulnerable areas; vulnerability of coastal areas
	Traditional risk adjustments as potential source of adaptation options against future climate change; tendency to add inferior options during strains
Indifference of public policies to greenhouse warming issues	Forced adaptations rather than planned or anticipatory adaptations to be the main approach against greenhouse warming impacts

and other services. Policies favoring this sector might be part of a strategy to ensure surplus production that partly protects the vulnerable areas and groups and partly supports rapidly growing urban populations (Mellor, 1976). This modern sector is less vulnerable, in general, to the stresses imposed by climatic variability, but adverse effects of changes induced by greenhouse warming on this sector would have very serious third-order impacts.

The vulnerability of agriculture in developing countries to possible future climate change is heightened by the indifference of public agencies in such countries to the issue of greenhouse warming—an indifference reflected in public policies and programs for agriculture. Developing and developed countries are not geared to provide protection against long-term changes in climate. In the developing countries the time horizon for public intervention is normally very short. Research and development efforts for agriculture tend to focus on current problems and short-term crises. Even those few efforts at long-term planning ignore such issues as climate change. The current climate and environment are taken as givens by most planners. If the past is a guide, at least a decade or more will elapse before the developing countries follow whatever public policies and programs the developed countries put forth to protect agriculture against possible impacts of greenhouse warming.

ABATEMENT AND ADAPTATION STRATEGIES

To abate further warming, the emissions of greenhouse gases into the atmosphere must be reduced. Adapting to the impacts of whatever warming is unavoidable involves other strategies. However, despite persistent alarm raised by concerned nongovernmental organizations about rising atmospheric pollution in developing countries (Center for Science and Environment, 1985), neither abatement nor adaptation is seriously debated or studied at this time. The indifference of the developing countries to these issues is largely due to a lack of awareness and to the preoccupation of governments with short-term issues, as noted above.

Prevention of Further Warming

There are divisive issues that stem from the prospect of greenhouse warming. The perception that some nations will be gainers and some losers, the inequalities in national contributions of greenhouse gases, as well as inequalities of national capacity to combat the problem (including differential rates for discounting the future), and uncertainties of scenarios all constrain the development of global consensus and action against greenhouse warming (Glantz, 1988). These same issues condition the approaches of developing countries to this problem.

Greenhouse warming is a problem of the global commons. If practiced only by certain countries, restraints on greenhouse gas emissions would be of very limited value to them. This would be especially true were such restraints to be followed only by the developing countries, since their contribution to the accumulation of greenhouse gases is, because of their limited industrialization, as yet quite small. Further, since global climate models indicate relatively small negative impacts of greenhouse warming on the tropics, the indifference of developing countries to the greenhouse warming issue continues, regardless of the possibly severe exposure of their substantial marginal areas to higher risks (Chen and Parry, 1987).

Other factors rooted in the present economic status and the poverty of developing countries help explain their lack of interest and involvement in greenhouse warming issues.

The much-discussed "logic of poverty" leads to a preference for low-cost options, such as using the fossil fuels that feed the greenhouse effect. Policymakers in these countries, even when aware of the long-term adverse consequences of certain options, may feel compelled by current economic pressures to ignore them. One case in point is the readiness of many developing countries to expose themselves to environmentally hazardous industries transferred from developed countries.

Second, because of policymakers' immediate concern with current problems of poverty and underdevelopment, greenhouse warming falls far down their lists of immediate priorities, and far beyond the time horizon governing policies and programs. When long-term planning is attempted at all, it may deal only with the future magnitude of current problems—in other words, how to increase food or water supplies for future populations. Potential new environmental problems and constraints are rarely recognized. Moreover, and not illogically, when climate is considered at all, there is more concern for short-term fluctuations in rainfall than there is for changes in precipitation patterns that may manifest themselves 50 or 100 years into the future. Incidentally, these generalizations apply equally to the approaches of international agencies and developed countries to aiding development efforts. Even a cursory look at such documents as the agriculture sectoral papers and world development reports by the World Bank, periodic world food surveys by the Food and Agriculture Organization, and a series of progress reports issued by international research centers sponsored by the Consultative Group on International Agricultural Research corroborate this finding.

Third, the recent history of economic development demonstrates a positive correlation between growth of per capita energy use and the rise of living standards in developing countries. In most cases, fossil fuels provide the cheapest sources of energy. In the absence of cheaper alternative energy sources, restriction of the use of fossil fuels in developing countries implies a willingness to accept current low standards of living (Reddy, 1985). Understandably, such an approach would be unacceptable to developing countries, especially to the poorest ones, still striving to reach respectable living standards. In view of these factors, the developing countries appear to have little to contribute to the prevention of greenhouse warming.

Yet there are some indications that indifference to greenhouse warming could be overcome. Despite poverty and the pressures of current needs, some developing countries have responded positively to long-term environmental problems once the consequences of their policies are clearly perceived—usually as the result of a particularly dramatic sign of environmental degradation—and when low-cost alternatives are presented. Several major and minor development projects in India have been halted following protests by environmental protection groups (Bandyopadhyay and Shiva, 1988). Perhaps this example offers hope that developing countries can be sensitized to greenhouse warming issues as well. Before this can happen, however, certain steps will be necessary: (1) more concrete and spatially disaggregated information on the climatic impacts of greenhouse warming must be made available; (2) a "lobby" to constantly convey information to various levels on the issues involved in greenhouse warming must be created; (3) simple, low-cost substitutes for fossil fuels must be provided, including energy conservation measures; (4) the developed countries must demonstrate their own serious intent to cope with greenhouse warming and their willingness to share technological and other means of abatement with the developing countries.

Perhaps the developing countries can contribute most to the abatement of further greenhouse warming through the conservation of tropical forests, since their destruction increases carbon dioxide in the atmosphere (Maini, 1988) and reduces the strength of the biospheric sink for atmospheric carbon dioxide. In developing countries the need for preventing forest loss is better understood than the need for restricting fossil fuel use. However, setting policies to deal with both problems is impeded by the same factors. Rising population pressures and poverty leave little room for environmentally preferable policies for dealing with either problem. To be politically feasible in a developing country, measures to counter greenhouse warming must be perceived as contributing to the general improvement in current economic well-being.

Adaptation Options

We have seen that substantial initiatives on the part of developing countries to abate greenhouse warming are not likely. We turn, then, to the matter of how developing countries may adapt to the impacts of whatever greenhouse warming is inevitable. Agriculture, as the major sector exposed to climatic change in developing countries, is the focus for this discussion. Adaptation options in developing countries are discussed in the framework of the following components: (1) uncertainties with respect to the impacts of greenhouse warming; (2) indifference of public policies and programs to greenhouse warming issues; and (3) potential agroclimatic changes following greenhouse warming. Information on these components is arrayed in table 11-5.

Uncertainties of Impact Scenarios

Crucial gaps in our knowledge of the impacts of greenhouse warming at regional and subregional levels persist, and the impacts to which society will have to respond are perceived only dimly. The absence of a concrete context makes it difficult to conceive of realistic responses. Hence the following discussion can, at best, suggest broad

Table 11-5. Framework for Assessing or Developing Potential Options for Adaptation to the Climatic Impacts of Greenhouse Warming (GHW)

Framework components	Implications/adaptation options
Uncertainty associated with climate impact scenarios	Vagueness of contexts to induce responses or adaptations
	Possibility of indicating broad approaches rather than suggesting an inventory of concrete measures
Indifference of public policies in developing countries to GHW issues	Search at farm level or public program level for potential adaptations which by nature would be forced adaptations rather than anticipatory adaptations
	Traditional farming systems and prior governmental crisis management strategies as potential sources of adaptations
	Sensitization of public policies to GHW issues as a form of potential adaptation
Nature of GHW-induced climatic changes:	
Changes likely to be gradual, not abrupt	Opportunity for gradual evolution of responses, using past experiences and modern science
Change, though slow and small in tropics, may accentuate short-term climatic risks in vulnerable and marginal areas	Lessons from existing farm strategies and public interventions for crisis management, unless changes are too rapid and unmanageable
GHW impacts likely to leave unchanged or to increase heterogeneity of land resources	Heterogeneities may facilitate identification of adaptation options by analog approach
	Need for preparation of inventory of existing farm practices from different regions to serve as a pool of potential adaptation measures
Possible similarities between GHW impacts and the impacts of other changes in production environment and resource base causing risk/crises	Analogous situations as a source of potential adaptations
	Need for preparation of inventory of past changes and responses thereto

approaches rather than offer an inventory of relatively concrete suggestions on potential adaptation strategies. Such broad categories of adaptation alternatives can be identified by searching for analogous situations in which factors other than greenhouse warming have adversely affected the agricultural resource base and production environment. Societal responses to these impacts can then be identified.

Indifference of Developing Countries to Greenhouse Warming

For reasons stated above, developing countries are likely to remain indifferent to greenhouse warming issues for a fairly long time to come, a situation that has a number of implications with respect to potential adaptations to the impacts of greenhouse warming. First, rather than waiting for and concentrating on public interventions and preparedness, the search for potential measures to manage the impacts of climatic change should be directed to the existing farming systems. These already represent adaptations to heterogeneous production environments in different regions of developing countries. An inventory of options may be developed to suit the different potential agroclimatic situations caused by climatic change. The main limitation of this approach is that such an inventory would be confined to the stock of traditional knowledge, knowledge that can complement, but is no substitute for, research on new technological and institutional options.

Second, to the extent that changes due to greenhouse warming present themselves as crises similar to those already faced by developing countries, experience in crisis management may provide ways to evolve potential strategies to combat the impacts of climatic change. Such an approach can help improve other adaptation options in the form of agricultural support systems, research, and technology.

The third and most important implication is that sensitization of developing countries to greenhouse warming issues should, in itself, be treated as an important form of potential adaptation. Sensitization is a prerequisite to any further adaptation measures, including those indicated above. Means and mechanisms to sensitize developing countries to greenhouse warming issues are described later.

Nature of Climatic Changes Due to Greenhouse Warming

Though the impacts of greenhouse warming cannot yet be modeled dependably, one may speculate on the broad nature of the potential changes. Some reasonable assumptions (Jaeger, 1988) are: (1) climatic changes will be gradual and not abrupt; (2) the impacts in the tropics, though relatively smaller and even positive in some cases, may aggravate short-term climatic crises in marginal areas; (3) climatic changes may not generally affect the existing spatial heterogeneity of land resources but may in some cases add to it; and (4) in terms of consequences, and speed in some cases, changes induced by greenhouse warming may be similar to other changes that have adversely affected the resource base and agricultural production environment in the developing world. Understanding the attributes of these past changes and their implications can facilitate systematic thinking about adaptation strategies in the face of future climatic changes.

Gradual Changes We assume for the purposes of this chapter that climate changes due to greenhouse warming, while rapid when compared to those of the past, will be gradual rather than abrupt. By implication the changes would permit enough time to search for and to evolve effective responses. The time permitted by a slow pace of climatic change can be used to improve predictions about the impacts of such changes at the regional and local level.

However, we need a better understanding of how much lead time is necessary in order to evolve adaptation options. Table 11-6 presents some examples of the approximate time required for different technological and institutional adjustments to changed circumstances. Most of the examples are based on Indian experience, and the first- and second-order impacts relate mainly to agriculture. The number of examples could be greatly increased by taking cases from other countries and from other sectors, such as science, medicine, and socioeconomics, in order to give a better idea of the time required for societal responses to change.

Table 11-6 shows that several crucial changes or adjustments have, in recent times, occurred within a period of ten to twenty years. This is a time frame much shorter than that envisioned for significant climate change due to greenhouse warming and raises the hope that, given appropriate resources and well-directed efforts, adaptations to climate change can be evolved in temporal step with the changes.

The greatest uncertainties remaining stem from a lack of precise information on the scale, speed, and nature of changes induced by greenhouse warming. Perhaps within another ten or fifteen years science will provide information to fill the gaps with respect to both climate change and adaptations to its impacts. However, even slow and small changes in climate could accentuate the frequent short-term climatic crises that occur in, say, drought prone, marginal areas in the semiarid tropics. This effect may not allow enough time to evolve adaptation strategies for such areas. In these cases, greater reliance at both the farm and the government levels must be placed on past and present experience with adjusting to crises.

Table 11-6. Approximate Time in Years for Societal Response to First-, Second-, and Third-Order Impacts on Agriculture

First-order impacts Changes related to the physical resource base and production environment (see table 11-1)	*Second-order impacts* Changes related to farming systems (see table 11-2)	*Third-order impacts* Changes in macrolevel activities related to agriculture (see table 11-3)
Development of new cultivars, and hybrids with required characteristics (7–12 years)	Crop rotation cycle (5–10 years)	Generation to implementations of new ideas (5–10 years)
Building resistance or tolerance in crops to insects, pests, disease, drought, soil salinity, etc. (10–15 years)	Diffusion of new technologies (5–10 years)	Conception to implementation of new institution (5–7 years)
Diffusion of new technologies, input use, and management practices (5–10 years)	Asset depletion-replenishment cycle in drought-prone areas (7–10 years)	Generation of effective lobbies and pressure groups on emerging issues (5–10 years)
Major shifts in cropping patterns following changes in resource base (5–10 years)	Productive life of farm assets (5–15 years)	Gestation period of most agricultural projects (7–20 years)
Translation of science fictions into scientific reality (20–50 years)	Life cycle of domestic animals (7–15 years)	Impact made by major public program (7–10 years)
Travel of relevant ideas and practices from developed to developing countries (10–15 years)	Felling cycle of farm trees (10–20 years)	Aid negotiations on specific major projects (5–7 years)
	Recovery from life-cycle risks (7–10 years)[a]	Rehabilitation of people and systems uprooted by major disasters (15–20 years)
	Complete enterprise substitution in response to changed circumstances (5–7 years)	
	Recovery from major disasters (7–10 years)[b]	

Note: Based largely on the experience of postindependence changes in India in different fields. Such examples could be multiplied and presented with more precise details about time involved.

[a]Life-cycle risks due to family accidents and misfortunes such as untimely death of bread earner or sudden loss of job.

[b]These disasters, like flood, drought, civil war, and earthquakes, are not family-specific.

Impacts in the Tropics Global climate models suggest that precipitation will probably increase, but in an uneven manner, throughout the tropics. A general increase would likely increase indifference to greenhouse warming in tropical developing countries. Yet even small negative impacts in already vulnerable areas there could become a major cause of concern. In the absence of planning for response to the impacts of climate change, the only alternative is to search for relevant adaptation measures in the body of experience gained by developing countries in dealing with droughts and floods.

Spatial Heterogeneities Interregional variability in climatic changes due to greenhouse warming could be high, perhaps as high as ± 50 percent for specific variables like precipitation and solar radiation. These spatial heterogeneities might further amplify existing regional differences due to soils, topography, and the like, as well as to the ways in which the natural resource base interacts with climate. Jodha et al. (1987) and Jodha (1986) have shown how, in India, great differences exist in such variables as length of growing season, runoff harvesting possibilities, groundwater potential, the extent of mixed crop and livestock farming, and many other factors within regions that receive broadly similar amounts of rainfall but have differences in soil, topography, and the like. In a sense, these differences reflect the structural and operational features of different farming systems that have evolved historically and that are adaptations to the heterogeneities of agricultural resource bases in different regions or subregions.

These different features of agriculture in different regions may themselves constitute a pool of measures and practices suitable to any increase in heterogeneity that might accompany climatic changes induced by greenhouse warming.

An increased heterogeneity of production environments following greenhouse warming could create several habitats where current experience and adaptation would apply. A major implication of this reasoning is the need for a focused and careful examination of different components of prevailing agricultural systems in different regions as a source of potential options for adapting to climatic changes. For instance, all else being equal, cultivars and farming practices in zones with short growing seasons could be adopted in zones where climatic change has worked to shorten the growing season. In the same way, research and technological strategies designed for one area could be applied in newly similar regions.

Comparable Risks and Crises Another aspect of climatic change due to greenhouse warming relates to the severity of risks or crises likely to be generated by these changes. Developing countries face a sufficient number of risks, and crises, even without climatic change. The causes of these crises may be different, but in terms of their consequences, and the pace of change in some cases, they may be similar to those that will be induced by climatic change. Measures adopted at the farm and public policy levels to manage these crises may offer lessons useful in evolving adaptation strategies in the face of climatic change. Rapid increases in land-use pressures, deforestation, the depletion of grazing resources, falling water tables, and recurrent drought are among current crises that illustrate the point.

Tables 11-7 through 11-10 list adaptations and responses

Table 11-7. Responses to Physical-Biological Changes (Some examples from India of adaptations to a declining agricultural resource base and deterioration of the production environment broadly comparable to the potential negative impacts of greenhouse warming)

Changes and their causes	Adaptations/responses: Farm level	Adaptations/responses: Public policy/program level
Increased soil salinity and water logging following faulty water management under new Chambal irrigation project (8–10 years)	Switched to salinity- and water-tolerant crops, syphoned off surplus water, abandoned cropping temporarily	Project on drainage improvement, research on soil amendments and salinity, and water-tolerant crops
Increased incidence of *striga* weed in sorghum crop in parts of Andhra Pradesh and Maharastra states (10–12 years)	Stopped growing sorghum, rotated sorghum with other crops	Research on *striga* problem
Rise of white grub in groundnut due to continuous planting of the crop (10–12 years)	Stopped planting groundnut for 5 to 7 years, started long-duration rotation sequence	Research on the white grub problem
Rapid increase in rodent populations in parts of arid region of Rajasthan (8–10 years)	Crop shift involving replacement of short-stem crops by long-stem crops (grain legumes by pearl millet)	Research on rodent control and public campaign for rodent control
Collapse of pearl millet-based green revolution due to emergence of downy mildews and ergot diseases (3–5 years)	Switch back from hybrid pearl millet to traditional varieties	Focused research for development of resistant varieties/hybrids

Table 11-8. Responses to Precipitation Changes (Some examples from India of adaptations to a declining agricultural resource base and deterioration of the production environment broadly comparable to the potential negative impacts of greenhouse warming)

	Adaptations/responses	
Changes and their causes	Farm level	Public policy/program level
Vulnerability of some areas to high risk of frequent droughts (historical record)	Adjustments through diversification based on agricultural activities with noncovariate production flows, flexibility of production and consumption activities involving recycling, asset depletion–replenishment cycles, seasonal migration	Permanent protection through provision of irrigation, links with surplus-producing areas for relief supplies, public works research and technology to strengthen capacity to withstand droughts, emphasis on activities suited to drought-prone areas
Reduced efficacy of farmers' adjustment strategies against drought due to institutional and technological changes (20–30 years)	Increased dependence on public relief, search for security through emphasis on nonagricultural activities, access to groundwater	Increased relief measures with emphasis on their production components
Crises caused by recurrence of severe droughts for 2 to 3 years in parts of Maharastra and Rajasthan (5–7 years)	Collapse of traditional measures like on-farm storage, product recycling, asset depletion–replenishment cycling, permanent loss of draft animals, peasants leased out or dispossessed from land to become agricultural laborers	Initiation of permanent rural employment program on year-round basis and not only during droughts, emphasis on high-potential areas to produce surplus, maintenance of high buffer stock to feed deficit areas
Short-term shifts in rainfall pattern perceived by farmers:		
Recurrence of midseason dryspell in groundnut-growing areas of Maharastra and Andhra Pradesh (8–10 years)	Shift from varieties with spreading root systems to those with deep roots	Research to offer multiple options
Recurrence of late-season drought in several parts of arid Rajasthan (8–10 years)	Increased emphasis on crops/varieties with high salvage potential (availability of fodder in case crop fails to mature)	Research for drought-resistant cultivars and agro-forestry systems
Recognition of slack potential of rainy season in high rainfall, deep vertisols areas of central India (20–30 years)	Introduction of short-duration crops of soya bean and mung bean in areas traditionally kept fallow during rainy season and cropped only after rainy season	Research and extension work and creation of marketing facilities to popularize new cropping systems

to changes in the agricultural resource base and deterioration in the production environment that have been made at the farm level and at the public policy and program level. These changes could be broadly comparable to some of the potentially negative impacts of greenhouse warming. Table 11-7 deals with physical and biological changes and responses thereto; table 11-8 with responses to precipitation-related changes; table 11-9 with responses to changes in the resource base; and table 11-10 with responses to changes related to marginal areas and marginal populations. The examples are all drawn from Indian agriculture. Since climate change and the situations described in the tables are not strictly comparable, many of the measures listed here may not serve well as guides for adaptation to climatic change. However, the main purpose in highlighting these measures is to demonstrate that even in poor countries communities can evolve workable responses to risks and crises. Furthermore, the measures listed indicate the broad areas where the search for potential adaptation measures could be focused—for example, on crops, structural aspects of agriculture, traditional wisdom, and conscious public interventions.

An important feature revealed by tables 11-7 through 11-10 is that some adaptations—extension of cropping to submarginal lands, lowering the standard of living, abandoning productive activities, raising costs of enterprises, and such—amount to the acceptance of inferior options. Such options are accepted only when the traditional wisdom provides none better. Experience and traditional wisdom represent an important dimension of resilience that characterizes traditional farming systems. With focused input from science and technology, not only could traditional farming systems be strengthened, but more productive and superior options could be generated as well.

Summary of Adaptation Options

The general indifference of the governments of developing countries to greenhouse warming issues, coupled with the lack of concrete information on spatially disaggregated climatic impacts, make it quite difficult even to conceptualize what kinds of adaptation strategies would be worth consideration. Nevertheless, if 2 × CO_2 climate change evolves over a period of, say, fifty years or more, we can be

Table 11-9. Responses to Changes in the Resource Base (Some examples from India of adaptations to a declining agricultural resource base and deterioration of the production environment broadly comparable to the potential negative impacts of greenhouse warming)

	Adaptations/responses	
Changes and their causes	Farm level	Public policy/program level
Significant decline in production base (individual land holding) following rapid population increase (30–40 years)	Intensification of land use by extension of cropping to submarginal areas, reduction of periodic fallowing of land, de-emphasis of long-duration crops, increased landlessness and poverty, extension of irrigation, and use of new crop technologies	Promotion of land-use intensification through provision of irrigation, new technology, and transfer of submarginal lands to crops, efforts to create off-farm jobs
Sinking of water table from 20 m to 100 m or more in parts of Gujarat, Tamil Nadu (15–20 years)	Abandonment of irrigation, further deepening of wells, reduced irrigation	Restrictions on digging new wells and tubewells, emphasis on groundwater recharge
Significant decline in the area of rural common property resources (CPRs) such as community pastures and forests due to privatization following population growth and public policies (20–30 years)	Overexploitation and degradation of remaining CPRs, change in composition of grazing animals (sheep and goats substituted for cattle), increased dependence on inferior products	Increased attention to conservation measures, afforestation, social forestry, wasteland development
Deforestation and degradation of pastures and forests (20–30 years)	Increased dependence on crop by-products for fodder and fuel, changes in feeding practices, change in composition of livestock favoring animals that can survive on degraded vegetation	Emphasis on afforestation, social forestry, pasture development programs, new usage regulations

optimistic that the agriculture of developing countries will respond adequately to emerging risks. This optimism is strengthened by the availability of scientific knowledge and capacities for at least rough quantification of environmental variables and their interaction with relevant agricultural variables. The exceptions to this optimistic assessment, which are many, fall in the marginal areas, particularly in the semiarid tropics, where even a slow, small change in climate can accentuate existing climatic risks. Moreover, it is in these areas particularly that we see little hope for planned or anticipatory adaptation strategies (Schneider and Thompson, 1985).

The alternative to planned or anticipatory adaptations is "forced adaptation strategies" (Jaeger, 1988). Accordingly, one way of identifying broad and potentially relevant options is to search out analogous situations and examine the transferability of knowledge and well-adapted components of agriculture from the current to the new situation. The rationale behind this reasoning is twofold: (1) the existing production environment at one location may be analogous to the production environment at other locations following climate change; (2) irrespective of their causes, several past changes unrelated to greenhouse warming might have had impacts on the production environment and resource base of a sort quite similar to those that will be caused by greenhouse warming. Societal responses to such changes may offer a pool of potential options for adaptation to climatic changes. The analog approach may have certain deficiencies when the situations considered are extremely location specific (Nix, 1985), but with precautions against this, the analog approach can serve as a useful first step in the search for adaptation options. Whether second and subsequent steps would be worthwhile would depend on whether greenhouse warming through multiple, simultaneous, and interacting changes to adaptation leads to a "no-analog" situation.

Impediments to Adaptation

Impediments to the evolution of potential adaptation options occur at two levels: first, at the level of specific adjustments of the type listed in tables 11-7 through 11-10 and, second, at the level of more general adjustments. Regarding the search for adaptations using an analog approach, the obstacle is the lack of a sufficiently large inventory of crisis or risk-oriented adjustment measures. Another impediment stems from the frequent failure of researchers and development planners to recognize that such measures can serve as a useful source of input for public strategies. This is largely a problem of a research culture that allocates sums for formal research to discover new options, but allots very little for identifying existing options well adapted to different parts of the world and for investigating how they might be transferable interregionally. Consequently, formal research reinvents the wheel now and again.

At a more general level, the major constraint on the evolution of adaptation measures stems from uncertainties

Table 11-10. Responses to Changes Related to Marginal Areas and Marginal Populations (Some examples from India of adaptations to a declining agricultural resource base and deterioration of the production environment broadly comparable to the potential negative impacts of greenhouse warming)

	Adaptations/responses	
Changes and their causes	Farm level	Public policy/program level
Intensification of desertification in arid areas; depletion of trees, shrubs, and perennial grasses (30–50 years)	Increased replacement of camels by cattle with disappearance of trees and shrubs, increased replacement of cattle by sheep and goats with depletion of perennial and rich grazing lands, substitution of rubber-tire carts for wooden-tire carts, increased seasonal migration	Provision of relief employment program, emphasis on irrigation
Disruption of traditional collective risk-sharing arrangements due to recent institutional changes (30–40 years)	Increased vulnerability to weather-induced crises, increased dependence on public relief, resort to migration	Public relief substituting for traditional self-help strategies
Disruption of self-provisioning systems and traditional cottage and handicraft industry (10–20 years)	Reduced diversity in occupational structure, overcrowding of agricultural land, migration to towns	Provision of off-farm employment

regarding greenhouse warming. These make the context for responses too vague to serve any practical purpose, even for conceiving of realistic adaptations. A problem of this great uncertainty is the indifference of developing countries to the whole debate regarding greenhouse warming. Uncertainty of impact scenarios, the normally short time-horizon of public policies and programs, and the overall scarcity of resources tend to dampen whatever initiative may exist in developing countries for searching out adaptation options useful against long-term climate changes.

A heavy responsibility falls on developed countries to carry developing ones along in an alliance against atmospheric pollution. Such a partnership, and the coordinated action it implies, may ultimately require more equitable sharing of global economic resources and technology. If the world community is to escape rapid warming of the atmosphere and its consequences, it will need more cohesion in the future than in the past. Relatively concrete steps to induce developing countries to act against greenhouse warming may include the following:

- A reduction in the information gap between developed and developing countries on all facets of the greenhouse warming issue, including new technological options to combat it.
- A demonstration by the developed countries of a commitment to dealing with the greenhouse warming problem (for example, by sacrificing lucrative economic opportunities in order to reduce their contributions to greenhouse gases or by transferring innovations against greenhouse warming internationally rather than by trading for them).
- A strengthening of the current information base on greenhouse warming and improvement of spatially disaggregated information about its possible impacts in order to provide concrete contexts for adaptations. This will go a long way toward sensitizing developing countries to greenhouse warming.
- A provision of direct help and a strengthening of the capacity of developing countries to understand and analyze issues related to greenhouse warming and to prepare themselves for effective responses. This might lead to focusing public attention on long-term planning, research, and development strategies. It might also help to reduce the gap between the perspectives of developing and developed countries on the greenhouse warming issue. Initiatives by developing countries to abate the growth of greenhouse warming and to prepare anticipatory adaptation strategies could be greatly facilitated by their working together with developed countries on the problem.
- A commitment to an integrated and comprehensive effort to prepare an inventory of current farming practices including those that have emerged in response to climatic and other crises, thereby generating a pool of potential adaptation options applicable in both the developed and the developing worlds.

REFERENCES

Bandyopadhyay, J., and V. Shiva. 1988. "Political Economy of Ecological Movements," *Economic and Political Weekly* (Bombay) vol. 23, pp. 1223–1232.

Center for Science and Environment. 1985. *The State of India's Environment, 1984–1985: The Second Citizens' Report* (New Delhi, Center for Science and Environment).

Chen, R. S., M. L. Parry. 1987. *Policy Oriented Impact Assessment of Climatic Variations* (Laxenburg, Austria, International Institute for Applied Systems Analysis).

Glantz, M. H. 1988. *Politics and the Air Around Us: International Policy Action on Atmospheric Pollution by Trace Gases*

(Boulder, Colo., National Center for Atmospheric Research, Environmental and Societal Impact Group).

Hobbs, J., J. Anderson, and H. Harris. 1987. "The Effects of Climatic Variations on Agriculture in the Australian Wheat Belt," in M. L. Parry, T. R. Carter, N. T. Konijn, eds., *The Impact of Climatic Variations on Agriculture*. Vol. 2, *Assessment in Semiarid Regions* (Dordrecht, The Netherlands, Kluwer).

Hrabovszky, J. P. 1985. "Agriculture: The Land Base," pp. 211–254 in R. Repetto, ed., *The Global Possible* (New Haven, Yale University Press).

Jaeger, J. 1988. *Developing Policies for Responding to Climate Change*. World Meteorological Organization/United Nations Environmental Programme, WMO/TD-No. 225.

Jodha, N. S. 1986. "Common Property Resources and Rural Poor in Dry Regions of India," *Economic and Political Weekly* (Bombay) vol. 21, pp. 1169–1181.

———, and A. C. Mascarenhas. 1985. "Adjustment in Self Provisioning Societies," pp. 437–468 in R. W. Kates, J. H. Ausubel, and M. Berberian, eds., *Climate Impact Assessment: Studies of the Interaction of Climate and Society* (New York, Wiley).

———, S. M. Virmani, A. K. S. Huda, S. Gadgil, and R. P. Singh. 1987. "The Effects of Climatic Variations on Agriculture in Dry Tropical Regions of India," in M. L. Parry, T. R. Carter, and N. T. Konijn, eds., *The Impact of Climatic Variations on Agriculture*. Vol. 2, *Assessment in Semiarid Regions* (Dordrecht, The Netherlands, Kluwer).

Kates, R. W. 1985. "The Interaction of Climate and Society," pp. 3–36 in R. W. Kates, J. H. Ausubel, and M. Berberian, eds., *Climate Impact Assessment: Studies of the Interaction of Climate and Society* (New York, Wiley).

Maini, J. S. 1988. *Forests and Atmospheric Change, Proceedings of the Conference on the Changing Atmosphere: Implications for Global Security*, Toronto, June 20–30.

Mellor, J. W. 1976. *The New Economics of Growth: A Strategy for India and the Developing World* (Ithaca, Cornell University Press).

Nix, H. A. 1985. "Biophysical Impacts: Agriculture," pp. 105–130 in R. W. Kates, J. H. Ausubel, and M. Berberian, eds., *Climate Impact Assessment: Studies of the Interaction of Climate and Society* (New York, Wiley).

Parry, M. L., T. R. Carter, and N. T. Konijn, eds. 1987. *The Impacts of Climatic Variations on Agriculture*. International Institute for Applied Systems Analysis/United Nations Environmental Programme.

Reddy, A. K. N. 1985. "Energy Issues and Opportunities," pp. 363–396 in R. Repetto, ed., *The Global Possible* (New Haven, Yale University Press).

Schneider, S. H., and S. L. Thompson. 1985. "Future Changes in the Atmosphere," pp. 397–430 in R. Repetto, ed., *The Global Possible* (New Haven, Yale University Press).

World Bank. 1987. *World Development Report 1987* (New York, Oxford University Press).

12

Human Dimensions of Global Change: Toward a Research Agenda

Ian Burton

The earth's environment is being transformed by human activity. Human activity, in turn, is being affected by these transformations. This interaction is being studied under the aegis of *global change* in the geosphere-biosphere (National Academy of Sciences, 1984). The purpose of this chapter is to explore the basis for and substance of a proposed research program focused on human dimensions of global change.

AN OVERVIEW DIAGNOSIS

If the received wisdom from atmospheric and other natural scientists is approximately correct—namely, that there will be a significant global warming accompanied by major shifts in climatic zones and a rise in sea level—then humankind does indeed face a great danger and a threat of an extremely complex and intractable kind. It is right to start with such a conditional statement because there remain several important areas in the geobiophysical models where knowledge is weak. The role of the oceans as a carbon sink is not well understood. Nor apparently is cloud cover, which if it increases, could enhance reflectance and moderate greenhouse warming to some extent.

There are a number of other areas where additional scientific knowledge, when it becomes available, may show that present estimates of greenhouse warming are too high or too low. This very uncertainty is an important complicating factor in any analysis of the global warming issue and in assessments of what, if anything, should be done about it.

Many senior-level government officials and elected leaders in various countries doubt the confident and dire predictions that are being made. There is, as yet, insufficient evidence to convince some skeptics that the climate is changing. Given the scientific uncertainty and the lack of clear evidence of damage beyond normal, expected fluctuations in climate, it is to be expected that policymakers will act minimally or not at all. Consider three other major environmental problems that affect entire regions of the planet and where the evidence of damage is already very strong: desertification, deforestation, and acid precipitation. All three are causes or effects of global atmospheric change, and yet—despite a great deal of science, many conferences, and a lot of public attention—progress in dealing with these problems has been extremely slow.

There are three further obstacles to prompt action on greenhouse warming. First, the perceived distribution of gains and losses; second, and related, the challenge of mobilizing a global response; and third, the costs of appropriate response strategies. Each will be discussed in turn.

Some think they may benefit from greenhouse warming. Others are tempted to take a free ride (see Darmstadter and Edmonds, chapter 3 in this volume). While the atmospheric scientists are seeking to persuade humankind that global warming constitutes a grave danger, some members of the public and government officials have an eye on the benefits and opportunities. An audience in southern Ontario, after hearing that their climate may some day resemble that of southern Kentucky today, could barely conceal their delight; only half in jest the listeners expressed impatience that the change would come too slowly! Even if such windfalls should prove illusory or a mixed blessing, there is no doubt that the losses and benefits from climate change would be distributed unevenly.

Actions to control or limit greenhouse warming require collective agreement at an international level. Unilateral action by the United States, or by members of the Organisation for Economic Cooperation and Development (OECD), will have limited effect if it is not part of an international response involving all major economies that produce carbon dioxide (CO_2). This includes countries like China and India that do not emit much CO_2 now, but intend to in the future. It has been proposed by some that the international community could follow up the Montreal Protocol on controlling emissions of chlorofluorocarbons (CFCs) with a similar agreement on CO_2 and perhaps other greenhouse gases. Discussions are beginning informally about a "Law of the Atmosphere." Countries lagging in development are not likely to accept curbs that threaten their development, and if countries that have been the major contributors of CO_2 to the atmosphere propose such curbs, they will carry little moral force (see Jodha, chapter 11 in this volume). To achieve agreement, some compensating mechanisms or cheap energy alternatives will be required.

A third inhibiting factor is the high cost of preventive, or ameliorative, measures. While considerable refinement of estimates is possible, and necessary, it is already quite clear that restraining CO_2 emissions is technically difficult and very costly. Unless and until some substantial and profitable use for recovered carbon is found, the prospects are not encouraging. The costs of inaction may prove to be very high, and delay now probably will increase the longer run costs substantially. Political and governmental structures of all kinds have a well-known predilection for short-term benefits over long-term costs. This is reinforced by some economic analyses that sharply discount the future by giving the longer run benefits and costs little or no weight in present-value terms.

Any realistic diagnosis of the greenhouse warming problem necessarily recognizes that the evidence is uncertain; that there will be an unequal distribution of risks, benefits, and costs; that present costs of action could be very high; and that in the absence of agreed collective action by all, now and into the future, single countries or groups of countries are likely to shirk unilateral or limited regional reaction.

The world scene is changing very rapidly under the influence of innovations in science and technology and with the associated reorganization of comparative economic advantage on a global scale. There have been extremely large shifts in agricultural patterns of production and trade in the past four decades. Under the impact of competitive trade policies, innovations in agriculture from applications of biotechnology, and many other factors, the pattern of rapid change will certainly continue, regardless of whether the climate changes (see Easterling, Parry, and Crosson, chapter 7 in this volume). Despite the potentially profound effect of climate change, it will nevertheless be only one of a number of factors in economic change and quite possibly a relatively minor factor in some sectors such as agriculture.

A favored research approach to assessing the impacts of climate change has been to argue by analogy from studies of the impact of past climatic events. From a recent study of a series of severe frosts in Florida, it has been inferred that a decline in orange juice production can be attributed to the frosts. It seems more likely that the frosts served mainly to trigger decisions not to replant orange groves that would have been made in any case on economic grounds. Brazilian orange juice is now considerably cheaper and has been gaining market share (Miller, 1988).

Similarly, the sequence of severe droughts in the Sahelian region of Africa has served to reveal a pattern of changing relationships between the economy of a region and the natural resources (population growth, overgrazing, soil erosion, effects of agricultural and trade policies, and so forth). The droughts have often triggered migration and emergency relief operations, but the drought itself is not the root cause of the problem (Garcia, 1981).

TOWARD PRESCRIPTION

Given this overview diagnosis, what should be done or proposed? The first, almost instinctive, reaction of research scientists in the natural and social sciences is to reach for a research grant! In this case they clearly should, even though this may add to the credibility gap between themselves and some decision makers. The risk should nevertheless be taken, and the efforts of the International Council of Scientific Unions (ICSU) to create the International Geosphere-Biosphere Program (IGBP), or "Global Change Program," are absolutely on target.

Knowledge of the functioning of the global environmental system is inadequate and needs to be strengthened as rapidly and as effectively as possible. Since the questions affect all humanity, it is proper that an international coordinating and stimulating mechanism be created. Given the political sensitivities of the issues raised, the research should be fostered, at least in part, by largely nongovernmental organizations like the ICSU (an association of international, discipline-based scientific unions such as the International Union of Biological Sciences, the International Union of Pure and Applied Chemistry, and the International Geographical Union). The ICSU organizes international, interdisciplinary studies and programs often through the mechanism of a specially created scientific committee—for example, the Scientific Committee on Problems of the Environment and the Scientific Committee on Oceanographic Research (ICSU, 1988).

The IGBP is being established with the following broadly defined objectives: "To describe and understand the interactive physical, chemical and biological processes that regulate the total earth system, the unique environment that it provides for life, the changes that are occurring in this system, and the manner in which they are influenced by human actions" (IGBP, 1988, p. 3).

"Influenced by human actions": that is a key point. It makes no sense to mount a major international natural science program under a scientific body like ICSU without considering at the same time how to describe and understand the social and economic processes that influence global change and will be affected by it. This is no small task. If knowledge and predictive capacity are lacking with regard to global geobiochemical processes, then how much more true is this of social and economic processes? (see Turner et al., forthcoming).

Humankind treads ever more heavily on the face of the earth. Sheer population growth and the economic expansion needed to support it guarantees this. The broad outlines of development are clear, and they point to a steady increase in the emissions of greenhouse gases from the use of fossil fuels and other activities, further spread of deserts, further deforestation, and further expansion of food production on less land. Very little is understood, however, about the relationship between environmental change and economic and population growth. A major hypothesis of *Our Common Future*, the report of the World Commission on Environment and Development (1987)—also known as the Brundtland Commission—is that environmental deterioration is not simply a result of economic development; environmental deterioration has become in many developing countries an impediment to further growth. This hypothesis and its variants should now be critically tested under a variety of environmental and economic conditions.

Little also is known from a goal-oriented perspective. Research could be much more productive in the relatively short run, and some of the leading questions are obvious. To what extent could changes in land use affect the pattern of global climate change? What changes in land use could be desirable, and how extensive would they need to be, how rapidly introduced, and at what cost? To what extent would alternative energy policies be able to influence the rate and extent of global warming? What energy policies are needed, and where, when, and how would they be introduced?

Clearly, many goal-oriented questions should be addressed and illuminated by research. For the reasons outlined earlier, such research would gain in credibility and in acceptability if it were organized and coordinated through an international mechanism. These claims are not meant to preclude more fundamental questioning of the relationships of people to planet, but rather to encourage internationally led research with a policy-oriented and normative goal. The fact that such research has been lacking and has been difficult to organize points to the relative lack of credible international research mechanisms and arrangements and adds to the sense of urgency. Some research is now being organized on an intergovernment level through the United Nations Environment Programme, the World Meteorological Organization and others, but there is no nongovernmental body on the social science side with the authority, status, or capacities of ICSU (World Meteorological Organization, with UNEP, 1988). Not only is internationally sponsored research needed, the institutional mechanism to bring that about also will have to be created.

HUMAN DIMENSIONS OF GLOBAL CHANGE (HDGC)

Three international organizations have attempted to create a social and economic science program that complements the International Geosphere-Biosphere Program. The International Federation of Institutes for Advanced Study (IFIAS), the International Social Science Council (ISSC), and the United Nations University (UNU) have jointly established a program entitled Human Dimensions of Global Change (HDGC). The overall objectives of the program were drafted and agreed to at an international symposium held in Tokyo on September 19–22, 1988. From the report of that symposium (The Human Dimensions of Global Change Programme, 1989, p. 4):

> Dramatic, potentially threatening, and possibly irreversible change in the global environment results from past and present human activities. Such change leads to serious impacts on human conditions, with the risk of compromising prospects for life in the future. This body has therefore resolved to initiate an international research program to investigate both the human causes and consequences of global environmental change. The objectives of the research programme are:
>
> - to improve scientific understanding and increase awareness of the complex dynamics governing human interaction with the total Earth system;
> - to strengthen efforts to study, explore, and anticipate social change affecting the global environment;
> - to identify broad social strategies to prevent or mitigate undesirable impacts of global change, or to adapt to changes that are already unavoidable;
> - to analyse policy options for dealing with global environmental change and promoting the goal of sustainable development.

The HDGC program is not limited to the greenhouse warming problem. It seeks to follow the lead of ICSU and IGBP in addressing a broader set of global change issues such as changes in species diversity, deforestation, and other land-use changes. Nevertheless, global climate warming is regarded as one of the most important global change issues, and one that undoubtedly is helping to drive the whole ICSU initiative and its companion program on human dimensions. Before discussing the possible research directions for HDGC, I want to examine the current initiative in its intellectual and institutional context.

THE INTELLECTUAL CONTEXT

Greenhouse warming and the broader issues of global change have come to the fore at a time when many scholars are consciously reevaluating their understanding of natural and human-made systems and the interactions between them. Expressed briefly and simply, they are making an effort to develop and deploy an evolutionary view of the global environment and of human society, especially the economic system.

The economic system deserves special attention for two reasons. First, it is the economic system that encompasses the major driving forces that are causing climate change, including the use of fossil fuels for energy and the cutting and burning of tropical rainforests. It is also in the economic system that the impacts of climate change will be felt and measured.

Second, both the economic system and the field of economics have been little related in the past to environmental change. Environmental matters have been relegated in economic theory to "externalities" and have been explained as "market failure." A corresponding effort was not made to alter the economy or economics—or, it may be added, economists. This is now beginning to change with the emergence of a new view (Clark and Juma, 1988).

According to this view, much scientific and social science thinking continues to be based on a Newtonian mechanical clockwork view of the universe. Models of this universe can be run forward or backward through time, and they suppose an equilibrium-seeking behavior in which perturbations to the system seem to bring about no more than temporary departures from equilibrium. Such a world view also encompasses processes of gradual incremental change and linear cause–effect relationships.

The alternative paradigm stresses the changing character of relationships, the irreversibilities that evolve through time and that cannot therefore be modeled in a time-independent fashion. As these systems evolve or are perturbed, they are liable to move into far-from-equilibrium states, and at such times their structure may change very dramatically. Students of social change and of biological systems have long understood that change is not necessarily gradual. Dramatic and catastrophic changes come about in human society. After such changes conditions do not return to normal or to the status quo antebellum. They are fundamentally and irrevocably changed into a new structure.

For example, the oil shocks of the 1970s have profoundly altered the structure of international economic relations at the macro level. In a sort of parody of an equilibrium-seeking economy, oil prices have fallen back to former levels as market forces have reasserted themselves. In the meantime, however, the position in the world economy of the United States, Japan, and the OPEC countries has been radically and permanently changed in part as a result of the oil crises and the responses that were chosen. The Third World debt problem also was largely created as an indirect result of the need to invest the sudden, huge surpluses of "petro-dollars."

Similarly, the technical innovations in communications and information systems (microchips, facsimile machines, fiber optics, to name a few) have dramatically changed business and trade relationships. In a very short space of time, a global stock market has been created that operates on a twenty-four-hour basis. The structure of world financial dealings has thereby been suddenly altered and will not return to the previous equilibrium.

One sees the new bravery of the social and biological scientists breaking out in the most unlikely places. It must be said that one of the most unlikely places is in neoclassical economics. Nowhere is it more needed, since the failure of current economic theory to describe adequately the macroeconomic system is almost daily becoming more evident. Particularly clear is the failure of neoclassical economics to take account of the far-reaching effects of scientific and technical innovations and the profound restructuring effects they are having on the global economy.

This much said, how should human interactions with global environmental change be studied? What sorts of policy-oriented research are appropriate in light of the uncertain future of the earth's natural systems as well as the world's socioeconomic systems—where national interests conflict and where a new intellectual paradigm identifies the possibility of rapid structural change in far-from-equilibrium situations? Recognition of such factors is crucial to understanding, and therefore to prediction and effective policy and management.

A general answer is that these ideas form the core of an emerging methodology called "complex systems analysis" and that it is in the development of such an approach that more intellectual capital might be usefully invested. "Complex systems" essentially are systems that incorporate a series of nonlinear relationships. Of course, students of systems analysis have always understood many relationships to be nonlinear, but for practical purposes they have had to assume them to be linear. Now the new mathematics of chaos is making possible the construction of models in which nonlinearities are included rather than assumed away (Allen, 1986). Quantitative, computer-run models of complex systems have been developed in some areas—fisheries, for example, where models developed by McGlade and Allen (1985) are designed to deal with the intricate interactions of fish population dynamics as well as the competitive behaviors of fishing fleets and the interventions of regulatory authorities. The adoption of such an approach need not require quantification, however. As Garcia (1981) in his work in Mexico and on the Sahelian drought has shown, a qualitative logic approach to the analysis of processes at local, national, and international levels, and their interaction, can reveal a great deal about the causes of environmental degradation and social impoverishment without resort to premature quantification.

INSTITUTIONAL CONTEXT

Organization of the International Council of Scientific Unions and the International Geosphere-Biosphere Program is a major institutional development that provides an excellent natural science base for the development of social and economic research on greenhouse warming and other elements of global change. Three other major developments in the past several years or so merit special attention. The first of these, and the one that should have profound and far-reaching significance, is the report of the World Commission on Environment and Development. Its report, *Our Common Future*, stresses the importance of considering environmental changes as an integral part of strategic economic thinking and links this closely to the state of the global economy, including such contentious issues as the Third World debt and the large and mounting level of military expenditures worldwide.

This report has set in motion a new international debate on how to harmonize the need for development to alleviate poverty (and to accommodate the next 5 billion people at a reasonable standard of living) with the need to protect the natural resources and environment of the planet. The development of an international program of research on climate warming is central to movement toward sustainable development. Given that the costs of controlling CO_2 emissions and of addressing the impacts of climate change are likely to be high, an acceptable path of development that offers hope for long-term sustainability must be sought.

The second institutional context for a HDGC program is a major shift in policy at the World Bank. The Bank has taken an approach to development activities that considers the natural resource and environmental endowments of a country as a vital element in its long-term strategic and macroeconomic planning. No longer does the Bank view environmental impacts of development as impediments to be considered at the project level; rather, it now sees the whole development activity as a mechanism for maintaining and restoring the natural capital of a country, and as vital to long-run sustainable development.

Increasingly, the macroeconomic and sectoral studies carried out for the World Bank (and bilateral aid agencies) will need to take into account future uncertainties with respect to climate and sea level. Any program of research on HDGC would clearly benefit from a close working relationship with the Bank, and vice versa.

The third institutional element of some significance is the multilateral Montreal Protocol of October 1987 on emissions of chlorofluorocarbons. That agreement demonstrates receptivity to the achievement of international consensus on the environment. The CO_2 issue is far more complex than the chlorofluorocarbon problem. Yet the CFC protocol provides an encouraging example that serves to show that the international system can achieve agreements of this sort.

Further evidence, if needed, may be seen in the report of the June 1988 Toronto Conference, *The Changing Atmosphere: Implications for Global Security*. A widely representative (but not officially or formally intergovernmental) group of participants agreed in principle to a final *Conference Statement* that inter alia proposed as a global target the reduction of CO_2 emissions of 20 percent of 1988 levels by 2005 (Canada, Atmospheric Environment Service, 1988).

RESEARCH DIRECTIONS

Broadly speaking, there are three sorts of actions that might be taken in relation to climate warming: control, adaptation, and rehabilitation. By *control* is meant all those actions that could reduce the human causes of climate change. It means modifying or finding alternatives to current practices in energy production, industrial processes, and agriculture and forestry that result in net increases in atmospheric CO_2 and other greenhouse gases. The scope of such a research agenda is extremely large. Its main elements have to be identified, and some sense of priority should be developed. Where are the greatest payoffs to research likely?

Some of the major candidates are very evident. Research and development on energy alternatives to the use of fossil fuels get a strong vote. So, too, do efforts at energy conservation and end-use efficiency. Responses to earlier energy crises have shown the potential here and provide the hope that much more could be accomplished. The storage of carbon in biomass—especially forests—is another potential area of policy-directed research. The potential to reduce CO_2 emissions by control of deforestation may be relatively small in physical terms. Its incremental value in economic terms as a demonstration of political will and capability is unknown.

Feasibility studies of alternative ways of reducing CO_2 emissions are required. Future technological options, costs and benefits, and likely political and public acceptability should be studied. The tasks are large, and the development of appropriate analytical frameworks is itself a challenge. There seems every reason to press ahead as vigorously as possible in this direction.

A second area of research is in modes of adaptation. There has already been significant attention to the impacts of climate warming on agriculture, forestry, water resources, and unmanaged ecosystems (see chapters 7, 8, 9, and 10 in this volume). Better knowledge of likely impacts is essential, and as the models of the global environment change and improve, the reliability of results of impact analysis should also improve.

Little attention has yet been given to the development of ways to reduce negative impacts by adaptation. Anticipatory adaptation sooner might achieve more at lower costs than forced adaptation later (see Jodha, chapter 11 in this volume). For example, a plan of land-use management taking future sea-level rise into account in the choice and design of current and near-future investments in coastal development and construction might reduce vulnerability and long-run costs of coastal protection. A research agenda for adaptation has not yet been developed, and it seems time that an initial attempt be made.

A third area of research relates to the possibility of atmospheric rehabilitation—that is, removing some of the greenhouse gases already added to the atmosphere. The most discussed option is reforestation, although preliminary indications of the magnitude of action required to achieve significant effects seem daunting (see Sedjo and Solomon, chapter 8 in this volume). The costs and benefits of marginal incremental change in forest cover are not known. But let us not write off reforestation too quickly. It could provide a much-needed breathing space by slowing down the rate of CO_2 accumulation in the atmosphere while longer term strategies are developed and put into place.

An early and pressing item on the agenda of HDGC must therefore be the development of a set of research priorities. These can be developed on the basis of curiosity-driven motivations of researchers in a laissez-faire mode. The choice of research agenda also should be influenced by some analysis of likely payoffs in terms of policy options that are technically possible, economically efficient, and socially acceptable. A research agenda that is driven in part by curiosity and in part by technical, economic, and social reliance may be quite small at present.

AN ENCOURAGING CONVERGENCE

Policy options for dealing with global warming will require a lot of scientific and technical research and development, a lot of economic analysis, and a strong element of political will backed by public support. Policies deserving of attention include

- Development of alternative energy technologies
- Increase in energy end-use efficiency
- Curtailment of use of fossil fuels and switching from richer to less carbon-rich energy sources
- Reassessment of the future of nuclear power
- Development of plants and crop varieties preadapted to different environmental conditions
- Coastal land-use planning and zoning
- Slowdown and reversal of deforestation and desertification

These and other policies related to climate warming will be difficult to develop and implement—especially when collective international agreement will be required to achieve sufficient effect. The realist will observe that there is a slim chance of all or any of these things being achieved in the name of response to global warming. It is here that luck and good judgment converge.

Many of the policy options that need to be developed for climate warming are also indicated for other reasons. The Brundtland report and the new World Bank policies are inspired by a common aim to bring about more environmentally harmonious and sustainable development. For all the reasons that have led these major forces to begin to incorporate environmental considerations into macroeconomic policy and strategic thinking, we need to find the basis for a long-term stable relationship between development and the atmosphere. The challenge has been recognized, there is a great expansion of scientific and intellectual resources directed to the problem, and the struggle has been joined.

INDUSTRIAL RESTRUCTURING

Most of the literature on policy responses to climate warming considers the prospects for international agreement and concludes that this will be difficult if not impossible to achieve. Aside from uncertainty about the winners and losers, there are other reasons for caution about the prospects for regulation at any level. We live in a time when dominant views are tending away from regulation, and there is a worldwide movement (transcending political and ideological differences) toward less government, less interference, and greater freedom of action. The time does not seem appropriate for a strongly guided approach to climate change.

There are, however, other ways in which desirable results may be obtained. Recent research on environment, energy, and material use in the OECD and European centrally-planned economies has shown that in the period 1973 to 1985 those countries that did the most to protect their environment, to conserve energy, and to reduce the materials content of economic production also achieved the highest rates of economic growth (Simonis, 1988).

This is not to imply a simple cause-and-effect relationship between environmental protection and economic growth, but it does suggest that common underlying causes—modernization of the economy and the shift to an information, high-technology, and service-oriented economy—contribute both to economic growth and to environmental protection.

Some of the most advanced economies, and those growing most rapidly, have been able to reduce the rate of growth of energy consumption—and of associated wastes including carbon dioxide (Simonis, 1988; World Resources Institute, 1985, p. 336). International agreements and regulation have a role to play in controlling greenhouse warming, but secular forces in the economy, if pushed forward more rapidly and spread more widely, might contribute significantly to the reduction of greenhouse gas emissions.

IMPOVERISHMENT

The developing countries, with few exceptions, are not yet concerned about the possible effects of climate change and sea-level rise (see Jodha, chapter 11 in this volume). When they are, what many natural and social scientists and economists in the developed countries now realize will be more strongly impressed upon the world. Climate change is likely to have the severest impacts on countries that are already poor. For countries as for people, to be poor is to be vulnerable. If individual nations and the collective global community are to respond well to greenhouse warming and other global change, steps taken to reduce vulnerability and to strengthen the resilience or response capacity of all economies constitute a move in the right direction. The developing countries struggle with debt rescheduling, adverse terms of trade, and lack of access to industrial-country markets. So long as this continues, they are likely to remain correspondingly vulnerable to climate change and resistant to international cooperation that does not provide for compensation from historical beneficiaries of the use of the global atmospheric commons as a CO_2 waste dump.

The idea of "compensation" in international environmental policies deserves much more attention. Nations have been required to pay damages (reparations) as part of the peace treaty signed at the end of a war. If a global fund could be created (for example, by a small tax on each barrel of oil that enters international commerce), it could be used as a means of compensating those countries asked to forgo the use of cheap fossil fuels in the interest of the global environment. Such schemes need to be explored and developed so that when international opinion catches up ideas will be ready for implementation.

Here again the prudent response to the greenhouse effect points in the same direction as other desirable policies. To bring about a satisfactory human response to climate warming, agreed in principle among the nations, and to avoid international conflict require no more than a set of policies that have been long advocated by those concerned with environmental protection and economic development—and global equity. Energy alternatives, energy-use efficiency, flexible and adaptive farming systems, sensible use of coastal lands, more equitable trade and macroeconomic relationships, easing of the debt burden, taxes and incentives for less gasoline consumption, and many other policy options need to be identified, properly assessed, and where appropriate, adopted. Much research is needed; if it is orchestrated internationally it is more likely to find acceptance.

REFERENCES

Allen, P. M. 1986. "Towards a New Science of Complex Systems," in *The Praxis and Management of Complexity* (Tokyo, United Nations University).

Canada, Atmospheric Environment Service. 1988. *The Changing Atmosphere: Implications for Global Security Conference Statement*. Toronto.

Clark, N., and C. Juma. 1988. "Evolutionary Theories in Economic Thought," pp. 197–218 in G. Dosi, C. Freeman, R. Nelson, G. Silvergert, and L. Soete, eds., *Technical Change and Economic Theory* (London and New York, Pinter Publishers).

Dosi, G., C. Freeman, R. Nelson, G. Silverberg, and L. Soete. 1988. *Technical Change and Economic Theory* (London and New York, Pinter Publishers).

Garcia, R. 1981. *Drought and Man* vol. 1, *Nature Pleads Not Guilty* (Oxford and New York, Pergamon Press).

———. 1984. *Food Systems and Society: A Conceptual and Methodological Challenge* (Geneva, United Nations Research Institute for Social Development).

The Human Dimensions of Global Change Programme. 1989. *Report of the Tokyo International Symposium on the Human Response to Global Change*. Report of a symposium held September 19–22, 1988, in Tokyo, Japan (Toronto, Ontario, Canada).

IGBP (International Geosphere-Biosphere Programme). 1988. *A Plan For Action*. Report 4 (Stockholm).

McGlade, J. M., and P. M. Allen. 1985. *The Fishery Industry as a Complex System*, Canada Technical Reports on Fisheries and Aquatic Science, no. 1347.

Miller, Kathleen A. 1988. "Public and Private Sector Responses to Florida Citrus Freezes," pp. 375–405 in Michael H. Glantz, ed., *Societal Responses to Regional Climate Change* (Boulder and London, Westview Press).

National Academy of Sciences. 1984. *Global Change in the Geosphere-Biosphere: Initial Priorities for an IGBP* (Washington, D.C., National Academy Press).

Prigogine, I. 1980. *From Being to Becoming* (San Francisco, Freeman).

Simonis, U. E. 1988. "Industrial Restructuring for Sustainable Development: Three Strategic Elements." Paper prepared for the International Symposium on Human Response to Global Change, Tokyo, September 19–22.

Turner, B. L., II, et al. Forthcoming. *The Earth as Transformed by Human Action* (New York, Cambridge University Press).

World Commission on Environment and Development. 1987. *Our Common Future* (London and New York, Oxford University Press).

World Meteorological Organization, with United Nations Environment Programme. 1988. *Intergovernmental Panel on Climate Change*. Report of the First Session (Geneva).

World Resources Institute. 1988. *World Resources 1988–89* (New York, Basic Books).

13

Thoughts on Abatement and Adaptation

Roger R. Revelle

I shall not attempt to summarize the many ideas and viewpoints presented by the participants at the June 1988 RFF Workshop on Controlling and Adapting to Greenhouse Warming. Instead I shall try to spell out some of the limits and some of the possible consequences of anticipated climate change during the next hundred years, and to suggest ways in which the changes could be mitigated. I shall discuss some aspects of the climate change problem that were not brought out in preceding chapters, but some coverage of the same ground is unavoidable.

As Schneider and Rosenberg emphasize in chapter 2, we are dealing with considerable uncertainty about the magnitude, and in some cases even the sign, of future climatic change, because we still do not understand many fundamental aspects of climate. There is disagreement on how sensitive the climate system may be to various burdens of radiatively active gases. We do not know nearly enough about clouds, which, depending on where they are and their characteristics, may either amplify or moderate the expected effects. Because of these and other uncertainties we must regard the general circulation models (GCMs) as indicators and not as forecasts. The recently recognized and totally unexpected "hole" in the ozone layer over the Antarctic makes clear that the atmosphere still has many surprises in store.

The role of the oceans, it seems to me, needs greater attention than it received in the workshop. The oceans and atmosphere are coupled in complex ways, and the flow of heat between the two fluids—one thick, one thin—seems to be one of the fundamental mechanisms of climate. Like the atmosphere, the oceans are slow to yield their secrets. We have much work ahead to achieve an adequate understanding of oceanic biogeochemistry and dynamics, and how the oceans may affect greenhouse forcing. Neither in regard to the atmosphere nor the oceans can we be sure that we have accounted fully for all of the factors involved, or that we have not ignored important feedbacks, positive or negative.

That the general circulation models are currently unable to predict the regional details of a future greenhouse climate is a fact that should receive even greater emphasis here. A recent comparison by Grotch (1988) of four numerical models of global climate—those of the National Center for Atmospheric Research in Boulder, Colorado; the Geophysical Fluid Dynamics Laboratory in Princeton, New Jersey; the Goddard Institute for Space Studies in New York City; and Oregon State University in Corvallis—makes clear that there is much work to be done before the models can tell us, for example, whether it will be wetter or dryer in any specific river basin. Until then, GCMs will not be of much use for regional impact studies.

With this as background, I turn to a number of questions having to do with the workshop themes of abatement and adaptation. Under the first rubric are the questions of future concentrations of radiatively active trace gases, the linkage of these gases with greenhouse warming, and other environmental problems. Also examined in the abatement context are opportunities to reduce fossil fuel use and therefore the emission of greenhouse gases, and the likelihood that natural forest expansion may provide an opportunity to control the rate of carbon dioxide (CO_2) accumulation in the atmosphere. Under the adaptation rubric I add some remarks to those of Easterling, Parry, and Crosson (chapter 7 in this volume) concerning the possible effects of greenhouse warming on agriculture in the United States and in the developing world. Finally, I offer some suggestions on capturing and retaining interest in greenhouse warming on the part of the decision-making public.

ABATEMENT

Future Atmospheric Concentrations of Greenhouse Gases

Prescriptions for abating greenhouse warming must be based upon believable estimates of future concentrations of radiatively active trace gases in the atmosphere. Not only is there considerable uncertainty about the workings of the climate system, but there is also great uncertainty about future concentrations of these gases and the future contributions of fossil fuel combustion and the biosphere to these concentrations. In 1860 the atmosphere's burden of carbon in the form of carbon dioxide was about 600 billion tons (gigatons [Gt]). We can expect that quantity to double by some time in the middle of the twenty-first century. For this to occur, about 1,200 Gt must be emitted from roughly the middle of the last century to the middle of the next, because approximately half of the total added carbon goes into the ocean and, perhaps, into the terrestrial biosphere. We can make a rough estimate of this partition between ocean and biosphere because we can compare the present rate of carbon dioxide production by the worldwide burning of fossil fuels with the measured rate of increase of atmospheric CO_2. (Part of the estimated increase of atmospheric CO_2 during the past century resulted from the cutting and burning of forests. Even today, perhaps one-sixth of the CO_2 added to the atmosphere comes from deforestation in the tropics.) To arrive at a quadrupling of the 1860s concentration—that is, at 2,400 Gt—would require combustion, since 1860, of nearly 3,600 Gt of fossil fuel carbon. The total estimated recoverable resources of coal, oil, and natural gas are of the order of 5,000 to 6,000 Gt; hence, a third doubling, to 4,800 Gt, which would require a total emission of 8,400 Gt, would be beyond the limit of recoverable resources, unless oil shales and other marginal carbon reserves could be economically utilized.

Pollution Problems Related to Climatic Change

That climate change would also exacerbate several other environmental problems was pointed out in the panel discussion (see chapter 14 in this volume), by Michael Oppenheimer of the Environmental Defense Fund. The cooling of the stratosphere that would accompany tropospheric or near-surface warming, for example, would probably intensify ozone depletion in the stratosphere, and thereby increase the flux of ultraviolet radiation to the Earth's surface. In his panel comments, Oppenheimer suggested that a 10 to 20 percent depletion of stratospheric ozone could occur within the next forty years, which in turn could increase the ultraviolet flux through the stratosphere by 30 to 50 percent.

Climate change would occur in a stressed world in which stratospheric ozone depletion is already occurring; in which tropical forests are being destroyed, together with a large fraction of their many plant and animal species; and in which freshwater lakes are becoming acidified and where there is widespread death of forest conifers in temperate latitudes.

In addition to the difficulties that the direct climatic effects of increased concentrations of greenhouse gases may hold for human beings and natural ecosystems, there will be important indirect effects—among them, intensification of other pollution problems. The nitrogen oxides and tropospheric ozone that contribute to the greenhouse effect may be among the principal causes of the current die-off of the Black Forest in Germany and the conifer forests of Austria. According to Smith and Tirpak (1988), global temperature increases would speed the reaction rates among chemical species in the atmosphere, causing increased ozone pollution near urban areas. The summer season would become longer, and it is in summer that air pollution levels are usually highest.

On the other hand, the fluxes of nitrogen oxides and other fertilizers into the surface layers of the ocean, carried by winds and runoff from the land, might very well increase ocean productivity and lead to more vigorous action by the biological pump—in other words, to the sequestering of larger quantities of organic carbon in the deep ocean waters. This would reduce, perhaps quite significantly, the buildup of carbon dioxide in the atmosphere. Again, the great uncertainty surrounding these questions dictates that we must intensify our effort to understand interactions between greenhouse warming and other environmental perturbations.

Controlling Fossil Fuel Use

Future rates of emission of CO_2 are estimated by Darmstadter and Edmonds (chapter 3 in this volume), who base their work on previous energy scenario computations by a number of researchers (Reilly et al., 1987; Edmonds et al., 1986). Darmstadter and Edmonds' base case shows emissions rising from an annual total of approximately 5 Gt in 1985 to 8 Gt in the year 2050—a cumulative volume of around 450 Gt over the period. Their analysis shows that these emissions depend principally on (1) world economic growth, (2) the average quantity of energy required to produce one unit of economic product, and (3) the amount of fossil fuel carbon that must be burned to produce the required amount of energy. Although Darmstadter and Edmonds confine their exposition to CO_2 alone, economic and social changes affecting the atmospheric concentrations of greenhouse gases other than carbon dioxide—primarily methane, nitrous oxide, and ozone in the lower atmosphere (Rasmussen and Khalil, 1986; Thompson and Cicerone, 1986)—must of course also be considered in order to assess the total dimensions of the greenhouse effect.

Growth in the world economic product depends on population growth (estimated to reach more than 9 billion by the year 2050) and on economic development in all countries, but especially in the poor developing countries, whose populations now make up about three-fourths of mankind and (for the most part) are growing rapidly. As these countries develop, their principal economic activities will shift from agriculture, which traditionally consumes comparatively little fossil fuel energy, to fossil fuel–intensive industry and transportation. Acting in the opposite direction could be significant increases in the efficiency of energy use and changes in the mix of economic activities in developed countries, which might move toward such relatively low-energy sectors as communications and information processing.

That population growth and rising living standards will require strenuous efforts to increase agricultural production may be self-evident, but it is worth underscoring in the context of energy and climate change. Most future increases in global food production are likely to come from technological changes that increase the yield of the land. Nevertheless, some clearing of forests is likely to continue for decades, mainly in Latin America, Africa, and the remaining forested regions of southeast Asia. Expansion of areas under cultivation, by clearing and burning forests, adds carbon dioxide, carbon monoxide, and methane to the atmosphere (Thompson and Cicerone, 1986). Increases in productivity will require more use of nitrogenous fertilizers, which add nitrous oxide to the atmosphere. Population growth and rising incomes will also result in an increase in the number of cattle and expansion of the areas of rice paddies; cattle and rice paddies are two major sources of methane.

The explosive growth of cities and towns in all developing countries means a movement away from rural areas. Such movement results in a shift toward more energy-intensive activities and away from traditional and agrarian ones, and a shift toward fossil fuels and away from traditional biomass fuels. Poor countries are determined to develop economically and, for every reason, they *must* do so. Must they pollute their way to prosperity?

The main question raised by several studies of future energy paths cited by Darmstadter and Edmonds is whether it is possible to drastically "decouple" economic growth and fossil fuel use. These studies propose the feasibility of a long-term energy strategy which slows down or reduces the emission of greenhouse gases *without penalizing human welfare*. Darmstadter and Edmonds question the assumption behind these "soft energy" scenarios, which treat energy and economic growth as virtually independent of each other. Are the technical feasibility and costs of a wide array of improvements in energy use technology, they ask, firmly enough established to support a zero-energy growth scenario? Persuasive evidence, pointing to one or another position on this important issue, remains to be presented; more study is clearly warranted.

One is tempted to apply the example set by the Montreal Protocol on substances that deplete the ozone layer that took effect on January 1, 1989. That protocol limits the future emissions of chlorofluorocarbons (CFCs). If commercial interests can join with governments to significantly curtail future emissions of CFC gases, might not a CO_2 emission protocol be equally possible and likely? Darmstadter and Edmonds point to a number of significant differences between the two classes of atmospheric threat. For example, the cost of switching to noncarbon-emitting energy sources and of enhancing fossil fuel conservation is apt be orders of magnitude higher than the cost of reducing chlorofluorocarbon emissions.

One factor that affects the quantity of fossil fuel carbon used to produce the required energy is the relative proportions of the natural gas, petroleum, and coal that are burned (Cheng et al., 1986). Combustion of natural gas produces only 57 percent as much CO_2 per unit of energy as does combustion of bituminous coal, and combustion of oil produces less that 70 percent as much.

At the present time, however, many countries appear to be committed to energy policies that assign a major future role to fossil fuels in general (Nordhaus and Yohe, 1983) and to coal in particular (National Coal Association, 1988). The USSR, the United States, and China possess perhaps 90 percent of the enormous volume of recoverable world coal reserves. It is easy to forget how relatively few years have passed since the World Coal Study (1980) urged that coal be an energy "bridge to the future" to ease the transition from oil dependence.

These facts underscore the importance of pursuing research and policies likely to lessen global dependence on coal without incurring a major economic penalty. Indeed, substitution of energy produced from sources other than fossil fuels—hydroelectric, geothermal, solar, biomass, or nuclear sources—would be even more effective than substituting natural gas for coal as a way of reducing carbon emissions to the atmosphere (Jaeger, 1988). But in the final analysis it seems clear that no single policy or technological thrust, be it supply expansion or demand management, is likely to be sufficient to ensure a climatologically harmless energy system. The dilemma justifies pursuit of a broad mix of strategies and options.

Capturing CO_2 by Afforestation

Another possibility for controlling the concentration of atmospheric CO_2 should be considered. All models of global climate change show the rise of temperature in high northern latitudes to be much larger than in mid-latitudes. Hence one solution for undesirable global warming might be to sequester very large quantities of carbon in the forest trees at high northern latitudes. Indeed, considerable sequestering could be done simply by letting boreal forests do what comes naturally as climate becomes warmer and atmospheric CO_2 content increases. For example, the 750 million hectares now covered by boreal forests might be greatly enlarged if their boundaries with the tundra were to expand northward and the forests were to cover a much larger area than might be lost to temperate forests and agriculture on the southern boundaries. Temperate forests also could play a part in preventing global warming if they respond to warmer temperatures and greater atmospheric carbon dioxide by increasing the rates and amounts of photosynthetic production of organic matter (Marland, 1988).

It is interesting to speculate on the quantity of carbon that could be sequestered in the boreal forests by expanding their areas and increasing their photosynthetic production. Sedjo and Solomon (chapter 8 in this volume) appear unconvinced that climate change and the CO_2 fertilization effect will lead to any net increase in CO_2 sequestering. However, in a recent study carried out by the International Institute for Applied Systems Analysis (IIASA) (Binkley, 1987), it has been suggested that the area of exploited boreal forest might increase by two-thirds, or about 500 million hectares, over the fifty-year period between 1980 and 2030. The total area covered by exploitable boreal forest would then be 1,250 million hectares. Present rates of growth in these forests, counting stem wood and roots, branches, and leaves, are about 0.8 metric tons per hectare per year. Presumably this could be increased by 35 percent when atmospheric carbon dioxide content doubles because of the fertilizing effects of the added CO_2. The IIASA study speculates that the effects of higher temperatures alone would lead to a doubling of boreal forest growth rates to about 1.60 metric tons per hectare per year. Combining this with the 35 percent CO_2 fertilizing effect, we obtain an average growth rate in the combined old and new exploitable boreal forest area of 2.16 metric tons per hectare per year, or 2.70 Gt of biomass. The incremental volume of carbon sequestered by the addition of 500 million hectares of boreal forest would be 2.10 Gt, which is 70 percent of the total carbon retained in the atmosphere attributable to the burning of fossil fuels and other human activities. Of course, the rate of carbon sequestering would be considerably less than this during the transition time before atmospheric CO_2 is doubled.

It is likely that the tendency to higher photosynthetic production would be maintained on the southern boundary of the present boreal forest area whether these forests remain boreal or are displaced by cool, temperate forests. Cool, temperate forests average 200 cubic meters of biomass per hectare, which compares to only 90 cubic meters for average boreal forests, and thus have the potential to sequester large amounts of carbon in the long run. However, existing forests in temperate and tropical latitudes presumably consist largely of mature trees which are no longer growing rapidly. Hence these forests might not contribute much to carbon sequestering.

Sedjo and Solomon devote considerable attention to purposeful afforestation. It certainly seems reasonable that forest growth could be expanded by various means if there were a widespread consensus that fuelwood or wood charcoal could be substituted for fossil fuel. This would have two advantages with regard to the carbon cycle and global warming. First, emissions of atmospheric carbon resulting from fossil fuel combustion would decline; and second, the standing stocks of forest trees, with their contained carbon, would expand, thereby removing carbon from the atmosphere.

Admittedly the above analysis is based on very crude estimates of the expansion of boreal forest areas and tree productivity that might be induced by increased atmospheric carbon dioxide and global warming. Factors that need to be considered in a more realistic analysis include: tree migration rates, as well as the effects of human settlement on these rates; better estimates of carbon dioxide fertilization effects in forest ecosystems; and rates of increase in heterotrophic respiration resulting from higher temperatures. Respiration is also affected by the availability of soil moisture (DaCosta et al., 1986); therefore changing patterns of precipitation will play a role. In particular, accelerated oxidation of peat in the present tundra areas could largely counteract the effects of increased photosynthesis resulting from the extension of the boreal forest. In addition, the economic costs of silvicultural practices intended to increase forest productivity in northern latitudes could amount to hundreds of billions of dollars. Finally, earth's global albedo (reflectivity for solar radiation) should diminish when present tundra areas become covered with forest trees, and a lower albedo would mean more absorption of solar radiation and therefore more heating.

ADAPTATION

Effects of Climate Change on the World Food System

From the standpoint of human welfare, the most serious impacts of climate change brought about by increasing atmospheric concentrations of greenhouse gases will be in the world food system—the complex of human activities that includes agricultural food production and food distribution and marketing. Since complete abatement of greenhouse warming is highly unlikely, adaptation will be necessary.

The concern here is with both the demand for and the supply of food. Generally accepted population and economic projections indicate that the demand for food and other agricultural products will increase by between 1 and 2 percent per year between 1980 and 2050, for total growth by the year 2050 of about 250 percent (see Easterling, Parry, and Crosson, in chapter 7). Nearly half of this growth will be caused by rising human numbers, particularly in the developing countries. But the larger part will result from increases in per capita food demands among present less developed countries as their economies improve and real incomes grow. (Human beings in the present developed countries, who will make up about 15 percent of the world's population in 2050, are, on average, already sufficiently well fed, and there probably will be little increase in their per capita food demands.) Economic models indicate that even rather large climate changes might have relatively little effect on the overall trends of world income. The implication is that the growth of world demand for food also would be little affected by climate change.

Both increasing temperatures and changing available water resources will affect future food supplies, as will increases in atmospheric carbon dioxide and the loss of low-lying farmland resulting from a rise in sea level (Easterling, Parry, and Crosson do not consider this last point in their analysis).

The direct effects of increases in CO_2 are to raise crop productivity, as Schneider and Rosenberg show in chapter 2. If effects noted under controlled climate experiments were to occur in the open air, and if climate were to remain fixed, the fertilization effect would result in an increase in production per unit area and per unit of water consumed of many important crop plants such as wheat, rice, and potatoes.

Fortunately, many of these crops are somewhat insensitive to differences in climate. A good example is hard red winter wheat, which has spread across large climatic gradients in the Great Plains of the United States—gradients that are larger than those implied by the greatest predicted greenhouse climatic changes (see chapter 7 in this volume). On the other hand, small changes in mean climatic conditions might result in large changes in the frequency of extreme climatic events, such as heat waves, floods, and droughts (examples of such possible changes in climatic variability are given in chapter 2). Were heat waves to increase in severity and duration as a result of greenhouse warming, there could be serious detrimental effects on the yields of maize and other summer crops, as well as on the health of livestock and poultry. An increase in the frequency of such extreme events would be more likely to be perceived by farmers as climate change than would a slow increase, over several decades, in mean global temperatures and other average climatic parameters.

What may all of this mean for the comparative advantage of United States agriculture? During the past forty years, natural, human, and institutional resources have favored agriculture in the United States. Now other countries are emerging as major competitors in global export markets. Consequently, it is very important for United States farmers to minimize production costs in order to stay competitive. The principal economic and social impacts of unfavorable climate change are unlikely to reduce farm production severely enough to cause a world food shortage. From the standpoint of U.S. farmers, the problem will be one of maintaining their comparative advantage.

An example will be useful here. Temperature increases resulting from an equivalent doubling of atmospheric CO_2 at middle and high latitudes (that is, in northern Japan, Canada, the Scandinavian countries, and the northern Soviet Union) would mean a substantial lengthening of the crop-growing season. At the same time, extreme high temperatures, if extended over several days during critical growth stages, and greater summer drying, should they occur, would reduce yields and increase stress on crops growing in the U.S. corn belt (Iowa, Missouri, Illinois, Indiana, and Ohio) and in the southern and southeastern United States (and probably also in the Ukraine, the breadbasket of the Soviet Union).

But even if considerations of comparative advantage are not taken into account, regional food production and farm incomes in the midwestern and southern United States would tend to decline, offsetting technical, managerial, and institutional adjustments. If these adjustments were to fail, the consequent decline in farm income could radiate through the economy of farming regions because of the lessened ability of farmers to purchase goods and services.

Loss of comparative advantage of farming in the United States could result in major shifts in the geographic location in the country of major crops, such as soybeans, corn, and wheat. However, various strategies are available to farmers to help them cope with losses of productivity induced by climate change (see Waggoner, 1983). These can be grouped as technical and socioeconomic policy responses. More information is needed on the range of potentially effective technical adjustments at the farm level, such as irrigation, crop selection, and fertilizer applications, and the range of potentially effective responses at regional, national, and international levels, such as plant breeding, improved agricultural extension services, reallocation of land use, and large-scale water transfers.

Appropriate policy responses to the loss of comparative advantage would be development of crop varieties having higher thermal requirements; changes from spring-sown to fall-sown varieties (a change already being implemented in the southern Canadian prairies); changes in the rate and times of fertilizer applications; and changes in crop management, particularly in the choice of area, location, and time of planting of crops that have different needs for moisture and warmth.

Irrigated Agriculture in the United States

Irrigation is required for most agricultural practice west of the one-hundredth meridian in the United States. This region includes the eleven arid western states of California, Nevada, Idaho, Colorado, Montana, Oregon, Washington, Wyoming, Arizona, Utah, and New Mexico, as well as the western portions of the six transition states of Texas, Oklahoma, Kansas, Nebraska, South and North Dakota. These seventeen states constitute the primary irrigation region of the United States. According to Peterson and Keller (forthcoming), 16.2 million of the total 19.8 million hectares of irrigated cropland in the United States are in the western states. This is about 15 percent of our cultivated farmland. Primary irrigation is also used for rice in Arkansas and

Louisiana, and for vegetables and fruit in Florida. Supplemental irrigation to prevent decreases in yield during droughts is practiced in the remaining twenty-eight contiguous states.

Under present climate conditions in the United States, droughts may affect one growing season or they may persist for decades. For example, a destructive drought began in 1925 in one irrigated area in west central Utah, and persisted for all but five years until 1982 (Israelson, 1935; Jennings and Peterson, 1935; Crafts, 1976). In 1983 a cycle of heavy precipitation began in Utah and continued until 1987, causing destructive floods in 1983 and 1984. Beginning in the late 1940s, groundwater was used to supplement river runoff and largely stabilized year-to-year fluctuations in water supplies.

Diminished water supplies are translated economically into less income (Peterson and Keller, forthcoming). But this is also true of decreased farm commodity prices, increased production costs, and the lowering of water tables because of development. The economic and social effects are indistinguishable. For farmers mining large aquifers such as the Ogallala in the Great Plains (an aquifer extending from Texas to South Dakota), accelerated pumping may make up for reduced rainfall resulting from climate change. But this can only hasten the day of reckoning when pumping costs will become too great to continue irrigation agriculture. In their study Peterson and Keller compute crop irrigation requirements (not including water-loss inefficiencies) for four different climatic scenarios: (1) the present agro-climatic condition; (2) 3°C warming; (3) 3°C warming plus a 10 percent increase in precipitation; and (4) 3°C warming with a 10 percent decrease in precipitation. In all three climatic change scenarios, irrigation increased because of a lengthened growing season, more multiple cropping, shifts in crops, and increases in the potential for evapotranspiration. Irrigation requirements increased 15 percent with 3°C warming and no change in precipitation; 7 percent with a 10 percent increase in precipitation; and 26 percent with a 10 percent decrease in precipitation. On average, potential evapotranspiration increased 8 percent, but changes in precipitation had only a small effect on irrigation requirements in the areas west of the one-hundredth meridian. The irrigation requirement was greater than 0.5 m for all but the northernmost tier of states of this western region.

With a warmer climate, reductions in cultivated area in most western states averaged 24 percent if there was no change in precipitation. The reductions averaged 15 percent with a 10 percent increase in precipitation, and 31 percent with a 10 percent decrease. The most severe impacts were in the Great Plains, and the least severe in the Pacific Northwest.

With the expected greenhouse warming, irrigators will be hard put to maintain even present levels of irrigation, and new irrigation supplies will be costly. But fortunately there are attractive and relatively inexpensive opportunities for increasing the efficiency of water use, including the improvement of storage and conveyance through better maintenance or replacement of structures, and, on the farm, laser-controlled land levelling, better choices among water-application technologies, and irrigation to maximize returns per unit of water rather than to maximize yield. In the biological area, there are promising opportunities for developing crop varieties having a higher harvest index (the ratio of harvested material to total plant biomass produced), which would not require increased water use.

Transfer of water would be increasingly important in a warmer climate because it would be necessary to supply more water for uses more pressing than irrigation, a matter dealt with in detail by Frederick and Gleick in chapter 10 in this volume.

Adapting Food Production Systems in Less Developed Countries

The effects of changing climate on agriculture in less developed countries will be different from those in the rich countries. Although changes in temperature are likely to be smaller, the impacts of climate change may be more severe. There are four principal reasons for this. (1) Agriculture is a much larger component of the national economies in the poor countries, and hence climatic impacts on agriculture would reverberate more strongly throughout the whole of those economies. (2) Because of rapid population growth and grossly limited resources of capital and technology, a large proportion of farmland in those countries is marginal. Not only does it exhibit low productivity, but it is highly vulnerable to erosion, salinization, and other environmental damages. (3) In less developed countries, other national problems of much more immediacy and potentially larger consequence tend to overwhelm the interest of national governments in climate change, as Jodha so effectively shows in chapter 11 in this volume. (4) Lack of capital, technology, and knowledge tend to greatly reduce the options available to poor countries for dealing with climate change problems. Lack of knowledge can be partly overcome by technical assistance from the rich countries. Yet one should take particular note of Jodha's view that farmers in the developing countries, having had much practical experience in ways of reducing or avoiding the risks of crop failure in the face of adverse environmental conditions such as droughts and floods, may have a unique store of knowledge that may help not only themselves, but the developed countries as well.

HOW CAN THE ATTENTION OF THE DECISION-MAKING PUBLIC BE CAPTURED AND RETAINED?

In a recent study, Ingram et al. (forthcoming) claim that for a scientific issue to be given a significant place on the political agenda of a government and its people, the problem should be "serious, certain, soon, and solvable, and it should have a villain." The climate change issue meets some of these criteria, but not others. There is a general consensus among earth scientists that climate change will *certainly* occur and that it will have *serious consequences*. But these consequences are *not likely to happen soon*. Moreover, there is no general agreement among experts about what should be done. It is clear that the *principal villains* are those who burn coal, oil, and natural gas. But unfortunately this includes almost everybody, at least almost everybody in the rich countries and regions, including Japan, Western Europe, and the United States and Canada, as well as in the Soviet Union and Eastern Europe.

The climate change problem, like the problem of ozone depletion in the stratosphere, is a natural one for governmental and nongovernmental international organizations. Many of these have already become involved, including the United Nations Environment Programme (UNEP), the World Meteorological Organization (WMO), the International Council of Scientific Unions (ICSU), the United Nations Educational, Scientific, and Cultural Organization (UNESCO), the Food and Agriculture Organization (FAO), the Organisation for Economic Cooperation and Development (OECD), and the International Institute for Applied Systems Analysis (IIASA). So far, however, these organizations have confined themselves to holding scientific conferences and sponsoring research. This is in contrast to the important role they played in forging an international agreement to reduce chlorofluorocarbon production as a means of lessening the threat to the stratospheric ozone layer. Coordinated international action on climate change may be very difficult because of the likelihood that climate change will have different effects on different countries, and because of differences in the fossil fuel resources of those countries. For example, the Soviet Union, with its enormous coal reserves and need for a longer growing season across its vast northern areas, could favor enhanced climatic warming caused by a large increase in coal combustion. Similarly, the People's Republic of China, with its very large coal reserves and a desperate need for economic development, might be willing to accept the consequences of climatic warming in return for industrial growth based on energy from coal.

There is a natural constituency for the climate change issue—those groups of environmentalists in many different countries who have repeatedly proven their effectiveness in political action. Such groups have concerned themselves with other international environmental problems like the destruction of tropical rain forests, extinction of species, export of dangerous pesticides, and adverse environmental impacts of international development projects. Unfortunately, they are not likely to be enthusiastic about many of the proposed devices for avoiding climate change, such as a huge expansion of nuclear power plants, or disruption of the ecology of far northern regions by converting vast areas to fast-growing boreal forests. However, environmentalists are likely to give strong support to drastic measures for energy conservation.

As Ingram et al. (forthcoming) point out, climate change may touch and affect a large number of issues without itself becoming a separate and singular issue. Decisions in transportation, siting of facilities, agriculture, water, energy development, fisheries, and many other areas may be affected and modified by considerations of climate change. This would be true, of course, even if we do not try to prevent or mitigate climate change. In general, in many of these problem areas the prospects for climate change can be used to reinforce policy changes that would be desirable on other grounds. For example, in agriculture, crop diversification and intensified crop-breeding programs are needed in any case, and prospective climatic change would simply provide a powerful added justification for these activities. In fisheries, better estimates of recruitment of new stocks, fish population sizes, and changing distributions are needed for better fisheries management, and would be even more essential under the impact of climatic change.

Especially in regard to energy and water resource problems, policies that promote conservation, efficiency of use, the establishment of priorities, and diversification of sources are highly desirable because these resources are increasingly limited, and, in the case of energy, exhaustible. Such policies may well be the most important means of delaying and abating the greenhouse effect, but they are needed in any case for a sustainable economy.

Until we know more about the magnitudes and rates of change of temperature and precipitation in specific regions, and about the probable frequency and intensity of extremely hot or dry conditions in the summertime and of tropical storms resulting from ocean surface heating, it will be hard to make a convincing case for more drastic measures of abatement or adaptation. Hence the most immediate and important policy decision facing us may be to support unstintedly research on and monitoring of the atmospheric and oceanic concentrations of greenhouse gases, the atmospheric, oceanic, and biological processes that determine the distribution and abundance of those gases, and the effects of greenhouse gases on atmospheric and oceanic temperatures, atmospheric moisture, and the large-scale circulation of the ocean and the atmosphere. Of almost equal urgency are studies of the socioeconomic impacts of anticipated climatic change, particularly as they affect the uses and sources of energy.

REFERENCES

Binkley, C. S. 1987. "A Case Study of the Effects of CO_2-induced Climatic Warming on Forest Growth and the Forest Sector: B. Economic Effects on the World's Forest Sector," in M. L. Parry, T. R. Carter, and N. T. Konijn, eds., *The*

Impact of Climatic Variations on Agriculture. Vol. 1, *Assessment in Cool Temperate and Cold Regions* (Dordrecht, The Netherlands, Kluwer) pp. 197–218.

Cheng, H. C., M. Steinberg, and M. Beller. 1986. *Effects of Energy Technology on Global CO_2 Emissions*. TR030 (Washington, D.C., Carbon Dioxide Research Division, U.S. Department of Energy).

Crafts, D. D. 1976. *History of Sevier Bridge Reservoir* (Delta, Utah, DuWil Publishing) pp. 95–96.

DaCosta, J. M. N., N. J. Rosenberg, and J. B. Verma. 1986. "Joint Influence of Air Temperature and Soil Moisture on CO_2 Release by a Soybean Crop," *Agricultural and Forest Meteorology* vol. 37, pp. 219–227.

Edmonds, J. A., J. M. Reilly, R. H. Gardner, and A. L. Brenkert. 1986. *Uncertainty in Future Global Energy Use and Fossil Fuel CO_2 Emissions 1975 to 2075*. TR036 (Washington, D.C., Carbon Dioxide Research Division, U.S. Department of Energy).

Grotch, S. L. 1988. *Regional Intercomparisons of General Circulation Model Predictions and Historical Climate Data*. TR041 (Washington, D.C., Carbon Dioxide Research Division, U.S. Department of Energy).

Ingram, H. M., H. J. Gortner, and M. K. Landy. Forthcoming. "The Politics of Climate and Water Resources," in P. E. Waggoner, ed., *Climate and Water* (New York, Wiley).

Israelsen, O. W. 1935. "Drainage, Irrigation, Soil, Economic, and Social Conditions, Delta Area, Utah: Drainage and Irrigation Conditions," Bulletin no. 255, Utah Agricultural Experiment Station, Logan.

Jaeger, J. 1988. "Developing Policies for Responding to Climatic Change," a summary of the discussion and recommendations of the workshops held in Villach, Austria (28 Sept.–2 Oct. 1987) and Bellagio, Italy (9–13 Nov. 1987), under the auspices of the Beijer Institute, Stockholm, Sweden. World Climate Programme Impact Studies, WMO/TD-No. 225.

Jennings, D. S., and J. Darrel Peterson. 1935. "Drainage, Irrigation, Soil, Economic, and Social Conditions, Delta Area, Utah: Drainage and Irrigation Conditions," Bulletin no. 256, Utah Agricultural Experiment Station, Logan.

Marland, G. 1988. *The Prospect of Solving the CO_2 Problem through Global Reforestation*. TR039 (Washington, D.C., Carbon Dioxide Research Division, U.S. Department of Energy).

National Coal Association. 1988. *Facts About Coal, 1988*. Washington, D.C.

Nordhaus, W. D., and G. W. Yohe. 1983. "Future Carbon Dioxide Emissions from Fossil Fuel," in *Changing Climate: Report of the Carbon Dioxide Assessment Committee*, National Research Council (Washington, D.C., National Academy Press) pp. 87–153.

Peterson, D. F., and A. D. Keller. Forthcoming. "Irrigated Agriculture", in P. E. Waggoner, ed. *Climate and Water* (New York, Wiley).

Rasmussen, R. A., and M. A. K. Khalil. 1986. "Atmospheric Trace Gases: Trends and Distributions Over the Last Decade," *Science* vol. 232, pp. 1623–1624.

Reilly, J. M., J. A. Edmonds, R. H. Gardner, and A. L. Brenkert. 1987. "Uncertainty Analysis of the IEA/ORAU CO_2 Emissions Model," *Energy Journal* vol. 8, no. 3.

Smith, J. B., and D. A. Tirpak. 1988. "The Potential Effects of Global Climate Change on the United States," draft report to Congress. Washington, D.C., Environmental Protection Agency.

Thompson, A. M., and R. J. Cicerone. 1986. "Atmospheric CH_4, CO and OH from 1860 to 1985," *Nature* vol. 321, pp. 148–150.

Waggoner, P. E. 1983. "Agriculture and a Climate Change Caused by More Carbon Dioxide," in *Changing Climate: Report of the Carbon Dioxide Assessment Committee*, National Research Council (Washington, D.C., National Academy Press) pp. 383–418.

World Coal Study. 1980. *Coal: Bridge to the Future*. Report of the World Coal Study (Cambridge, Mass., Ballinger).

14

Epilogue

Chester L. Cooper

When All Is Said—

However rich and varied the intellectual feast provided at the RFF Workshop on Controlling and Adapting to Greenhouse Warming, there may still remain a sense of undernourishment on the part of many readers. What does the compilation of papers add up to? What were the principal points of agreement? What major insights were revealed? What should the policymakers be expected to do or to eschew? In short, when all is said—so what?

This question was addressed, sometimes directly, sometimes inferentially, in a two-hour panel discussion that followed the formal presentations at the workshop. The panelists—Robert W. Fri, president of Resources for the Future; D. Gale Johnson, professor of economics, University of Chicago; Michael Oppenheimer, senior scientist, Environmental Defense Fund; Milton Russell, professor of economics and senior fellow, Energy, Environment, and Resources Center, University of Tennessee, and senior economist, Oak Ridge National Laboratory—sought to place key issues raised at the workshop in a policy framework. What follows here is drawn from the panel discussion, from audience interventions engendered by the panel, and from selected policy-related issues raised in the formal presentations.

A disclaimer at this point is prudent: any attempt to provide, within a limited compass, the gist and flavor of a two-day meeting involving a dozen formal presentations and several hours of informal discussion requires much picking and choosing. Such an enterprise is bound to involve generous amounts of subjectivity and arbitrariness.

THE ISSUE

Ten years or so ago greenhouse warming was only beginning to be recognized as worthy of attention by serious investigators in a broad range of disciplines. The subject of increasing accumulations of atmospheric carbon dioxide (CO_2) and the ecological consequences thereof was, until then, the province of a handful of scientists engaged in a lonely endeavor. While there was general agreement that the consequences of growing levels of CO_2 in the atmosphere warranted serious study, there was considerable uncertainty as to what additional CO_2 emissions meant in terms of local and regional climatc changes, when such changes would be evident, and what the consequences would be when (if?) they occurred.

Now, a decade or so later, the uncertainties (some with regard to matters not recognized earlier) remain. Nevertheless, there appears to be an emerging consensus, as the contributions to this volume attest, that discussion of policy aspects of the climate change issue can fruitfully be organized around a number of fundamental propositions:

1. The greenhouse effect is real; the greenhouse gases that have accumulated in the atmosphere as a result of human activities are sufficient to cause a significant warming. Models of atmospheric behavior predict that even if emissions were stabilized at today's rates they would lead to continued accumulation of greenhouse gases in the atmosphere. If emissions continue at current rates, global mean temperature by about the middle of the twenty-first century may be higher than at any time in the past million years.
2. Such warming could continue indefinitely unless emissions are reduced to the capacity of their sinks to absorb them. Getting international agreement to reduce emissions by the necessary amount will be difficult because nations will have different views about what constitutes fair international, intra- and inter-generational sharing of the costs of halting or not halting global warming.
3. Stabilization or reduction of emissions presupposes fundamental changes in the amount and types of fuel required by a growing world economy. This proposition

points up the need to give concentrated attention to three areas of research: first, the feasibility of substituting nonfossil for fossil fuels in large quantities; second, the magnitude of energy conservation potentials; and, third, the policy and consensual processes needed to effect desirable changes in energy production and usage.

4. Should warming occur as predicted by the atmospheric models, climate throughout the world will be altered, but projections of specific climatic changes at the scale of specific regions are still conjectural.

5. Carbon dioxide, the dominant greenhouse gas, is an essential building block in photosynthesis. Rates of photosynthesis increase in most plants with increasing concentration of CO_2 in the ambient air. Whether these effects, observed in controlled environments, will occur in the open field is uncertain. Higher rates of respiration due to the higher temperatures of a greenhouse-warmed world would likely offset increases in photosynthesis to some degree.

6. Transpiration (water use) decreases in most plants when they are exposed to elevated levels of CO_2. This effect, if it occurs in the open field, will moderate the increase in evapotranspiration that is due to higher temperatures.

7. The efficiency of water use (photosynthate produced per unit of water consumed) could increase in most plants.

8. The climate change resulting from an effective doubling of atmospheric CO_2 would not appear likely to seriously threaten the ability of world agricultural capacity to expand in step with the growth in world agricultural demand. However, an increased frequency of extreme events such as drought could cause short-term shocks to world agricultural production and prices. Moreover, climate change resulting from unlimited increases in greenhouse gases beyond an effective doubling of CO_2 would be likely, in time, to severely impede the growth of global agricultural capacity.

9. Farm-level adjustments to climatic change (such as altering planting dates and switching crop varieties) and long-term investments in adaptive plant breeding could partially offset the negative effects and even, perhaps, manifest the positive effects of climate change.

10. Agriculture may expand in regions where it is now limited by low temperatures and short growing seasons. It may suffer in regions where rains become deficient, or prosper where rains increase.

11. In contrast to agriculture and forestry, where some benefits of warming are possible, the prospect of sea-level rise offers few apparent benefits and is likely to cause considerable disruption throughout the world. Coastal ecologies will be affected, low-lying agricultural lands may have to be abandoned and their populations moved to higher ground if higher ground is available, and cities and harbors may have to be rebuilt on higher ground or protected by expensive and complicated defense structures.

12. Forests of the world may expand into the high latitudes as a result of climate change, although it is equally or more likely that the growth and species composition of some existing forests will be negatively affected.

13. Massive afforestation offers the technical possibility of providing a major sink for the amounts of CO_2 currently accumulating in the atmosphere, but the areas that would have to be planted would be immense (on the order of 4.5 million square kilometers). The initial investment, could such land be found, might reach half a trillion dollars (which is less than the superpowers spend annually for defense). The ultimate harvestable yield of such forests would exceed current annual wood production by a factor of four, with possibly disruptive effects on world lumber markets.

14. The wood so produced might be used to fuel power plants, especially in tropical countries, making a potentially significant and ecologically acceptable substitution for fossil fuels.

The emerging consensus in favor of these propositions provides a useful focus for discussions within the research community of the consequences of climate change. As for the policy community, the agnosticism, evasion, and indifference which were endemic not long ago have pretty much disappeared; many policymakers in most industrialized countries now appear to be aware that CO_2 and trace gas accumulations warrant concern, and that at some point in the next century the effects—economic, ecological, social, and political—of these accumulations could be traumatic for much of the world and disastrous for some of it. Recognition of this ran through all the formal presentations at the RFF workshop and through the panel discussion as well. Some of the points made focused on national (that is, United States) approaches to policy, while others dealt with multinational or international initiatives. Some of the participants advanced broad guidelines; others suggested specific measures. Some ideas entailed considerable economic, political, or social costs; some appeared to be virtually cost-free. What follows is a synthesis of some of the policy-related discussions and a culling of specific policy-related propositions.

SOME GENERAL POLICY CONSIDERATIONS

Virtually every participant in the RFF workshop argued that steps should now be taken both to reduce—and eventually halt—greenhouse warming and to adapt to it. A necessary condition for both strategies is the achievement of consensus on the nature of the problem and on its scale, seriousness, and urgency. The considerable progress made by the research community in increasing public and official awareness of this issue has been reinforced, at least for the moment, by increased public sensitivity to the issue as a consequence of recent extreme climatic events in the United States—particularly the severe drought of 1988. Indeed, panelist Robert Fri noted that the subject of greenhouse warming is becoming a fad in the media and in public discussion. Milton Russell agreed, and warned, "The global warming situation is too serious to be treated as a fad. It is a condition that must be dealt with in a sustained, long-term fashion and on many fronts. Such events as the hot, dry summer of 1988 may fuel public clamor for hasty, ill-conceived government action." Stephen Schneider reminded participants that "bad decisions are likely to flow from such public hysteria."

Sober research, together with popular concern, has put the global warming issue on the policy agenda for the United States, and some other countries as well. Many workshop participants, however, appeared to be ambivalent about the effects of having this on the policy agenda. There was general recognition that the broad issue of greenhouse warming warrants serious national and international attention, but as Fri reminded the participants, uncertainty with respect to critical specific questions makes it "very important to have the right policy, aimed at the right thing, for the right reasons." And so, while it is well that the global warming issue has become a matter of public and official notice, there was a widely held view, if not a consensus, among workshop participants that thoughtful national and international examination of realistic and promising choices will accomplish more in the end than hasty, costly, and highly uncertain quick fixes.

This is not to say that scientists need be wary of forwarding their views to the policy community until all uncertainties are resolved. The research community, after all, recognizes the need for specifying levels and areas of uncertainty and, over the years, has become more sensitive to political realities.

There was also general agreement with the admonition from workshop participant Assistant Secretary of Agriculture Orville Bentley that "large concepts and philosophies and policy positions [with regard to climate change] will be hard to sell," and that measures should be taken "on an incremental basis." These, Bentley suggested, should be concentrated in areas where "dual benefits can be achieved." Russell made much the same point: "There are a lot of things (e.g., increasing energy efficiency) that should be done in any case, but global warming increases the incentive to do them now." Fri agreed: "When government has a research program that is good for other reasons and is even better when looked at in terms of climate change, the program should get high-priority funding."

As for policymakers, they need not remain idle or aloof. When confronting a choice between two alternatives of roughly equal cost in such policy realms as energy, environment, economic development, or agriculture, it would be the better part of prudence to opt for the alternative least likely to exacerbate greenhouse warming. And agencies dealing with development in Third World countries—the U.S. Agency for International Development, the World Bank, and other national and international economic assistance institutions—should explore with those countries the prospects for relying on more benign energy sources than coal.

CONTROL OF? ADAPTATION TO?

Each of the two strategies for dealing with greenhouse warming—control and adaptation—has its own philosophical basis, involves different cost considerations, operates on a different time scale, responds to different national needs and capabilities, and presents different policy challenges. Yet the two basic approaches overlap and are not mutually exclusive. As Paul Portney pointed out in his formal paper, "The optimal policy will combine both control and adaptive strategies."

In the recent past, discussions of the strategy of controlling global warming included the prevention of additional warming from present levels, as well as ideas about delay or mitigation of warming. The concept of total prevention has now been abandoned. "We are already committed to a warming due to already-emitted gases," warned panelist Michael Oppenheimer, and Milton Russell presented as his "first working hypothesis" the proposition that "the globe is going to get warmer." In their formal presentation Stephen Schneider and Norman Rosenberg asserted, "There is no doubt that the concentration of carbon dioxide in the atmosphere has been rising—when the carbon dioxide concentration rises, the temperature at the earth's surface must rise, too."

Agreements for international action to prevent some greenhouse warming may take years or decades to develop. The immediate challenge of any control strategy, then, is to dampen the rate of CO_2 increase sufficiently to provide time for global society to cope with the effects of warming. But in the view of workshop participants, even this relatively modest, short-term goal is unlikely to be achieved soon, since concerted, effective action would have to be taken by virtually all of the world's industrialized countries and by a significant number of the developing nations. This, panel member D. Gale Johnson pointed out, "is extremely difficult because we are dealing with a problem of the global 'commons,'" which poses hard enough challenges on a national scale. Roger Revelle was more specific: "In the long run coordinated action may be very difficult because of the probably different effects of climate change on different countries, and also their different fossil fuel resources."

Although virtually all workshop participants recognized the obstacles in the way of an international agreement to halt global warming, some felt that the 1987 Montreal Protocol for protection of the ozone layer provided a ray of hope. They cautioned, however, that the actual rates of future emissions will hinge on the success of the protocol's implementation and success in finding acceptable substitutes for the chlorofluorocarbons (CFCs) now in use. Moreover, as Joel Darmstadter and Jae Edmonds noted in their formal paper, even the present, relatively permissive version of the protocol "has already changed anticipated rates of (CFC) emissions."

Others expressed the view that the protocol did offer a model for cooperative international action in reducing fossil fuel consumption. The Ozone Convention, according to Oppenheimer, "established a mechanism for workshops, information exchange and increasing the availability and dissemination of information. There is a possibility that a global policy may emerge. After all, no one is proposing that we rush into an international agreement for major cuts in CO_2 emissions by 1995." (This point was soon to be undercut: a few weeks after the RFF workshop, the Conference on the Changing Atmosphere, held in Toronto, called for a reduction of CO_2 emissions by approximately 20 percent of 1988 levels by the year 2005. This was only the initial goal; additional reductions were to be achieved in subsequent years.)

The Montreal Protocol aside, workshop participants and panel members strongly supported, as a matter of prudence, certain policies falling under the general rubric of "Control" as a complement to an "Adaptation" strategy. As Oppenheimer observed: "No matter how much research we accumulate, I don't believe a specific decision on averting climate changes versus adapting to it can ever be made."

While recognizing the difficulty of achieving international agreement on reducing CO_2 emissions, many participants expressed sympathy with the proposition that industrially advanced countries such as the United States could make a useful contribution by simply setting an example. Schneider, in a comment from the floor, referred to a "global survival compromise" that would involve wealthy countries "making some sacrifice in terms of improving their own [energy] efficiency and helping developing countries to find acceptable alternatives" to fossil fuel. Moreover, many steps to reduce fossil fuel consumption could be taken unilaterally by the United States or by some other industrially advanced country acting strictly in its own national interest and for reasons largely independent of global warming. Although these initiatives alone would do little to slow warming, they could strengthen the position of the industrialized countries in seeking to persuade Third World countries to pursue development strategies less dependent on fossil fuel.

These aspects of the discussion of control and adaptation strategies could be summed up in two statements:

1. An incremental approach is more likely to accomplish sought-after objectives than draconian lurches and lunges.
2. Proposals that make sense for other economic and environmental reasons are more likely to be accepted and implemented than proposals directed only at greenhouse warming.

There are obvious reasons why such admonitions make sense for policy advisers and policy planners. But one reason that may not come readily to mind has to do with costs, especially intergenerational costs. As a general conceptual proposition, Pierre Crosson and other economists at the workshop suggested that policymakers should opt to halt greenhouse warming at a point where the costs of warming, in terms of lost income, broadly defined, would equal the costs of halting it. Since the cost of halting warming, if we begin now, would fall mainly on the present generation while the cost of not halting it would fall mainly on future generations, a key policy issue would be to find a fair intergenerational sharing of these costs. These notions aroused considerable discussion, and much disagreement, particularly on the part of noneconomists. Crosson's proposition, they felt, relied on a cost-benefit approach under circumstances in which both costs and benefits would be difficult to quantify. There was general agreement among economists and noneconomists alike, however, that the two precepts noted above—incrementalism and duality of purpose—would make the cost of both control and adaptation strategies more manageable.

Control Strategies

The workshop participants generally accepted the proposition that, although international agreement on measures to halt global warming soon are highly unlikely, steps to slow the warming trend by a few decades are feasible and would be worth serious consideration. According to Schneider and Rosenberg, such a course would gain valuable time to develop adaptation approaches and ease adjustment problems. It was explicitly or implicitly within this context that control strategies were discussed.

Energy Efficiency

High on the list of steps that panel members felt could contribute to the delay of greenhouse warming was the improvement of energy efficiency. According to Oppenheimer: "Energy efficiency would bring us benefits on a wide variety of fronts." Russell said: "There are great global benefits associated with improving energy efficiency and helping less-advanced countries [he referred especially to China and its dependence on coal] to develop more modern fuel-using technologies." Fri commented that one of the most important steps to be taken is "to increase the productive use of energy in the economy."

The steps toward improving energy efficiency that were discussed at the workshop meet the two criteria noted earlier: they are incremental—that is, no one proposed that they be accomplished through sudden and sweeping government dicta (an unrealistic objective in any case); and they have duality of purpose—there are compelling economic and technological reasons to seek greater energy efficiency, aside from problems of global warming.

Fuel Switching

Fuel switching, as a way of reducing the rate of CO_2 emissions, was discussed by several participants, in particular Darmstadter and Edmonds in their formal paper, and Johnson in his panel remarks. But here, unlike the case for energy efficiency, apparent difficulties cloud the prospects for significant results. To achieve switching on a scale large enough to make a difference would involve a massive, worldwide movement away from fossil fuels (particularly coal) toward such CO_2-free energy sources as hydro, solar, and nuclear power. But, according to Darmstadter and Edmonds, "A large part of the world appears to be committed to energy policies which assign a major future role to coal in particular and fossil fuels in general."

If industrially advanced and developing nations are to be weaned away from fossil fuels early enough and to an extent that would make a significant contribution to delaying global warming, a practical alternative sourcc of energy will have to be available. This would place the principal burden on nuclear energy, although recent progress in solar energy technology holds great promise. Wind and hydro power are likely to make only marginal contributions.

In retrospect, too little time at the workshop was devoted to the feasibility of nonfossil alternatives. Johnson was one of the few workshop participants to address the nuclear issue: "Nuclear energy could make a great contribution toward averting global warming, but there is a disposition to believe that the risks of nuclear energy exceed those of global warming. Do we really know this, or is it something we just believe on faith?"

Capturing CO_2 Through Forest Management

Although much of the discussion concerning control strategies concentrated on approaches to reducing CO_2 emissions, one line of inquiry took another tack—the possibility of increasing the capacity of the global carbon sink. According to Roger Sedjo and Allen Solomon in their formal paper, among the most promising ways to accomplish this would be through "artificially established plantation forests." Johnson noted in this connection that the additional forestation "might be responsible for absorbing 10 percent of CO_2 emissions." But to have any important impact on slowing the rate of global warming, such an effort would have to be on a vast scale and at an enormous cost. According to Sedjo and Solomon, "The land area of new manmade plantations required to stabilize CO_2 emissions at current levels was estimated at a minimum of 465 million hectares," and "start-up costs would be at least $372 billion." Expansion on this scale would greatly increase potential wood supplies, with possibly disruptive effects on world lumber markets. Sedjo and Solomon suggested that some of the additional wood might substitute for fossil fuel in power generation, a use consistent with the carbon sequestering effect so long as the forest biomass remained permanently larger than it is now.

In his paper, N. S. Jodha gave some modest support to the forest-as-carbon-sink idea, noting that while developing countries are likely to be indifferent to international efforts to reduce fossil fuel consumption, "the need for preventing loss of forest is better understood."

Conclusion on Control Strategies

At first blush, the prospects for developing an effective strategy to halt or even delay greenhouse warming would seem unpromising; major difficulties stand in the way of obtaining the necessary international agreement. After all, substantial, near-term national sacrifice would be entailed to accomplish a goal that is at once remote (at least in terms of typical political timetables) and not yet clearly understood. Moreover, hovering over all negotiations on this thorny matter is the specter of unequal sacrifice and unequal gain among nations involved. Finally, there is the dilemma of dealing with the global commons problem.

And yet, reflecting on the formal papers and informal discussions surrounding control strategies, one senses that the outlook for an international accord may not be altogether bleak. The main ground for hope is the now-accepted proposition that improved energy efficiency is a desirable and urgent objective. Technologies have already been developed that are significantly more efficient than those now in general use. If it can be demonstrated that the new technologies are economically competitive, especially in the case of countries where coal is now the fuel of choice, there would be a real chance of pushing back the onset of significant greenhouse warming. Exploration of the energy option should clearly be high on the policy agenda of any nation concerned about climate change. Yet another option worthy of consideration is a robust program for increasing global forest cover. The fact that each of these options has an inner economic logic of its own increases the odds that, with patience and foresight (and no small amount of luck), a planetary bargain can be struck.

Adaptation Strategies

The costs of the control and adaptation strategies are highly uncertain, although there is every reason to believe they will be high in both cases. There are important differences, however, in the nature and perception of the two kinds of cost. As Russell explained, "Preventive measures carry present costs. These measures involve collective goods, truly global in nature, and they will have uncertain and distant benefits which will vary, and be perceived to vary, among countries and interest groups." Adaptation measures, on the other hand, "tend to be location-specific and their benefits will likely be closer in time and space and more immediate in terms of cause and effect to those who pay the cost."

For the rich, then, it is feasible to take out insurance against the high probability that the global community will be unable or unwilling to take measures soon to stem the trend of greenhouse warming. But while the United States and Japan may find the costs of defending their shores against rising sea levels acceptable, for Bangladesh and Egypt the costs may be crushing. And while Germany and Australia may have the luxury of choosing the extent to which they will adapt their agriculture to climatic change, the range of adaptive options available to Ethiopia and Guatemala are likely to be much more limited unless they receive outside assistance. In short, adaptation strategies may have more potential for the rich than for the poor countries; at the same time, unilateral initiatives for control strategies by the latter would appear even less promising for them than adaptive strategies. The different positions of the rich and poor countries with respect to adaptation strategies thus may be a festering source of future international ill will, if not outright conflict, over how to deal with climate change.

Adaptation in Agriculture

The implications for agriculture of increasing atmospheric accumulations of CO_2 are complicated and uncertain. On the one hand, crop growth and yields of most plants may be increased by CO_2-enrichment of the atmosphere. On the other hand, CO_2-induced changes in patterns of precipitation and temperature will affect, for good or ill, agricultural output in various parts of the world. But which locales will be affected, how they will be affected, and the degree to which they will be affected are not yet clear. In the light of these uncertainties it is difficult, and would, in any case be unwise, to rush policymakers into decisions about adaptive responses.

An effective policy for agricultural adaptation of greenhouse warming also requires concern for "incrementalism." As Easterling, Parry, and Crosson noted in their paper, "Climate changes are generally expected to occur continuously (rather than instantaneously) and several agroclimatic thresholds are likely to be crossed at different points along the way—research will produce more realistic insights by viewing climatic change as a transient process."

Agricultural adaptation in the face of global warming is not a matter for conjecture or debate, according to several experts. Easterling, Parry, and Crosson stressed that the issue "is not Can agriculture adapt? but *How best* can agriculture adapt?" They addressed this question in terms of the American Midwest. Most of the global climate models suggest that, with global warming, the climate in that region will become hotter and drier, with unfavorable implications for the region's agriculture. The severity of the effects would depend very much on the adaptive options available to midwestern farmers under gradually worsening climatic conditions. Easterling, Parry, and Crosson stressed that current farm management practices offer only limited insights into these options because, by the time the climate changes significantly, many new practices will have been developed. They emphasized the responsibility of the agricultural research establishment to develop such new practices with an eye to their usefulness to farmers in adapting to the changing climate regime. They were confident that much could be done to promote adaptation, but were uncertain whether it would be enough to avoid some loss of competitive position for midwestern agriculture.

In his paper, N. S. Jodha emphasized that because governments in developing countries face so many urgent problems of development, "public policies and programmes are likely to remain indifferent [to climatic change] for a fairly long time to come." Thus, "rather than looking for planned public interventions and preparedness, the search for potential measures to manage impacts of climatic changes should be directed to the existing farming systems." The "main limitation of this approach is that [the] inventory of such measures is confined to the stock of traditional knowledge." Since little official attention is being paid to "planned or anticipatory adaptations," the agricultural sectors in the developing countries will tend to rely on "forced adaptation strategies." Nevertheless, if changes in climate are gradual, "say, over fifty years or more, we can be optimistic that responses in developing countries' agriculture to emerging risks will be adequate."

Jodha's concern that governments in developing countries are not sufficiently sensitive to the risk of global warming was echoed by other participants with regard not only to the developing world, but to the advanced as well. "Until there is some large-scale, worldwide consciousness [of greenhouse warming], and until everybody is making some degree of sacrifice [to reduce greenhouse risks]," Schneider warned, "we are probably not going to get anywhere. . . ."

As for more research, Johnson expressed the hope that "American efforts to finance international research centers and bilateral research aid to developing countries would give additional attention to research that would help poor farmers everywhere adapt more rapidly to global warming."

Adaptation in Water Management

Agricultural vulnerability may be heightened when one confronts what Kenneth Frederick and Peter Gleick described in their presentation as the "increased variability—and thus more frequent extremes" of precipitation. This, they warned, could lead to "higher peak flood flows, more persistent and severe droughts, greater uncertainty about the timing of the rainy season. Few such changes would be beneficial."

Frederick and Gleick emphasized "the importance of nonstructural alternatives for coping with future water scarcity and variability." Among these, they cited "improving the management of the existing infrastructure, encouraging conservation, marketing water—in response to changing supply and demand conditions—and technologies that increase the options for and decrease the costs of responding to changes in water supplies."

Adapting to Sea-Level Rise

One of the more traumatic effects of greenhouse warming may be a significant rise (that is, 0.5–1.0 m or more over the next century) in sea levels throughout the world. This would entail profound consequences: coastal flooding, soil salination, the degradation of freshwater aquifers, damage to coastal ecosystems, and a host of other unfavorable effects. Sea-level rise would vary even within fairly small geographic areas throughout the world, but the effects would be generally grave and, in some places, disastrous.

Local communities and even individuals can take modest steps to ensure against or adapt to the threat of rising sea levels (for example, by moving to higher ground, building breakwaters or sea walls, and erecting houses on stilts). But major resources would have to be committed by large jurisdictions if effective long-term, large-scale measures were to be taken (for example, massive dike construction, or relocation of population centers). The costs of sea-level rise, Pierre Crosson stressed, "may have more effect in forging a global consensus to halt warming than any other set of climate change costs [including the cost of adapting agriculture to climate change]."

In his paper Gjerrit P. Hekstra argued, in essence, that because effective measures to control global warming are not likely to be taken soon enough to avoid some rise in sea levels, policy responses to the likely consequences must be addressed. The work in train at the Delft (Netherlands) Hydraulics Laboratory is an example of what could be done. The object of that work is "to enhance national and regional awareness of increased sea-level rise and to develop strategies for responding to the impacts." The work "aims to build up a global network of experts in national agencies who can develop and use software and data bases with which to model the policy options appropriate for each case study area. Results are to be presented through workshops for policymakers."

Hekstra noted that the course of climatic change, especially as it is manifested in sea-level rise, "is certain to cause problems for many of the world's coastal countries," and he urged that environment ministries assure that "the present momentum of concern about the future of climate and the biosphere" should be translated into the "appropriate will to action."

IN SUM

Few participants in the RFF workshop would have predicted, even just a few years ago, that the CO_2 issue would by now have worked its way to such a high place in the political agenda of the United States and numerous other nations. But emissions of CO_2 and other trace gases, as well as climate change, are now no longer the exclusive province of technical journals, scientific meetings, and scholarly debate. They have become the stuff of the popular press, political discussion, congressional hearings, and proposed legislation. In short, greenhouse warming has become an "in" subject in the media and the political arena. Indeed, the problem is in danger of becoming a chic intellectual toy for instant experts, and a "fun" challenge for policy buffs. (A retrospective look at the life cycle of the limits-to-growth issue of the 1970s would be instructive in this connection.)

Current American popular and political concern with regard to greenhouse warming bemuses some scientists. It may have occurred not so much as a consequence of articles and speeches by serious members of the research community, but more as a consequence of the conjunction of the hot, dry summer of 1988 with on-going scholarly research. While there are obvious advantages to having public attention focused on the greenhouse issue, even if for the wrong reasons, there is also a danger that this concern will disappear with the return of more apparently "normal" weather.

As for the policymaking community, prudence would dictate a stance removed from the zigs of popular panic and the zags of popular indifference. There is evidence at hand—not overwhelming, but sufficient; not perfect, but good enough—to warrant careful official attention and thoughtful policy planning. Steps should be taken to reduce the rate of increases in emissions of CO_2 and chlorofluorocarbons; investigations should be funded to narrow uncertainties and increase policy options; research should be encouraged to increase the resilience and robustness of agricultural systems, initiatives should be put in train to improve water management. Each of these measures promises multiple benefits. Even if the risk of greenhouse warming turns out to be grossly exaggerated as a consequence of some dreadful scientific error or of divine grace, such objectives would still be worth pursuing.